Synergies of 6G and IoT

Synergies of 6G and IoT

Edited by

Himanshi Babbar
Chitkara University Institute of Engineering and Technology,
Chitkara University, Punjab, India

IEEE Press
WILEY

Published by John Wiley & Sons, Inc., Hoboken, New Jersey.

Library of Congress Cataloging-in-Publication Data:

Names: Babbar, Himanshi editor
Title: Synergies of 6G and IoT / edited by Himanshi Babbar.
Description: First edition. | Hoboken, New Jersey : Wiley-IEEE Press, [2026] | Includes bibliographical references and index.
Identifiers: LCCN 2026006092 (print) | LCCN 2026006093 (ebook) | ISBN 9781394370672 Hardback | ISBN 9781394370696 adobe pdf | ISBN 9781394370689 epub
Subjects: LCSH: 6G mobile communication systems–Industrial applications | Internet of things–Industrial applications
Classification: LCC TK5103.252 .S96 2026 (print) | LCC TK5103.252 (ebook)
LC record available at https://lccn.loc.gov/2026006092
LC ebook record available at https://lccn.loc.gov/2026006093

Cover Design: Wiley
Cover Image: © lowball-jack/Getty Images

Set in 9.5/12.5pt STIXTwoText by Straive, Pondicherry, India

Contents

About the Editor *xix*
List of Contributors *xxi*
Preface *xxv*

1 Introduction to 6G and IoT *1*
Sridevi Sridhar, Thilagavathi Narayanamoorthi, Akila Krishna Moorthy, Durgadevi Palani, and Priyatharshini Rajaram
1.1 Introduction *1*
1.1.1 Understanding the Internet of Things *3*
1.1.1.1 Definition and Genesis *3*
1.1.1.2 The Evolution of IoT *5*
1.1.1.3 Key Components of IoT *6*
1.1.2 Overview of 6G Networks *8*
1.1.2.1 Advances in 6G *8*
1.1.2.2 Key Features of 6G *9*
1.1.3 Convergence of 6G and IoT *10*
1.2 Empowering IoT with 6G *11*
1.3 Next-Gen IoT Paradigms *12*
1.3.1 Visual Internet of Things (VIoT) *12*
1.3.2 Social Internet of Things (SIoT) *14*
1.3.3 Artificial Intelligence of Things (AIoT) *15*
1.3.4 The Synergistic Ecosystem: See, Interact, Think *16*
1.4 Research Advances and Applications *17*
1.4.1 6G and IoT Research Trends *17*
1.4.2 Application Areas *17*
1.4.2.1 Blockchain for IoT Security *17*
1.4.2.2 Integrated Sensing and Communication (ISAC) *18*
1.4.2.3 Human-Centric and Bio-IoT *18*
1.4.2.4 Sustainability and Green 6G/IoT *18*
1.5 Conclusion *19*
References *20*

2 **Foundations of 6G and IoT: Technologies, Challenges, and Future Directions** *23*
B. Navaneethi, A. Priyadharshini, and Saju Mathew
2.1 Introduction *23*
2.2 Evolution of Mobile Networks (1G to 6G) *24*
2.2.1 What Will 6G Bring? *26*
2.3 6G Network Architecture for IoT *28*
2.3.1 6G-Enabled IoT: Vision, Techniques, and Objectives of 6G *28*
2.3.1.1 Toward 6G IoT: Advances and Challenges *28*
2.3.1.2 Open Challenges and Future Directions *29*
2.3.1.3 Key Advantages of 6G and IoT Integration *29*
2.3.2 Synergy of 6G and IoT Networks *30*
2.3.3 Applications and Use Cases *37*
2.3.3.1 Seamless Integration of Physical and Digital Worlds *40*
2.3.4 Key Technologies Enabling 6G *40*
2.3.5 IoT in the 6G Era *43*
2.3.6 Edge and Fog Computing in 6G IoT *44*
2.3.6.1 Four-Dimensional Communication *45*
2.3.7 Issues with Safety and Confidentiality in the Internet of Things Ecosystems Powered by 6G *45*
2.3.7.1 Expanding Threat Surface and Vulnerabilities *46*
2.3.7.2 Privacy Risks in Data-Driven Environments *46*
2.3.8 Emerging Solutions and Technologies *46*
2.3.9 Key Research and Standardization Challenges *47*
2.3.10 Collaborative Approach to Security Innovation *47*
2.3.10.1 Energy Efficiency and Sustainability in 6G Networks *48*
2.3.10.2 Sustainability Challenges in Future Networks *48*
2.3.10.3 Green Enabling Technologies *48*
2.3.10.4 Design Techniques and Architectural Aspects of 6G in IoT *49*
2.3.11 Toward a Sustainable 6G Environment *49*
2.3.11.1 Future Roadmap *50*
2.4 Conclusion *52*
Further Reading *53*

3 **Green and Intelligent IoT in 6G Era: Machine Learning for Sustainable Smart Cities** *57*
Priyanka Datta, Neha Garg, Amitesh Aggarwal, Amanpreet Kaur, and Yonis Gulzar
3.1 Introduction *57*
3.1.1 Overview of IoT and Data Analytics *57*
3.1.1.1 Importance of Data Analytics in IoT *58*

3.1.2 Role of Machine Learning in IoT *58*
3.1.2.1 Data Processing and Pattern Recognition *59*
3.1.2.2 Predictive Maintenance *59*
3.1.2.3 Anomaly Detection *59*
3.1.2.4 Optimization and Automation *59*
3.1.2.5 Personalization *60*
3.1.2.6 Security Enhancements *60*
3.1.2.7 Edge Computing and Real-Time Decision-Making *60*
3.1.2.8 Improving User Experience *60*
3.2 IoT Data Characteristics *61*
3.2.1 Types of IoT Data *61*
3.3 Data Challenges *62*
3.4 Data Preprocessing Techniques *63*
3.5 Machine Learning Techniques for IoT *64*
3.5.1 Supervised Learning *64*
3.5.1.1 Decision Trees *64*
3.5.1.2 Support Vector Machines (SVMs) *65*
3.5.2 Unsupervised Learning *65*
3.5.2.1 K-means *66*
3.5.2.2 Density-Based Spatial Clustering of Applications with Noise (DBSCAN) *66*
3.5.3 Reinforcement Learning *67*
3.5.3.1 Q-learning *67*
3.5.3.2 Deep Q-Networks (DQN) *67*
3.5.4 Deep Learning *67*
3.5.4.1 Convolutional Neural Networks (CNNs) *68*
3.5.4.2 Recurrent Neural Networks (RNNs) *68*
3.5.5 Federated Learning *68*
3.5.6 Transfer Learning *69*
3.6 Case Studies *69*
3.6.1 Predictive Maintenance in Manufacturing *69*
3.6.1.1 Background *69*
3.6.1.2 Implementation *69*
3.6.1.3 Example *69*
3.6.1.4 Conclusion *70*
3.6.2 Smart Cities and Environmental Monitoring *70*
3.6.3 Healthcare and Wearable Devices *71*
3.6.3.1 Background *72*
3.6.3.2 Applications in Patient Monitoring *73*
3.6.3.3 Applications in Health Data Analysis *73*
3.6.3.4 Challenges and Ethical Considerations *74*

3.6.3.5 Conclusion *75*
3.7 Challenges and Future Directions *75*
3.8 Conclusion *78*
References *78*

4 Synergies of 6G and IoT *83*
Harpreet Kaur, Kritika, Jasmine, and Vanshika Gupta
4.1 6G Technology Foundations *83*
4.1.1 Terahertz Communication *83*
4.1.2 Reconfigurable Intelligent Surfaces (RISs) *84*
4.1.3 Massive MIMO and Beamforming in 6G *85*
4.1.4 Integrated Space-Air-Ground-Sea Communication *87*
4.2 IoT Evolution and its Integration with 6G *87*
4.2.1 IoT Architecture Evolution *88*
4.2.2 IoT Requirements for 6G *88*
4.2.2.1 Latency *88*
4.2.2.2 Coverage *89*
4.2.2.3 Energy *89*
4.2.3 AI, Blockchain and Edge Computing for Secure IoT *90*
4.2.4 Sustainable IoT Design and Green Networking *91*
4.3 Network Architecture and Management *92*
4.3.1 Network Slicing and QoS in 6G IoT *92*
4.3.2 Edge Intelligence and Federated Learning *94*
4.3.3 AI-Native Network Automation *96*
4.3.4 Protocols and Standardization Roadmaps *96*
4.4 Key Use Cases of 6G-IoT Synergy *97*
4.4.1 Smart Healthcare and Internet of Medical Things *98*
4.4.2 Autonomous Vehicles and V2X *98*
4.4.3 Industry 4.0 and Robotics *99*
4.4.4 Smart Cities and Environmental Sustainability *99*
4.4.5 Remote Connectivity via Satellites and UAVs *100*
4.5 Threat Landscape in 6G-IoT *101*
4.5.1 AI-Powered Intrusion Detection *101*
4.5.2 Blockchain for Trust and Data Integrity *101*
4.5.3 Privacy by Design in 6G-enabled IoT *102*
4.6 Economic, Market, and Ecosystem Insights *102*
4.6.1 Market Forecasts and Investment Trends *102*
4.6.2 Ecosystem Collaboration and Policy Frameworks *103*
4.6.3 Monetization Strategies and Business Models *104*
4.7 Societal Impacts and Ethical Considerations *104*
4.8 Conclusion *104*
References *105*

5 Societal Implications of Synergizing 6G and IoT *109*
V. Vidhya
5.1 Introduction *109*
5.2 Bridging the Digital Divide *110*
5.3 Transforming Healthcare Delivery *111*
5.4 Revolutionizing Education and Learning *112*
5.5 Smart Cities and Urban Development *113*
5.6 Workforce Evolution and Employment *114*
5.7 Environmental Sustainability and Climate Action *115*
5.8 Smart Cities and Infrastructure: The 6G-IoT Revolution *116*
5.8.1 Introduction *116*
5.8.2 A New Urban Command System *116*
5.8.3 Automated Infrastructure and Proactive Maintenance *117*
5.8.4 Intelligent Urban Mobility and Traffic Systems *117*
5.8.5 Waste Management and Environmental Oversight *117*
5.8.6 Energy Management and Smart Grids *118*
5.8.7 Safety, Surveillance, and Emergency Response *118*
5.8.8 Citizen Interaction and Digital Governance *119*
5.8.9 Privacy and Ethical Dilemmas *119*
5.8.10 Economic Growth and Urban Innovation *119*
5.8.11 Implementation Hurdles *119*
5.9 Agricultural Innovation and Rural Empowerment Through 6G-IoT Integration *120*
5.9.1 Introduction *120*
5.9.2 Precision Agriculture and Smart Farming *120*
5.9.3 Climate Adaptability and Crop Prediction *120*
5.9.4 Livestock Management *121*
5.9.5 Water and Resource Management *121*
5.9.6 Market Access and Supply Chain Clarity *121*
5.9.7 Empowerment of Rural Communities *122*
5.10 Environmental Sustainability and Ecosystem Monitoring *122*
5.10.1 Introduction *122*
5.10.2 Real-Time Environmental Monitoring *122*
5.10.3 Forest and Wildlife Conservation *122*
5.10.4 Smart Waste Management Systems *123*
5.10.5 Renewable Energy and Grid Optimization *123*
5.10.6 Disaster Prevention and Ecosystem Restoration *123*
5.11 Ethical and Privacy Considerations *123*
5.12 Conclusion and Future Directions *124*
References *125*

6 **Enabling Technologies in 6G for VIoT and Autonomous Driving** *127*
Ishan Kumar
6.1 Introduction *127*
6.1.1 Background and Problem Statement *128*
6.1.2 Research Gap *128*
6.2 Literature Review *128*
6.2.1 Novelty of the Research *129*
6.3 Methodology *129*
6.3.1 Problems Faced by Current Vehicular Systems *129*
6.4 Characteristics of 6G Networks for Vehicular Communication *130*
6.4.1 Ultrareliable Low-Latency Communication (URLLC) *131*
6.4.2 High-Precision Positioning and Localization *132*
6.5 Edge Computing and Fog Computing for Vehicular Applications *134*
6.5.1 Edge Intelligence and Distributed Data Processing *134*
6.5.2 Fog-Based Decision-Making for Autonomous Vehicles *135*
6.6 Conclusion *139*
References *139*

7 **Trends and Forecasts in 6G and IoT** *143*
Betshrine R. Jibinsingh and Leema Nelson
7.1 Introduction: The Dawn of Pervasive Interconnectivity *143*
7.1.1 Charting the Course for Tomorrow: Anticipating Key Developments *145*
7.1.1.1 Identifying Emergent Patterns and Future Projections in 6G and IoT *145*
7.1.1.2 Envisioning Future Applications and Transformative Use Cases *148*
7.2 Peering into the Future: Innovations on the Technological Horizon *151*
7.2.1 Emerging Next-Generation Technologies: Quantum Communication and Advanced AI Integration *151*
7.2.1.1 The Advent of Quantum-Secure Networks and Novel Encryption Methodologies *151*
7.2.1.2 Embedding AI at the Network Core and Extending It to the Extreme Edge *152*
7.2.1.3 Exploring Neuromorphic and Bio-inspired Computing Paradigms *152*
7.2.1.4 Pioneering Brain–Computer Interfaces and the Cognitive Internet of Things *153*
7.2.2 Harnessing Cross-Domain Synergies and Convergence with Allied Technologies *153*
7.2.2.1 Leveraging Blockchain for Enhanced IoT Security, Transparency, and Decentralization *153*

7.2.2.2 Fostering Synergies with Renewable Energy Systems and Intelligent Smart Grids *154*
7.2.2.3 Integrating IoT with Advanced Satellite Communication Networks *154*
7.2.2.4 Interfacing with Biotechnology: The Rise of the Bio-Internet of Things *155*
7.3 Actively Shaping the Future: Strategies and Policies for a Connected World *155*
7.3.1 Crafting Strategies for Sustainable Development and Inclusive Innovation *155*
7.3.1.1 Championing Green Networking Principles and Low-Power IoT Solutions *156*
7.3.1.2 Systematically Bridging the Global Digital Divide for Equitable Access *156*
7.3.1.3 Establishing Robust Ethical Frameworks and Promoting Responsible AI *157*
7.3.1.4 Fostering Inclusivity in Design Processes and Deployment Strategies *157*
7.3.2 Formulating Policy Recommendations and Fostering Collaborative Initiatives *157*
7.3.2.1 Harmonizing Regulatory Approaches and Advancing Spectrum Governance *158*
7.3.2.2 Strengthening Public–Private Partnerships in Research and Development *158*
7.3.2.3 Promoting Global Standards and Interoperability Frameworks *158*
7.3.2.4 Investing in Education, Skills Development, and Workforce Readiness *159*
7.4 Conclusion: Embracing a Hyper-connected, Intelligent Future *159*
References *160*

8 Advancing Connectivity: The Role of Satellite IoT in 6G Networks *163*
Garima Chopra
8.1 Introduction *163*
8.1.1 Applications and Use Cases *166*
8.1.2 Motivation *166*
8.1.3 Organization *167*
8.2 Hybrid Network Architectures for Seamless Connectivity *167*
8.2.1 Architecture of Sat-IoT *168*
8.2.2 Interoperability and Standard Protocols *169*
8.2.2.1 Importance of Interoperability in Satellite IoT *169*
8.2.2.2 Problems for Achieving Interoperability *169*
8.2.2.3 Standard Protocol in Sat-IoT *170*

8.2.2.4 Standardization Bodies and Initiatives *170*
8.2.2.5 Research Areas *171*
8.2.2.6 Role of Edge Computing for Sat-IoT *172*
8.3 6G Connectivity Solutions for Remote and Rural Areas *173*
8.3.1 Connectivity Limitations in Underserved Areas *173*
8.3.2 Satellite-IoT Integration for Ubiquitous Coverage in 6G Networks *173*
8.3.3 MEC-Enabled Satellite Networks: Edge Intelligence at Scale *174*
8.3.4 Bridging the Digital Divide with Satellite Connectivity *174*
8.3.5 Use Cases in Remote Monitoring, Agriculture, and Disaster Management *175*
8.3.5.1 Remote Environmental Monitoring *176*
8.3.5.2 Smart and Precision Agriculture *176*
8.3.5.3 Disaster Management and Emergency Response *176*
8.4 Future Research Directions *177*
8.4.1 AI-Driven Satellite Routing and Network Orchestration *177*
8.4.2 Green and Energy-Efficient Sat-IoT Protocols *178*
8.4.3 Security and Privacy for Non-Terrestrial IoT Networks *178*
8.5 Conclusion *178*
References *179*

9 Emerging IoT Scenarios Enhanced by Machine Learning in 6G *181*
Renu Popli, Deepanshi Batra, Rajeev Kumar, and Ashutosh Sharma
9.1 Introduction *181*
9.2 Emerging IoT Scenarios Enhanced by Machine Learning in 6G *182*
9.2.1 IoT Evolution with 6G *182*
9.2.2 Machine Learning Framework for 6G-Enabled IoT *183*
9.3 AI/ML Approaches for 6G Networks *183*
9.3.1 Edge Computing and Distributed Intelligence *184*
9.3.2 LLM-Empowered IoT Architecture *185*
9.4 Smart Cities and Urban Infrastructure *186*
9.4.1 Intelligent Transportation Systems *186*
9.4.2 Environmental Monitoring and Management *187*
9.4.3 Public Safety and Emergency Response *187*
9.5 Healthcare and Telemedicine *188*
9.5.1 Remote Patient Monitoring *188*
9.5.2 Smart Hospitals and Telesurgery *189*
9.5.3 Personalized Medicine and AI Diagnostics *189*
9.6 Industrial Automation and Smart Factories *190*
9.6.1 Intelligent Manufacturing Systems *190*
9.6.2 Predictive Maintenance and Quality Control *190*
9.6.3 Digital Twins and Simulation *191*

9.7 Agricultural Applications and Food Production *191*
9.7.1 Autonomous Farming Equipment and Drones *192*
9.7.2 Livestock Monitoring and Management *193*
9.7.3 Indoor and Vertical Farming *193*
9.8 Extended Reality and Digital Twins *194*
9.8.1 Comprehensive Digital Twins *194*
9.8.2 Internet of Senses and Embodied AI *195*
9.9 Challenges and Future Research Directions *195*
9.9.1 Technical Challenges *195*
9.9.2 Ethical and Societal Implications *196*
9.9.3 Future Research Directions *197*
9.10 Conclusion *197*
References *198*

10 Cybersecurity and Privacy in 6G-IoT Ecosystems *201*
Aman Jandwani, Amit K. Arora, Dadas A. Bhanudas, Ekta Singh, Madhur, and Paras Jandwani
10.1 Introduction *201*
10.1.1 Significance of Cybersecurity and Privacy in Hyper-connected 6G-IoT Environments *202*
10.1.2 Objectives of the Chapter *202*
10.1.3 Scope of the Chapter *203*
10.1.4 Foundational Motivation *203*
10.2 Advanced Encryption Mechanisms in 6G-IoT *204*
10.2.1 Current Encryption Technologies *204*
10.2.2 Symmetric Encryption (AES) *204*
10.2.3 Asymmetric Encryption (RSA) *204*
10.2.4 Limitations in 6G Scenarios *204*
10.2.5 Emerging Encryption Solutions *205*
10.2.6 Post-Quantum Cryptography (PQC) *205*
10.2.7 Lightweight Cryptography for IoT *205*
10.2.8 Homomorphic Encryption (HE) *206*
10.2.9 Secure Multiparty Computation (SMPC) *206*
10.3 Decentralized Security Models: Blockchain Integration *207*
10.3.1 Current Blockchain Applications in IoT *207*
10.3.2 Smart Cities and BFLIoT *207*
10.3.3 Innovations in Blockchain for 6G-IoT *208*
10.3.4 Scalability Through Sharding and Off-Chain Transactions *208*
10.3.5 Integration with Edge Computing and AI *208*
10.3.6 Energy-Efficient Consensus Mechanisms *209*
10.4 AI-Driven Threat Detection Systems in 6G-IoT *209*

10.4.1 Emerging AI Techniques *210*
10.4.1.1 Deep Learning and Reinforcement Learning *210*
10.4.2 Adversarial Training and Metaheuristic Algorithms *211*
10.4.3 Collaborative Intrusion Detection Systems (CIDS) *212*
10.5 Converging Technologies: Encryption, Blockchain, and AI in 6G-IoT *212*
10.5.1 Introduction: The Convergence Imperative *212*
10.5.2 Synergistic Integration: How These Technologies Interact *213*
10.5.3 End-to-End Secure Communication Pipelines in 6G-IoT *214*
10.5.4 Federated Learning and Privacy Preservation *215*
10.5.4.1 Federated Learning in 6G-IoT *216*
10.5.5 Blockchain as the Trust Anchor for Decentralized AI *217*
10.5.5.1 Verifiable AI Collaboration and Model Integrity *217*
10.5.5.2 Smart Contracts for Autonomous AI Governance *218*
10.5.5.3 Consensus and Trust Management in Multi-agent Systems *218*
10.6 Compliance and Privacy-Preserving Technologies in 6G-IoT *221*
10.7 Conclusion *221*
References *226*

11 Sustainable Development and Green IoT in 6G Era *231*
Parmeshwar Yadav, Aman Jandwani, Ekta Singh, Priyanka Darekar, Amit K. Arora, Dadas A. Bhanudas, and Paras Jandwani
11.1 Energy-Efficient Network Architectures *231*
11.1.1 AI-Assisted Dynamic Resource Allocation *232*
11.1.2 Green Communication Protocols *232*
11.1.3 Energy-Efficient Hardware and Transceiver Design *233*
11.1.4 Impact on Carbon Emissions and Environmental Goals *233*
11.2 Sustainable Internet of Things (IoT) Devices *234*
11.2.1 Optimization of Wireless Sensor Networks (WSNs) *234*
11.2.2 Edge-AI Integration for Real-Time Processing *235*
11.2.3 Quality of Service (QoS) Adaptation and AI-Aware Scheduling *235*
11.2.4 Recyclable, Modular, and Batteryless Device Design *236*
11.3 Advanced Protocols and Hardware for 6G *237*
11.3.1 Protocol-Level Sustainability: Smart, Adaptive, and Lightweight Communication *237*
11.3.2 MIMO, URLLC, and AI-Aided Protocol Stacks *238*
11.3.3 Circular Economy in Hardware Fabrication *238*
11.3.4 Security, Efficiency, and Blockchain-Verified Protocols *239*
11.3.5 AI-Driven Hardware Lifecycle Management *239*
11.4 Applications in Smart Cities and Agriculture *240*
11.4.1 Smart City Infrastructure: Connectivity, Control, and Intelligence *241*

11.4.2 Precision Agriculture: AI, Sensors, and Sustainability *242*
11.4.3 Blockchain and Data Transparency in Smart Systems *242*
11.4.4 Integration of Circular Economy Models *242*
11.4.5 Sustainable Supply Chains and Blockchain Logistics *243*
11.5 Circular Economy and Intelligent Resource Allocation in 6G *243*
11.5.1 Recycling of Electronic Components *243*
11.5.2 Sustainable Supply Chains and Blockchain Integration *244*
11.5.3 Dynamic Bandwidth Allocation *244*
11.5.4 Energy-Aware Grid Interaction *245*
11.5.5 Blockchain for Energy Trust and Automation *246*
11.5.6 AI-Powered Lifecycle and Predictive Maintenance *246*
11.6 Conclusion *246*
References *251*

12 Technological Advancements in Healthcare: AI-Powered Smart Health Evolution with Real-Time Predictive Analytics *253*
Anamika Larhgotra, Aditya Ahirwal, and Avinash Pandey
12.1 Introduction *253*
12.2 Literature Review *257*
12.3 System Architecture, Design, and Context *258*
12.3.1 System Architecture Overview *258*
12.3.2 Real-Time Processing and Feedback *260*
12.3.3 Advantages and Disadvantages *260*
12.3.3.1 Advantages *260*
12.3.3.2 Disadvantages *261*
12.3.4 Technologies and Techniques *261*
12.3.4.1 Technologies Used *261*
12.3.4.2 Techniques Used *262*
12.4 Methodology *263*
12.4.1 Data Collection *264*
12.4.2 Data Preprocessing *264*
12.4.3 Data Integration *265*
12.4.4 Model Development *265*
12.5 Results and Discussion *266*
12.5.1 Accuracy of Medical Predictions *266*
12.5.2 Precision, Recall, and F1-Score *267*
12.5.3 Real-Time Data Processing Performance *267*
12.6 Case Studies *272*
12.7 Challenges and Future Exploration *275*
12.7.1 Challenges *275*
12.8 Future Exploration *276*

12.8.1 Social Development and Women's Security 277
12.9 Conclusion 277
References 278

13 6G and IoT Integration for Industrial Innovation 281
Gurinder Kaur Sodhi and Reetu Malhotra
13.1 Introduction 281
13.2 Overview of 6G Technologies 282
13.2.1 Key Performance Targets of 6G 282
13.2.2 Enabling Technologies 282
13.3 IoT in Industrial Contexts 283
13.3.1 Definition and Scope of Industrial IoT (IIoT) 283
13.3.2 The IIoT Scope Encompasses 284
13.3.3 Limitations with Current Technologies (5G, Wi-Fi) 284
13.3.4 Types of IoT Devices in Industrial Use 284
13.4 Synergies Between 6G and IoT 285
13.4.1 How 6G Breaks Through Current IIoT Limitations 286
13.4.1.1 Edge AI Real-Time Decision-Making 286
13.4.2 Interoperability and Scalability Through Network Slicing 286
13.4.2.1 Advantage 287
13.4.2.2 Autonomous Systems and Swarm Intelligence with 6G-Enabled IoT 287
13.5 Industrial Applications Enabled by 6G-IoT Integration 287
13.5.1 Smart Manufacturing Real-Time Process Monitoring and Control 287
13.5.2 Smart Logistics and Warehousing Autonomous Drones and AGVs Based on EdgeAI 289
13.5.3 Automated Supply Chains Through Smart Contracts 289
13.5.4 Energy and Utilities 289
13.5.5 Hazardous Environment Remote Control 289
13.5.6 IoT + 6G in Wind/Solar Asset Monitoring 289
13.5.7 Agriculture and Agri-Tech Real-Time Soil Monitoring and Precision Irrigation 290
13.5.8 6G Swarm Robotics for Harvesting and Monitoring 290
13.5.9 Nano-IoT and 6G Drone Remote Sensing 290
13.5.10 Healthcare and Industrial Safety 290
13.5.10.1 Industrial Wearables and Biometrics for Worker Health 290
13.5.11 Real-Time Hazard Detection and Evacuation Systems 290
13.5.12 Emergency Robotics and Response Coordination 290
13.6 Challenges and Considerations 291
13.6.1 Spectrum Availability and Terahertz Deployment Challenges 291
13.6.2 Privacy, Security, and Ethical Concerns 291

13.6.3 Standardization and Regulatory Frameworks 292
13.6.4 Legacy System Integration 292
13.6.5 Digital Twins, Metaverse, and Extended Reality (XR) Role 293
13.6.6 AI Co-Designed Networks and Federated Learning in IIoT 293
13.6.7 Green and Sustainable 6G-IoT Ecosystems 294
13.7 Conclusion 294
13.7.1 Summary of Key Synergies and Transformative Potential 294
13.7.1.1 Strategic Implications for Industry Leaders and Policymakers 294
13.7.1.2 Roadmap for Adoption and Implementation 294
References 295

Index 297

About the Editor

Dr. Himanshi Babbar is Assistant Professor in CSE, working in the Centre for Cybersecurity Research and Education (C3R), Chitkara University, Rajpura, Punjab, India. She has seven years of teaching experience, CGC, Landran, Mohali, Punjab. She received an MCA (Master's in Computer Applications) degree from Chitkara University, Punjab Campus in 2015 and completed her PhD and Postdoctoral Fellowship in Computer Applications from Chitkara University, Punjab Campus and the UAE in 2021 and 2022, respectively. Her areas of research are Software-Defined Networking, Load Balancing, Deep Learning, Intrusion Detection, and Internet of Things. She has served as a reviewer in many conferences and journals of the International Conference on Intelligent and Innovative Technologies in Computing, Electrical and Electronics (ICIITCEE 2023), IEEE Access, Scientific Reports, Peer-to-Peer Networking and Applications, Ad Hoc Networks, Elsevier, Springer's Journal of Network and Systems Management, etc. She has published/accepted/presented many papers in national and international conferences, published more than 40 papers in SCI indexed journals, 20+ papers in Scopus Indexed conferences, and filed/published/granted more than 15 patents. She received the Young Researcher Award from the Institute of Scholars (INSC) in April 2023. She has won various Research Excellence Awards in 2024 and 2025 at Chitkara University, Punjab, India.

List of Contributors

Amitesh Aggarwal
Chitkara University Institute of Engineering and Technology
Rajpura, Punjab
India

Aditya Ahirwal
Department of Computer Science & Engineering
Chandigarh University
Mohali, Punjab
India

Amit K. Arora
Computer science and Engineering
IIT Madras
Chennai, Tamil Nadu
India

Deepanshi Batra
Department of Computer Science and Engineering
Ambala College of Engineering and Applied Research
Devsthali, Kurukshetra University, Kurukshetra
Ambala, Haryana
India

Dadas A. Bhanudas
Department of Management
Neville Wadia Institute of Management Studies & Research
Pune, Maharashtra
India

Garima Chopra
Department of Electronics and Communication
Chitkara University Institute of Engineering and Technology, Chitkara University
Rajpura, Punjab
India

Priyanka Darekar
Department of Management
Indira Global School of Business
Pune, Maharashtra
India

Priyanka Datta
Department of Computer Science and Engineering
Laxmi Institute of Technology
Sarigam, Gujarat
India

Neha Garg
Chitkara University Institute of Engineering and Technology
Rajpura, Punjab
India

Yonis Gulzar
Department of Management Information Systems, College of Business Administration
King Faisal University
Al-Ahsa
Saudi Arabia

Vanshika Gupta
Chitkara School of Engineering and Technology
Chitkara University
Baddi, Himachal Pradesh
India

Aman Jandwani
Department of Management
Neville Wadia Institute of Management Studies & Research
Pune, Maharashtra
India

Paras Jandwani
Department of Electronics
School of VLSI & Embedded (SVE)
NIT Kurukshetra
Kurukshetra, Haryana
India
and
Department of Management
Accenture LLP
Chicago
USA

Jasmine
Chitkara School of Engineering and Technology
Chitkara University
Baddi, Himachal Pradesh
India

Betshrine R. Jibinsingh
Ramanujan Computing Centre
College of Engineering Guindy, Anna University
Chennai
India

Amanpreet Kaur
Chitkara University Institute of Engineering and Technology
Rajpura, Punjab
India

Harpreet Kaur
Chitkara School of Engineering and Technology
Chitkara University
Baddi, Himachal Pradesh
India

Akila Krishna Moorthy
Department of Computer Science and Engineering
SRM Institute of Science and Technology
Chennai, Tamil Nadu
India

Kritika
Chitkara School of Engineering and Technology
Chitkara University
Baddi, Himachal Pradesh
India

Ishan Kumar
Department of Computer Science and Engineering
Chitkara University
Rajpura, Punjab
India

Rajeev Kumar
Chitkara University Research Innovation and Network
Institute of Engineering and Technology
Chitkara University
Rajpura, Punjab
India

Anamika Larhgotra
Department of Computer Science & Engineering
Chandigarh University
Mohali, Punjab
India

Madhur
Department of Commerce
Global School of Finance & Accounting
Chandigarh University
Mohali, Punjab
India

Reetu Malhotra
Applied Sciences Department
Chitkara University Institute of Engineering and Technology
Chitkara University
Rajpura, Punjab
India

Saju Mathew
Department of Computer Science with Data Analytics
SNMV College of Arts and Science
Coimbatore, Tamil Nadu
India

Thilagavathi Narayanamoorthi
Department of Computer Science and Engineering
S.A. Engineering College
Chennai, Tamil Nadu
India

B. Navaneethi
Department of Computer Application
SNMV College of Arts and Science
Coimbatore, Tamil Nadu
India

Leema Nelson
Centre for Cybersecurity Research and Education
Chitkara University
Rajpura, Punjab
India

Durgadevi Palani
Department of Computer Science and Engineering
SRM Institute of Science and Technology
Chennai, Tamil Nadu
India

Avinash Pandey
Department of Computer Science & Engineering
Chandigarh University
Mohali, Punjab
India

Renu Popli
Department of Computer Science and Engineering
Institute of Engineering and Technology
Chitkara University
Rajpura, Punjab
India

A. Priyadharshini
Department of Computer Application
SNMV College of Arts and Science
Coimbatore, Tamil Nadu
India

Priyatharshini Rajaram
Cardiff School of Technologies
Cardiff Metropolitan University
Cardiff
United Kingdom

Ashutosh Sharma
Department of Computer Science
Henan University of Science and Technology
Luoyang, Henan
China

Ekta Singh
Department of Management
Neville Wadia Institute of Management Studies & Research
Pune, Maharashtra
India
and
Department of Commerce Global School of Finance & Accounting
Chandigarh University
Mohali, Punjab
India

Gurinder Kaur Sodhi
Electronics and Communication Department
Desh Bhagat University
Mandi Gobindgarh, Punjab
India

Sridevi Sridhar
Department of Computer Science and Engineering
SRM Institute of Science and Technology
Chennai, Tamil Nadu
India

V. Vidhya
Department of Artificial Intelligence & Data Science
Easwari Engineering College
Ramapuram, Chennai
India

Parmeshwar Yadav
Department of Management
Indira School of Business Studies (PGDM)
Pune, Maharashtra
India

Preface

The dawn of the sixth generation (6G) of wireless communication marks a defining moment in the evolution of connectivity—one that transcends speed and bandwidth to embrace intelligence, sustainability, and human-centric design. As the digital and physical worlds converge, the Internet of Things (IoT) stands at the heart of this transformation, driving innovation across domains such as healthcare, transportation, industry, and smart cities. The synergy between 6G and IoT is not merely an incremental advancement; it is a paradigm shift that will reshape how societies connect, communicate, and collaborate.

This book, *Beyond Connectivity: Exploring the Synergies of 6G and IoT*, emerged from the recognition of a growing need for a comprehensive, interdisciplinary exploration of these converging technologies. While extensive literature exists on 5G and IoT individually, there remains a paucity of integrated frameworks addressing their coevolution in the 6G era. This book aims to bridge that gap by presenting both foundational and forward-looking perspectives that blend theory, technology, and practice.

Each chapter of this volume has been meticulously designed to provide readers with a layered understanding of the 6G–IoT ecosystem—from fundamental principles and enabling technologies to industry applications and ethical implications. We delve into the mechanisms driving ultrareliable low-latency communication (URLLC), massive machine-type communication (mMTC), and the role of artificial intelligence (AI), edge computing, and digital twins in shaping next-generation-connected environments. The discussion extends to real-world implementations across smart cities, healthcare, and autonomous systems, highlighting both opportunities and challenges.

In addition to its technical depth, this book underscores the societal and economic transformations that the fusion of 6G and IoT will bring. It emphasizes inclusivity, security, sustainability, and ethics—factors that will define the responsible evolution of intelligent connectivity in the coming decade.

We envision this book as a resource for researchers, students, engineers, industry professionals, and policymakers who seek not only to understand the technology but also to shape its future. By combining analytical rigor with accessibility, *Beyond Connectivity* aspires to inspire innovation and interdisciplinary collaboration.

We would like to extend our sincere gratitude to our colleagues, reviewers, and research collaborators whose insights have enriched this work. Special appreciation goes to my husband Mr. Ashish Wadhwa, my lovely children, Viven Wadhwa and Bhavisha Wadhwa, and specially our institution, **Chitkara University, Punjab**, for providing an enabling environment that fosters research and innovation at the intersection of technology and human advancement.

We hope this book serves as both a reference and a catalyst—guiding readers to envision the limitless possibilities of a hyperconnected, intelligent world driven by the synergies of 6G and IoT.

Dr. Himanshi Babbar
Assistant Professor
Chitkara University,
Punjab, India
31 January 2026

1

Introduction to 6G and IoT

Sridevi Sridhar[1], *Thilagavathi Narayanamoorthi*[2], *Akila Krishna Moorthy*[1], *Durgadevi Palani*[1], *and Priyatharshini Rajaram*[3]

[1] *Department of Computer Science and Engineering, SRM Institute of Science and Technology, Chennai, Tamil Nadu, India*
[2] *Department of Computer Science and Engineering, S.A. Engineering College, Chennai, Tamil Nadu, India*
[3] *Cardiff School of Technologies, Cardiff Metropolitan University, Cardiff, United Kingdom*

1.1 Introduction

Sixth generation (6G) is the upcoming generation of wireless communication expected to offer ultrafast speeds, near-zero latency, and smart connectivity. Combined with the Internet of Things (IoT), it will power smart cities, connected vehicles, and intelligent healthcare systems. This chapter gives a basic overview of what 6G and IoT are. This includes what 6G and IoT mean, why they matter, their expected impact, and how they might be used in the future. 6G wireless communication combined with IoT technologies creates an essential shift in digital technology development. The digital world has embraced IoT as its foundational component by allowing the networking of billions of physical objects, starting from household devices to industrial sensors up to self-driving vehicles over the last 20 years [1]. The intersection between operational devices enables continuous data processing and automated action steps while driving operational improvements and user-centric experiences in healthcare facilities and agricultural farms as well as smart cities and manufacturing plants.

According to Tong and Zhu [3], 6G envisions a paradigm shift from "connected people and things" to "connected intelligence." In this vision, devices, networks, and applications operate synergistically to offer predictive, context-aware, and adaptive services. 6G will integrate cutting-edge technologies such as artificial intelligence (AI), terahertz (THz) communication, reconfigurable intelligent

Synergies of 6G and IoT, First Edition. Edited by Himanshi Babbar.

Published 2026 by John Wiley & Sons, Inc.

surfaces (RIS), quantum communication, and edge computing into its core architecture [4]. These advancements will expand the potential of IoT ecosystems, enabling previously unfeasible applications such as autonomous drone fleets, real-time digital twins, immersive holographic communication, and fully automated industrial systems, as in Qadir et al. [5].

As shown in Figure 1.1, the evolution of wireless communications—from 1G in the 1980s to the anticipated full deployment of 6G in the 2030s—demonstrates a consistent trend toward higher capacity, broader coverage, and more intelligent services. Each generation brought forth new capabilities, and 6G is set to encompass all previous functionalities while introducing groundbreaking advancements such as [1]:

Full global network coverage,
Massive-scale IoT and AI-based mobile applications,
Enhanced satellite communications,
Seamless integration of autonomous systems.

In this broader context, 6G represents more than a communications upgrade; it signifies a technological convergence. It combines existing and emerging innovations across all network layers—from the physical infrastructure to the application layer—to establish a digitally unified, intelligent interface among users, services, sensors, machines, and computing platforms. As described by Lin and Widmer [6], this

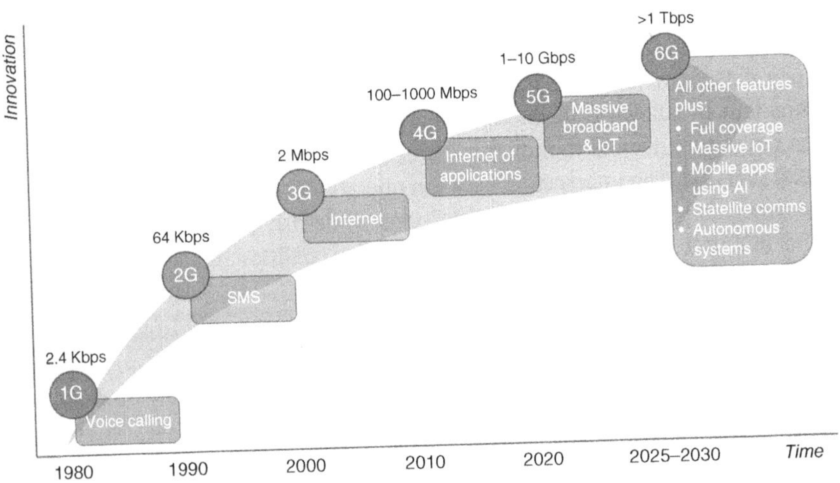

Figure 1.1 Evolution of wireless communications toward 6G networks and applications. *Source:* Polymeni et al. [2] / MDPI / CC BY 4.0.

interface directly addresses the pressing demands of a hyper-connected, data-centric world detailed in Hersent et al. [7]. New possibilities are emerging in areas such as:

Ultra-mobile broadband, offering unprecedented speeds,
Ultrareliable low-latency communication (URLLC) for mission-critical tasks,
Global geographic coverage, including rural and remote areas,
Massive machine-type communication (mMTC), enabling the dense deployment of IoT devices. These capabilities lay the technological foundation for what is now being termed the sixth-generation Internet of Things (6G-IoT).

The 6G-IoT framework is expected to overcome several long-standing limitations in current IoT systems by ensuring:

Scalable and reliable connectivity,
Optimized communication protocols for massive networks,
Geographic expansion of intelligent services beyond urban areas.

Moreover, the growing integration of AI into 6G systems will empower devices to move beyond passive data transmission. Devices will evolve into intelligent agents, capable of interpreting, learning from, and actively responding to environmental data in real time, as quoted in Lin and Widmer [6]. This transformation is especially critical in high-impact domains such as industrial IoT, smart cities, and remote monitoring—with particularly promising applications in next-generation smart agriculture.

1.1.1 Understanding the Internet of Things

The IoT refers to a network of interconnected physical objects—or "things"—embedded with sensors, actuators, software, and communication capabilities that enable them to collect, transmit, and respond to data. These "things" can range from everyday household items like refrigerators and thermostats to complex industrial systems such as machinery, vehicles, and environmental sensors.

1.1.1.1 Definition and Genesis

Researchers have worked on IoT development through a steady genesis, which started with digital communication expansion, as shown in Figure 1.2. The IoT concept rose from multiple decades of development throughout networking, computing, and human–computer interaction. The diagram "Evolution of IoT" presents an innovative model that demonstrates how the internet progressed from basic human-to-human interaction into sophisticated machine-to-machine total systems used for present-day IoT systems.

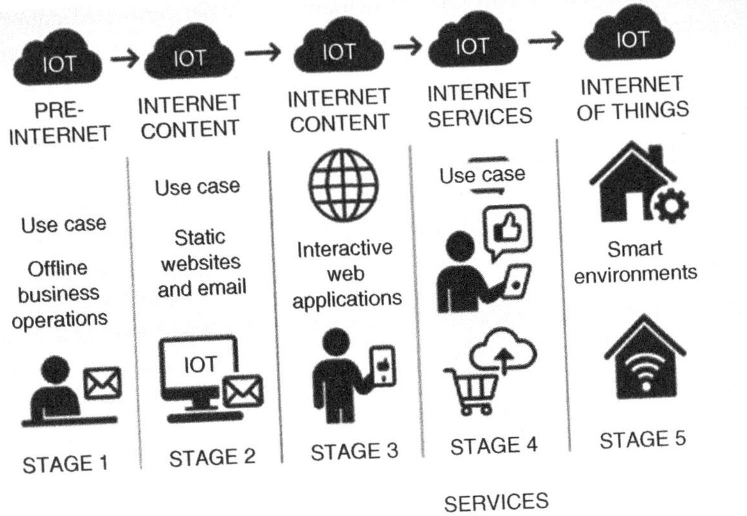

Figure 1.2 Genesis of the internet of things.

Pre-Internet (Human-to-Human)
The introductory phase of human technology predates the advancement of present-day internet connectivity. Human communication relied principally on personal dialogues transmitted by mobile telephony and fixed telephony systems alongside SMS (short message service). The concept of digitized inter-object communication did not exist at that time. Throughout this time, fundamental networking technologies alongside smart networking research started developing.

Internet of Content (The "WWW" Era)
The World Wide Web (WWW) creation transformed the internet into an advanced medium that allowed both data distribution and content access. Web browsing along with email service and entertainment portals started becoming regularly used services during this time. The contemporary era developed the fundamental structure of digital technology through its support of smart IT solutions and web-based operations.

Internet of Services (Web 2.0)
Interactive materials and user-generated content replaced static information as the primary focus of web development. The era known as Web 2.0 brought attention to e-commerce alongside productivity tools and dynamic program applications. The technical systems evolved to become more open for collaboration while also emphasizing app-driven functionalities, which eventually led to integrated

information networks. Users experienced their first encounter with web-based services, which operated directly between themselves and their devices.

Internet of People (Social Media Era)
Social internet emerged through the establishment of platforms, including Facebook, Skype, Twitter, and YouTube. People joined the connected network through this expansion, which bridged services to human interaction, thus creating a global human connection. Smart devices transformed into individual platforms for online communication as smartphones and tablets appeared on the market. These platforms paved the way for continuous data generation and mobile computing—key prerequisites for IoT.

Internet of Things (Machine-to-Machine Communication)
The IoT represents the highest development stage where machine-to-machine (M2M) communication occurs. Through data analysis, devices have the capacity to execute autonomous identification tasks and tracking while monitoring systems and autonomous decision-making capabilities. Computers now connect people and services with every object in existence. IoT enables billions of smart devices to exchange ambient data in a structured semantic manner for creating real-time applications across various sectors, including smart cities, agriculture, logistics, and healthcare.

1.1.1.2 The Evolution of IoT

The historical timeline of the "Internet of Things" demonstrates technical advancements, which eventually shaped today's IoT systems. The vision of a hyper-connected world received its fundamental development through each individual advancement.

1974—The Beginning of TCP/IP
Devices across the internet began communicating through the basic language established by Transmission Control Protocol/Internet Protocol (TCP/IP). The establishment of communication standards through TCP/IP represented a significant first advance in worldwide digital network development.

1990—The First IoT Device
John Romkey established the first internet-connected toaster, which represented the practical evolution of IoT technology in 1990. Scientists tried to link physical matter to internet functionality for the first time using this device.

1991—The First Webpage
Tim Berners-Lee established the WWW through its first webpage release, which gave people worldwide access to digital content and defined the basis of networked applications.

1999–"IoT" Is Coined
Kevin Ashton established the definition of "Internet of Things" at the Auto-ID Center during his research on RFID technology for supply chain monitoring. The conceptual milestone introduced the future of smart objects by showing how intelligence should be embedded into physical elements.

2013–Official Recognition of IoT
The scientific community started recognizing IoT as the transformative digital paradigm, which led to wide-scale investments from governments and industries for IoT-based innovations. This year started to observe the mass implementation of IoT technologies.

2016–The Mirai Botnet Attack
Through the Mirai malware, millions of IoT devices without proper security were compromised to launch an enormous DDoS attack, which targeted prominent websites. Security measures for IoT systems became indispensably important after this incident.

2020–IoT Devices Exceed 20 Billion
IoT technologies reached a turning point when interconnected devices worldwide exceeded 20 billion units in quantity for the first time. Buyya and Dastjerdi [1] defined IoT as an extension of internet capabilities that surpasses computer-based devices to manage and process physical world data at extensive scales. IoT evolution results from three major developments, which include wireless communication systems, electronic miniaturization, cloud technology expansion, and edge intelligence advancement.

Initially, IoT developed as a system for remote monitoring alongside basic automation through proprietary communication standards. The concept developed across time into a worldwide network through which billions of devices function simultaneously in real time. Exclusive digital transformation efforts across smart homes, precision agriculture, healthcare systems, urban planning, logistics, and manufacturing have positioned IoT as a key modern initiative. The development of IoT has created a foundation for enterprising wireless systems named 6G-IoT, which will power the next level of intelligent, automated, interconnected devices.

1.1.1.3 Key Components of IoT

A functional IoT system is composed of several interdependent components (Figure 1.3) that enable data sensing, transmission, analysis, and decision-making. The core components include:

Sensors and Actuators: Sensors are responsible for capturing real-world data such as temperature, pressure, motion, light, humidity, and location. Actuators

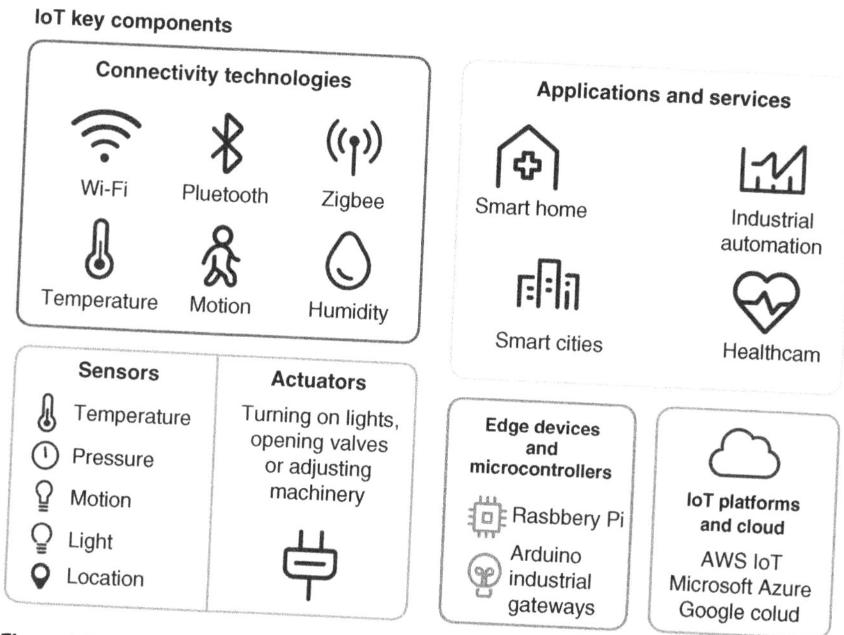

Figure 1.3 Key components of IoT.

enable devices to perform physical actions based on data insights—such as turning on lights, opening valves, or adjusting machinery. These elements form the interface between the physical and digital worlds.

Edge Devices and Microcontrollers: Edge devices are embedded computing platforms that connect to sensors and manage local data processing before relaying it to higher levels (e.g., cloud). Examples include Raspberry Pi, Arduino, and industrial gateways. They support real-time analytics, reduce latency, and improve privacy by minimizing data transmission.

Connectivity Technologies and Protocols: These technologies facilitate communication between devices and the network. Common IoT communication options include Wi-Fi and Bluetooth (short-range), Zigbee, Z-Wave, and LoRaWAN (low-power, long-range), 4G/5G cellular (for wide-area, high-throughput applications). Future integration will include 6G and Terahertz communication for ultrahigh-speed data transfer. Efficient communication is at the heart of every IoT deployment. Due to the constraints in power, bandwidth, and computational capacity of IoT devices, lightweight and scalable communication protocols are essential. As outlined by Hersent et al. [7], the following protocols are widely adopted in IoT networks: (1) MQTT (Message Queuing Telemetry Transport), a publish-subscribe protocol designed for constrained devices and low-bandwidth, high-latency networks; (2) CoAP (Constrained Application Protocol), a RESTful protocol similar to HTTP but optimized for lightweight devices; (3) 6LoWPAN (IPv6 over Low-Power Wireless Personal Area Networks), which enables IPv6 communication over IEEE 802.15.4-based networks; and other emerging protocols such as LPWAN and DDS (Data Distribution Service). These protocols are fundamental to ensuring energy-efficient, reliable, and scalable communication, especially as IoT scales into billions of devices connected over 6G networks.

IoT Platforms and Cloud Infrastructure: Platforms like AWS IoT, Microsoft Azure IoT Hub, and Google Cloud IoT Core offer tools for device management, data ingestion, analytics, and integration. These platforms play a critical role in aggregating and processing data, applying machine learning algorithms, and providing actionable insights through dashboards and Application Programming Interfaces (APIs).

Applications and Services: The value of IoT is realized through its diverse applications: Smart homes (energy management and security systems), healthcare (remote monitoring and wearable devices), smart cities (traffic control and pollution monitoring), industrial IoT (IIoT) (predictive maintenance and robotics), and agriculture (soil monitoring and automated irrigation). Each application requires tailored configurations of the above components to meet specific performance, scalability, and security requirements.

Understanding the architecture and functionality of IoT is essential to grasp its potential and limitations—especially in the context of next-generation networks like 6G. From sensing and actuation to protocol design and application delivery, each layer of the IoT ecosystem plays a crucial role in building a connected, intelligent, and responsive environment. As the convergence with 6G unfolds, IoT is expected to evolve into an even more context-aware, autonomous, and resilient infrastructure, redefining digital interaction in both everyday life and industries.

1.1.2 Overview of 6G Networks

As we approach the limits of what 5G can deliver, 6G wireless networks emerge as the next revolutionary phase in mobile communication systems. Beyond incremental improvements in speed and capacity, 6G is poised to fundamentally transform how networks interact with the world around them—ushering in a new era of ubiquitous intelligence, autonomy, and immersive digital experiences. This section explores the evolution, key capabilities, and enabling technologies that define 6G.

1.1.2.1 Advances in 6G

6G development goes beyond speed improvements because it transforms the entire communication network infrastructure. The research of Tong and Zhu [3] shows that 6G represents a paradigm shift from "connected things" to "connected intelligence" because all network components become self-directing through smart, adaptive, and independent operations. 6G advances by integrating network intelligence into core and edge systems through a combination of AI and Blockchain technology, as well as Quantum Communication, as discussed in Javeed et al. [4] and Alsharif et al. [8]. The newly developed technologies allow for instant context-based decisions and network self-repair mechanisms, as well as secure device-to-device connection capabilities. The development of 6G technology pursues emerging applications, which need extraordinary performance levels, including holographic telepresence and tactile internet functions, along with autonomous drone coordination and real-time digital twin solutions for factories, and the potential integration of brain–computer interfaces. 6G networks must provide both improved bandwidth and latency reduction, along with the additional capacity to understand and react to complicated environments automatically.

1.1.2.2 Key Features of 6G

The promise of 6G is underpinned by several defining characteristics that set it apart from previous generations as in Figure 1.4.

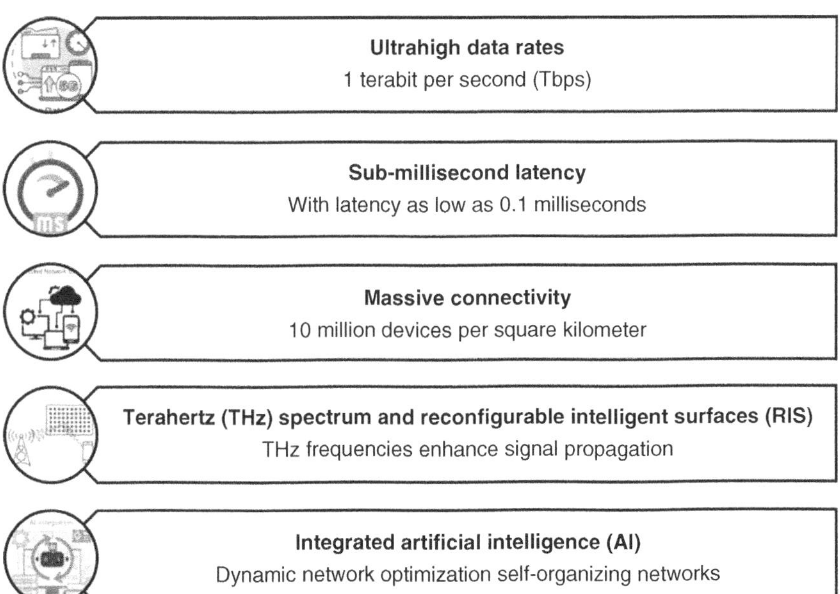

Figure 1.4 Key features of 6G.

Together, these features form the backbone of a future where communication networks are no longer passive conduits of data but active participants in shaping intelligent, sustainable, and autonomous ecosystems [9]. In the sections that follow, we will explore how these capabilities translate into real-world applications and how 6G synergizes with the IoT to unlock new possibilities.

1.1.3 Convergence of 6G and IoT

The convergence of 6G wireless communication and the IoT signifies a paradigm shift in how digital systems are designed, deployed, and operated.

Through traditional IoT development, vast connectivity occurs when billions of sensors collect and exchange data through centralized cloud platforms (Figure 1.5). Massive IoT comprises widely used connected devices, including smart locks, light bulbs, padlocks, smart plugs, laptops, security cameras, as well as factories and homes. The devices create vast amounts of data that move to central processing and analytical centers for further examination. Cognitive IoT represents the new technological frontier, which 6G technologies have brought into existence. Cognitive IoT achieves functions that its previous system did not because it advances data collection and basic connectivity. The system adds intelligent functions to physical objects so they can determine predictions and respond with localized awareness while avoiding dependence on centralized platforms.

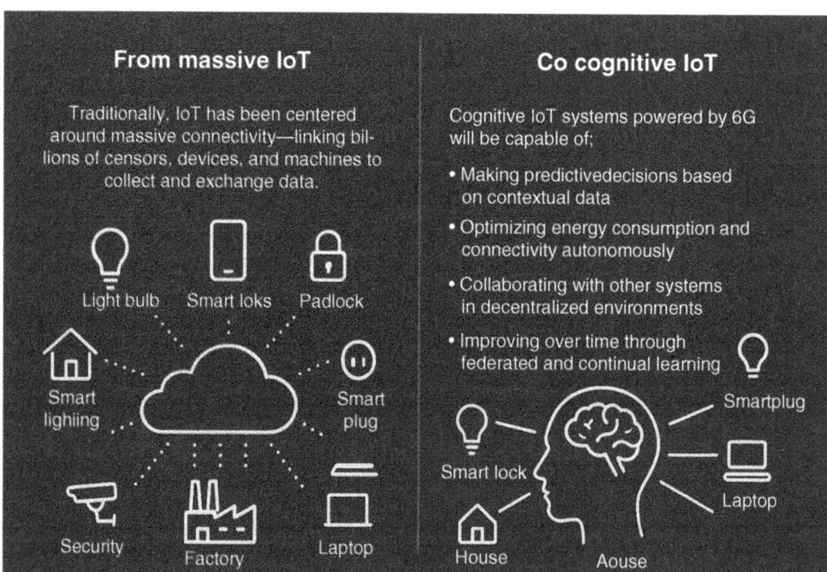

Figure 1.5 From massive IoT to cognitive IoT.

Cognitive IoT systems are capable of (1) making predictive decisions based on contextual data, (2) optimizing energy consumption and connectivity autonomously, (3) collaborating with other systems in decentralized environments, and (4) improving over time through federated and continual learning.

This transformation empowers smart devices—such as homes, smart locks, laptops, smart plugs, and lighting systems—to think, adapt, and learn from their environments. As a result, cognitive IoT represents a shift from passive data collection to intelligent, self-optimizing networks that can operate more efficiently and autonomously.

1.2 Empowering IoT with 6G

Although 5G networks deliver peak data rates of up to 10 Gbps—nearly 100 times faster than 4G LTE, which typically offers speeds around 100 Mbps—and reduce latency from about 50 ms (4G) to as low as 1 ms, they still face limitations in addressing the rapidly expanding and diverse requirements of the IoT ecosystem [10]. Despite its advancements, 5G faces limitations in supporting the growing scale and complexity of IoT. While it can connect millions of devices, future IoT ecosystems will require support for billions or even trillions of devices. As IoT expands into critical domains like smart cities, autonomous vehicles, and telemedicine, networks must offer ultralow-latency, high reliability, and seamless scalability [11]. These limitations highlight the need for 6G to enable a truly intelligent, hyper-connected world. The integration of 6G and IoT opens up a world of possibilities, but it comes with its fair share of opportunities to work on.

1) **Hyper-connectivity and Seamless Interactions:** 6G will enable the creation of a truly connected world where billions of devices can communicate seamlessly, as discussed by Zheng et al. [12]. This will lead to the development of smart cities, industries, and homes that function more efficiently, reducing energy consumption, improving safety, and enhancing the quality of life for people around the globe.
2) **Intelligent Systems:** With the power of AI and edge computing, 6G will allow IoT devices to make smarter, real-time decisions. The work by Khalid et al. [13] outline that this will be crucial in applications like autonomous vehicles, where immediate responses are needed, or healthcare, where predictive analytics can help prevent health emergencies.
3) **New IoT Applications:** 6G's low-latency and high-speed communication will enable IoT applications that require ultrafast responses, such as remote surgery, immersive virtual and augmented reality (AR) experiences, and real-time collaborative robotics, as in Ndiaye et al. [14].

The tandem development of 6G and IoT is a unique prospect to engineer the future of technology that could change nearly everything in society. Moving off the capabilities of 5G, we can expect 6G to deliver an unparalleled level of speed, connectivity, and intelligence. Thanks to the ability to handle massive numbers of devices, ultralow latency, and the incorporation of sophisticated technologies such as AI and edge computing, 6G will make IoT realize its true potential. The outcome will be a hyper-connected world with devices easily talking to each other and making real-time intelligent decisions, leading to smarter cities, efficient industries, and improved healthcare, as discussed in Ahad and Tahir [15].

However, this change does not come without its fair share of issues. The increase in the pace of development of the IoT devices creates a cause for concern in security, privacy, and cyber threats as explained in Sicari et al. [16]. The infrastructure needed for 6G networks will entail huge outlays of investments, but energy efficiency and sustainability will play important roles in curtailing the effects on the environment. In addition, it will take a lot of time and effort for the global cooperation needed for standardization and regulation of 6G technology. Although there are these challenges, the combination of 6G and IoT has great potential in the future. If we are able to overcome these obstacles, new avenues for innovation and enhancement of life quality, as well as an improvement of our world in terms of its being connected, efficient, and sustainable, will open up before us.

1.3 Next-Gen IoT Paradigms

The evolution from 5G to 6G marks a transformative leap in the realm of the IoT. While 5G introduced enhancements in speed, connectivity, and latency, 6G is set to revolutionize IoT by enabling ultrahigh data rates, terahertz communication, and seamless integration with AI. This convergence gives rise to advanced IoT paradigms: Visual IoT (VIoT), Social IoT (SIoT), and Artificial Intelligence of Things (AIoT). Each of these paradigms leverages the unique capabilities of 6G to unlock new applications and efficiencies across various sectors.

1.3.1 Visual Internet of Things (VIoT)

VIoT refers to the integration of visual data—such as images and video feeds—into IoT systems, enabling devices to "see" and interpret their environment (Figure 1.6). With 6G's high-speed data transmission and low latency, real-time processing of high-resolution visual data becomes feasible, facilitating applications that require immediate visual analysis.

Key Enablers with 6G: Ultrafast image and video transmission: 6G's terabit-per-second data rates enable the seamless transmission of high-resolution video

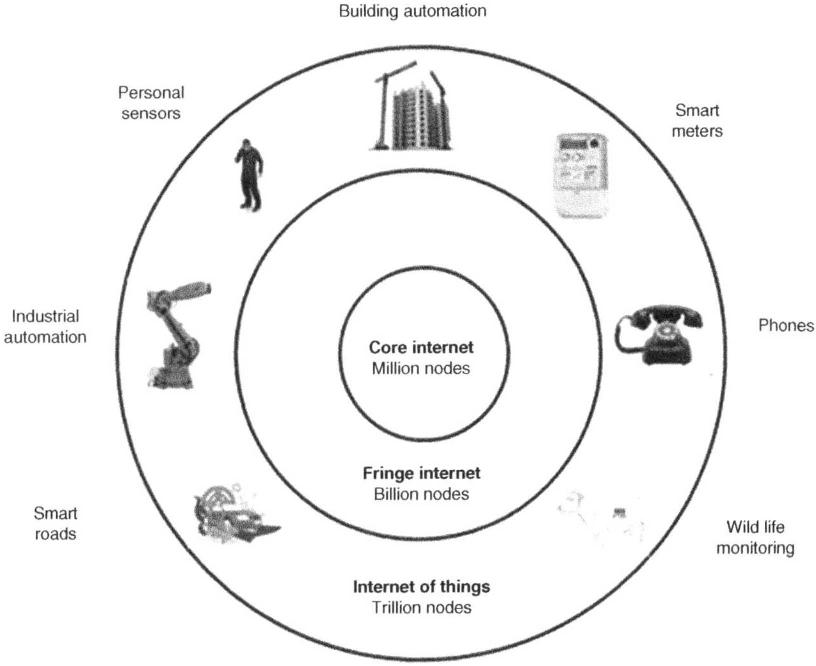

Figure 1.6 Visual Internet of Things (VIoT) architecture. *Source:* rimglow / Adobe Stock Images.

feeds, which is crucial for applications like real-time surveillance and autonomous vehicles.

1) **On-Device AI for Edge-Based Visual Analytics:** By embedding AI capabilities directly into IoT devices, VIoT systems can process visual data locally, reducing latency and bandwidth usage.
2) **Enhanced Augmented Reality (AR) and Virtual Reality (VR) Integration:** 6G's capabilities support immersive AR and VR experiences, which can be utilized in sectors like education, healthcare, and entertainment. Example Applications:

   Smart Surveillance Systems: VIoT enables real-time monitoring and analysis of video feeds from security cameras, enhancing public safety and enabling rapid response to incidents.

   Autonomous Vehicles: Vehicles equipped with VIoT capabilities can process visual data from cameras and sensors to navigate and respond to their environment autonomously.

   Remote Healthcare Diagnostics: Medical professionals can utilize VIoT systems to analyze live video feeds from patients, facilitating remote consultations and diagnostics.

1.3.2 Social Internet of Things (SIoT)

SIoT introduces a paradigm in which IoT devices can establish social relationships autonomously, based on predefined rules and shared interests (Figure 1.7). With 6G's enhanced connectivity, devices can form complex, dynamic networks that mimic human social structures, improving service discovery, trust, and collaboration.

Key Features Enabled by 6G:

1) **Scalable Device-to-Device Communication:** 6G's advanced networking capabilities allow for efficient direct communication between devices, facilitating the formation of decentralized networks.
2) **Secure and Trusted Interactions Among Heterogeneous Devices:** Enhanced security protocols in 6G ensure that devices can interact safely, even when they originate from different manufacturers or ecosystems.
3) **Context-aware Service Personalization:** SIoT systems can leverage contextual information to offer personalized services, improving user experience and satisfaction.

Example applications:

1) **Smart Homes:** Devices within a smart home can communicate and collaborate to optimize energy usage, enhance security, and provide personalized experiences for residents which is briefed by Malik et al. [17].

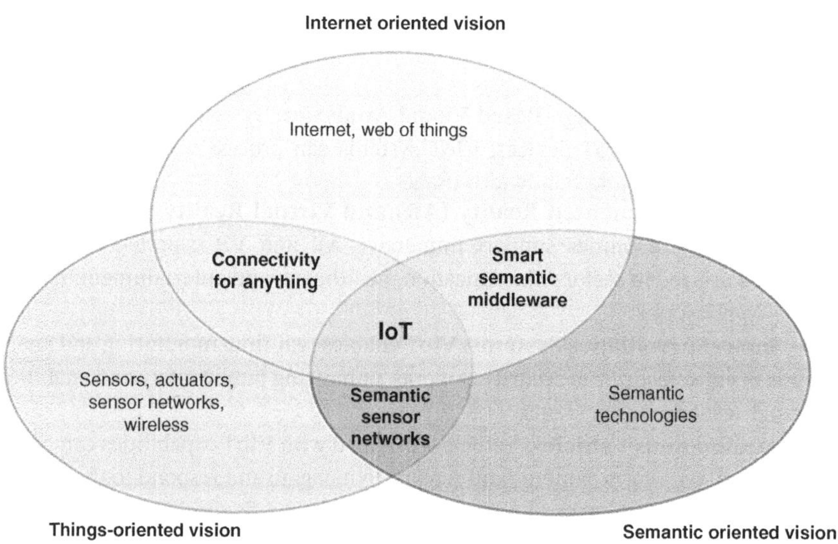

Figure 1.7 Social Internet of Things (SIoT) framework.

2) **Collaborative Robotics:** In industrial settings, robots equipped with SIoT capabilities can share information and coordinate tasks, improving efficiency and flexibility.
3) **Healthcare Networks:** Medical devices can form networks to share patient data and collaborate on diagnostics, leading to more accurate and timely medical interventions.

1.3.3 Artificial Intelligence of Things (AIoT)

AIoT represents the fusion of AI and IoT, where intelligent analytics and decision-making are embedded into IoT systems (Figure 1.8). 6G supports this evolution by enabling faster, more efficient processing of massive data streams through edge computing and AI model distribution across the network. Key Advantages with 6G:

1) **Real-Time Decision-Making at the Edge:** AIoT systems can process data locally, allowing for immediate responses to changing conditions without relying on centralized cloud services.
2) **Improved Network Efficiency Through AI-Driven Optimization:** AI algorithms can optimize network resources and traffic management, enhancing overall system performance.
3) **Autonomous Systems with Self-Learning Capabilities:** AIoT devices can learn from their environment and experiences, adapting their behavior to improve performance over time.

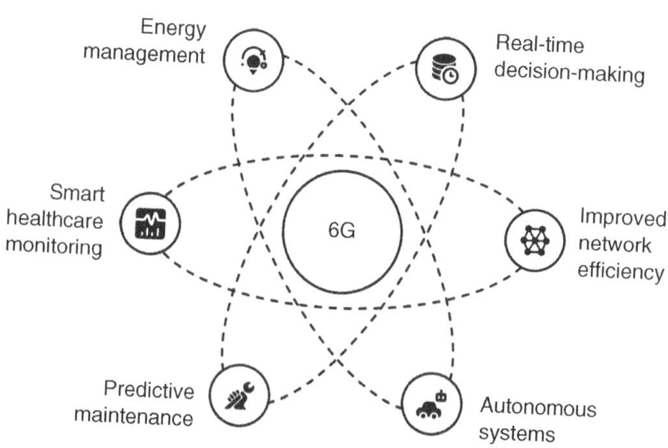

Figure 1.8 AIoT model with 6G cellular technologies.

Example applications:

1) **Predictive Maintenance in Manufacturing:** AIoT systems can analyze data from machinery to predict failures before they occur, reducing downtime and maintenance costs.
2) **Smart Healthcare Monitoring:** Wearable devices can monitor patient vitals and use AI to detect anomalies, alerting healthcare providers to potential issues.
3) **Energy Management in Smart Grids:** AIoT systems can analyze energy consumption patterns and optimize distribution, leading to more efficient energy use.

1.3.4 The Synergistic Ecosystem: See, Interact, Think

The convergence of VIoT, SIoT, and AIoT in 6G networks represents a critical step toward realizing fully autonomous, intelligent, and context-aware IoT ecosystems. Each of these paradigms contributes unique capabilities—visual perception, social collaboration, and intelligent decision-making—which, when combined and supported by 6G infrastructure, create a rich and dynamic cyber-physical world.

At the core of this integration is the transformation of IoT devices from simple data collectors into intelligent agents that can see (VIoT), interact socially (SIoT), and reason autonomously (AIoT). Supported by 6G's high-speed, low-latency communication and distributed intelligence, this ecosystem becomes far greater than the sum of its parts. The synergy gives rise to certain challenges, and thereby relevant mitigations are tabulated in Table 1.1. Key synergies enabled by this integration of 6G and IoT are (1) enhanced situational awareness, (2) collaborative intelligence, and (3) autonomous operations.

Table 1.1 Challenges and Mitigations

Challenge	Description	Potential Mitigation
Data overload	Massive streams of video and sensor data	Edge filtering, compression, and AI summarization
Interoperability	Devices from diverse manufacturers and platforms	Standardized protocols and interfaces
Privacy and ethics	Use of visual and social data for decision-making	Differential privacy, explainable AI, and secure access control
Energy consumption	High-performance processing demands	Green AI, energy-efficient hardware, and optimized algorithms

1.4 Research Advances and Applications

Telecommunications and computer science are fields in which research on the development of the 6G wireless technology and its integration with the IoT are prominent topics of investigation. In the discussion below, there is an analysis of the main developments and research agendas in these areas, with much attention paid to the intersections.

1.4.1 6G and IoT Research Trends

Semantic and goal-oriented communication in 6G and IoT research focuses on transmitting only relevant and meaningful data, which enhances efficiency and reduces bandwidth use—ideal for resource-constrained IoT devices. By prioritizing essential information, these systems become more energy-efficient, responsive, and reliable, especially in real-time applications like smart cities and industrial automation. This approach minimizes unnecessary data transmission, conserves network resources, and supports more effective decision-making. Additionally, since the transmitted data is filtered, labeled, and contextualized, it becomes easier to analyze using AI and machine learning, enabling smarter downstream processing. As a result, IoT ecosystems are simplified and can derive deeper insights from collected data. This method not only improves overall performance but also boosts scalability, allowing networks to support a growing number of devices without sacrificing efficiency. By focusing on critical, goal-driven data, networks can ensure timely delivery and optimal functioning even under high traffic conditions.

1.4.2 Application Areas

1.4.2.1 Blockchain for IoT Security

Blockchain technology presents a promising solution to enhance the security of IoT systems due to its decentralized architecture and robust access control. By utilizing distributed ledger technology, blockchain offers a tamper-proof and transparent platform for monitoring IoT devices, reducing single points of failure and large-scale attack risks [18]. Its cryptographic features ensure secure communication and transactions. Blockchain also strengthens trust management in large IoT ecosystems through consensus mechanisms and smart contracts, enabling automated and reliable interactions [19]. Reputational systems based on blockchain can evaluate device behavior and credibility. Additionally, blockchain's distributed nature supports scalability, addressing IoT's big data storage and processing needs. This integration has transformative potential across sectors like supply chains, smart cities, and healthcare. However, challenges such as high energy consumption, slow transaction speed, and lack of interoperability between blockchain platforms must be addressed to fully harness its advantages in IoT environments.

1.4.2.2 Integrated Sensing and Communication (ISAC)

Integrated Sensing and Communication (ISAC) is a significant improvement of the wireless technology because sensing and data-transfer features are combined into a single system. This integration helps improve the resource utilization that will also simplify hardware and improve the overall system performance. By using the same signal for sensing and communication, the ISAC will allow the devices to sense their environment while simultaneously sending data, meaning context-aware and ultrareliable systems can be achieved. For autonomous systems, like self-driving cars, ISAC can do real-time environment awareness and communications between vehicles and infrastructure. In the healthcare industry, ISAC technology can be used in wearable gadgets that can be used to monitor vital signs and relay the data to the healthcare providers with ease. Moreover, ISAC has potential uses in smart home appliances, industrial automation, and military activity, where there is a need for a sense of the situation and safe communication. Given that this technology will continue to advance, it is expected to contribute significantly to the creation of intelligent, effective, and responsive wireless systems in the different sections of society.

1.4.2.3 Human-Centric and Bio-IoT

Human-centric and Bio-IoT technologies are transforming healthcare by enabling continuous monitoring of patients through wearable and implantable sensors. These systems provide real-time physiological data, allowing early detection of health risks and personalized treatment based on individual needs and genetics. Applications range from remote monitoring to precision medicine, improving patient outcomes and reducing healthcare costs. However, ethical concerns such as data privacy, informed consent, and patient autonomy must be addressed. To ensure responsible implementation, strong regulatory frameworks and ethical guidelines are essential. This integration also supports advanced decision-making tools, enhancing diagnosis and treatment accuracy for healthcare professionals.

1.4.2.4 Sustainability and Green 6G/IoT

Sustainability and green technologies are becoming central to 6G and IoT development, as emphasized by Dang et al. [20]. Researchers are focusing on energy-efficient protocols that reduce power usage through intelligent data transmission, sleep-mode activation, and recyclable hardware using biodegradable materials. Efforts also include low-carbon designs that consider the full product lifecycle, from manufacturing to disposal. Eco-friendly network architectures aim to reduce reliance on energy-intensive data centers by leveraging edge computing and local processing. Additionally, integrating renewable energy sources like solar and wind and exploring energy harvesting from radio waves or thermal gradients are key to building self-sustaining systems. These initiatives are critical for lowering the environmental impact of digital technologies and supporting a sustainable, low-carbon 6G/IoT ecosystem.

1.5 Conclusion

With the advent of 6G networks, whose blazing rates, ultralow latency, overwhelming connectivity, and native AI, the IoT paradigm will undergo fundamental shifts—from the current network of connected things to a cognitively aware, intelligent, and autonomous system of systems. In the near future, IoT devices will move on from being passive data transmitters and become active cognitive agents in their own right, capable of sensing their environment, learning from it, and making intelligent decisions in the real time–space continuum. The processing power of 6G networks at the edge of processing huge amounts of data, adapting to changing conditions, and providing ultrareliable communication will fuel the next generations of IoT applications—ranging from autonomous vehicles, remote surgical systems, and smart agriculture to immersive digital twins and human–robot collaboration in Industry 5.0. However, this transformative vision brings forth huge challenges. Matter of security, energy efficiency, standardization, and data governance have to be approached in a proactive manner. The scale, complexity, and societal implications of the 6G-enabled IoT systems require interdisciplinary work with engineers, computer scientists, policymakers, and industry executives. Ethical issues such as algorithmic transparency, digital inclusion, and sustainability should be incorporated into the construction and implementation of future systems.

As discussed in this chapter, the path to this intelligent, connected future will be driven by constant innovation, smart regulation, and a collaborative desire to ensure that the 6G and IoT are available from all corners of the society. In this new age of the connected intelligence, the lines which separate the physical and digital will become diffuse, and new opportunities, which are still in their infancy, will be born. Combined, 6G and IoT create the building blocks of a future where machines, humans, and environments communicate immediately with each other through lightning-fast data transfer, insignificant latency, and smart automation. On trying to understand the fundamentals of 6G and IoT, we can discuss how the potent mixture of the two is going to transform industries, societies, and the concept of how we live.

References

1 Buyya, R. and Dastjerdi, A.V. (2016). *Internet of Things: Principles and Paradigms*. Morgan Kaufmann.

2 Polymeni, S., Plastras, S., Skoutas, D.N. et al. (2023). The impact of 6G-IoT technologies on the development of agriculture 5.0: a review. *Electronics* 12 (12): 2651. https://doi.org/10.3390/electronics12122651.

3 Tong, W. and Zhu, P. (2021). *6G: The Next Horizon – From Connected People and Things to Connected Intelligence*. Springer.

4 Javeed, D., Saeed, M.S., Ahmad, I. et al. (2024). Quantum-empowered federated learning and 6G wireless networks for IoT security: concept, challenges and future directions. *Future Generation Computer Systems* 160: 577–597. https://doi.org/10.1016/j.future.2024.06.023.

5 Qadir, Z., Le, K.N., Saeed, N., and Munawar, H.S. (2023). Towards 6G internet of things: recent advances, use cases, and open challenges. *ICT Express* 9 (3): 296–312. https://doi.org/10.1016/j.icte.2022.06.006.

6 Lin, X. and Widmer, A.V. (2022). *Fundamentals of 6G Communications and Networking*. Wiley.

7 Hersent, O., Boswarthick, D., and Elloumi, O. (2016). *The Internet of Things: Key Applications and Protocols*, 2e. Wiley.

8 Jahid, A., Alsharif, M. H., & Hall, T. J. (2021). *The convergence of blockchain, IoT and 6G: Potential, opportunities, challenges and research roadmap* (arXiv:2109.03184). arXiv. https://arxiv.org/abs/2109.03184

9 Khan, L.U., Yaqoob, I., Imran, M. et al. (2020). 6G wireless systems: a vision, architectural elements, and future directions. *IEEE Access* 8: 147029–147044. https://doi.org/10.1109/ACCESS.2020.3015289.

10 Kim, J.H. (2021). 6G and internet of things: a survey. *Journal of Management Analytics* 8 (2): 316–332. https://doi.org/10.1080/23270012.2021.1882350.

11 Lu, Y. and Zheng, X. (2020). 6G: a survey on technologies, scenarios, challenges, and the related issues. *Journal of Industrial Information Integration* 19: 100158. https://doi.org/10.1016/j.jii.2020.100158.

12 Zheng, K., Xu, R., Mei, J. et al. (2024). *Ambient IoT towards 6G: Standardization, potentials, and challenges. IEEE Access* 12: 146668–146677. https://doi.org/10.1109/ACCESS.2024.3474298.

13 Khalid, W., Rehman, M.A.U., Van Chien, T. et al. (2024). Reconfigurable intelligent surface for physical layer security in 6G-IoT: designs, issues, and advances. *IEEE Internet of Things Journal* 11 (2): 3599–3613. https://doi.org/10.1109/JIOT.2023.3297241.

14 Ndiaye, M., Niane, K., Saley, A.M., and Raimy, A. (2022). Future 6G communication networks: typical IoT network topology and terahertz frequency

challenges and research issues. 2022 *International Conference on Intelligent Robots and Systems and Emerging Technologies (IRASET)*, pp. 1–5. https://doi.org/10.1109/iraset52964.2022.9744327.

15 Ahad, A. and Tahir, M. (2023). Perspective—6G and IoT for intelligent healthcare: challenges and future research directions. *ECS Sensors Plus* 2 (1): 011601. https://doi.org/10.1149/2754-2726/acabd4.

16 Sicari, S., Rizzardi, A., Grieco, L.A., and Coen-Porisini, A. (2015). Security, privacy and trust in internet of things: the road ahead. *Computer Networks* 76: 146–164. https://doi.org/10.1016/j.comnet.2014.11.008.

17 Malik, U.M., Javed, M.A., Islam, S.U., and Zeadally, S. (2021). Energy-efficient fog computing for 6G-enabled massive IoT: recent trends and future opportunities. *IEEE Internet of Things Journal* 9 (16): 14572–14594. https://doi.org/10.1109/jiot.2021.3068056.

18 Son, B.D., Hoa, N.T., Van Chien, T. et al. (2024). Adversarial attacks and defenses in 6G network-assisted IoT systems. *IEEE Internet of Things Journal* 11: 19168–19187.

19 Ansari, S., Nyamasvisva, T.E., and Ansari, A. (2024). 6G vision, IoT technologies, challenges and emerging technology solutions—a survey. *Journal of Independent Studies and Research Computing* 22 (1): https://doi.org/10.31645/jisrc.24.22.1.5.

20 Dang, S., Amin, O., Shihada, B., and Alouini, M.-S. (2019). What should 6G be? *IEEE Access* 7: 99729–99747. https://doi.org/10.1109/ACCESS.2019.2920934.

2

Foundations of 6G and IoT: Technologies, Challenges, and Future Directions

B. Navaneethi[1], *A. Priyadharshini*[1], *and Saju Mathew*[2]

[1] *Department of Computer Application, SNMV College of Arts and Science, Coimbatore, Tamil Nadu, India*
[2] *Department of Computer Science with Data Analytics, SNMV College of Arts and Science, Coimbatore, Tamil Nadu, India*

2.1 Introduction

The revolution motivated by the combination of fifth-generation (5G) network technologies and Internet of Things (IoT) applications, sixth generation (6G) is anticipated to perform better than its predecessor in numerous aspects, enhancing both business efficiency and our everyday lives. From the introduction of first generation (1G) in the 1980s to the expected completion of 6G in the 2030s, wireless communications have evolved over time. 6G is expected to offer every feature of earlier network generations in addition to a few new ones, including complete coverage of the network, the creation of extensive IoT and mobile applications powered by artificial intelligence (AI), enhanced satellite communication, and the development of fully self-sufficient systems. Every ten years, mobile communication systems are often upgraded to a more sophisticated generation. As is customary, mobile communication networks, which have existed since the 1980s, transitioned to 5G in 2020. Many people consider 5G to be the apex of mobile communication technology. The swift development of mobile communication technologies has completely changed how enterprises and society run. From 1G's analog voice services to 5G's improved broadband experience, each generation has brought new advancements in connectivity, speed, and latency. However, the need for a more powerful networking architecture is indicated by the rise of immersive experiences, data-intensive applications, and hyper-connected intelligent systems 6G. In parallel, real-time, dependable, and scalable communication

Synergies of 6G and IoT, First Edition. Edited by Himanshi Babbar.

Published 2026 by John Wiley & Sons, Inc.

frameworks are required as the IoT continues to transform industry sectors, including manufacturing, healthcare, and transportation. In order to support the next generation of digital ecosystems, this study explores the synergistic evolution of 6G and IoT, looking at important enabling technologies, obstacles, and future research prospects. It is well known that 5G and its predecessor, fourth generation (4G), sometimes referred to as LTE-advanced, will create an application-oriented ecosystem and intelligent services enabled by the IoT. More importantly, 5G has three features that were specifically designed to get beyond the drawbacks of 4G enhanced mobile broadband (eMBB), ultrareliable low-latency communications (uRLLC), and massive machine-type communications (mMTC). 5G networks, for example, are expected to provide a 20 Gbps peak data rate, 100 times more energy efficiency, three times spectral efficiency, and a Gbps user experience with a 1 ms end-to-end latency compared to 4G networks. Additionally, 5G would enable smooth connectivity for devices with 500 km/h mobility, 1 million devices/km connection density, and 10 Mbps/m^2 area traffic capacity. Despite expectations that 5G networks will enable a wide range of intelligent Internet of Everything (IoE)-related services, they won't be sufficient to meet the needs of smart communities in the future. As smart cities enable the digital layer on top of the existing infrastructure, they automate our surroundings, causing a dramatic increase in stakeholder demand. As a result, it is becoming increasingly crucial to manage this digitalization properly in order to provide a pervasive solution for smart cities, disaster management, and other services.

2.2 Evolution of Mobile Networks (1G to 6G)

To address the intricate challenges and transformative requirements of today's age, 6G wireless networking is being developed as an all-encompassing framework that surpasses the abilities of earlier generations. Its goal is to combine both current and new technologies throughout all levels of communication infrastructure—from the physical and radio access layers to the application and service levels. The outcome will be a completely digital, smart, and integrated interface that effortlessly links users, services, devices, sensors, autonomous machines, and smart environments. This comprehensive fusion will establish the foundation of a period characterized by immediate, engaging, and context-sensitive interactions.

New opportunities are opening up in crucial areas like mMTC, uRLLC, eMBB, and wide geographic coverage, including underserved and remote areas, as society shifts toward ubiquitous computing and pervasive connectivity. These characteristics are essential for the development of cyber-physical systems, immersive virtual worlds, autonomous systems, and next-generation IoT applications.

To completely comprehend the transformative possibilities of 6G, it is essential to recognize the evolutionary path of mobile communication technologies—from the analog voice systems of 1G to the smart, data-centric networks of 5G. Every generation has arisen in reaction to evolving societal needs and technological possibilities. While 1G concentrated on facilitating mobile voice communication, second generation (2G) brought in digital transmission and messaging features. Third generation (3G) introduced mobile internet connectivity, 4G facilitated fast broadband and multimedia services, and 5G established the groundwork for real-time machine and device communication due to its low-latency, high-bandwidth architecture.

Despite the substantial progress of 5G, new applications and services are testing the capabilities of existing networks. The growing need for faster data rates, reduced latency, increased device density, improved security, and context-aware intelligence is propelling the creation of a new generation of networks. 6G goes beyond mere speed—it signifies a transition to intelligent, adaptable, and self-sufficient communication systems that can learn from, respond to, and anticipate user and environmental actions.

The journey to 6G is influenced by multiple intersecting trends: the emergence of AI-powered systems, the investigation of terahertz (THz) frequency ranges, the incorporation of non-terrestrial networks (NTNs) (such as satellites and drones), and the emphasis on sustainability and digital equality. Collectively, these developments showcase a perspective of a world where connectivity is smart, worldwide, robust, and eco-friendly—facilitating innovative services in education, healthcare, transportation, manufacturing, and more.

Progression of Wireless Generations

- 1G brought analog voice communication in the 1980s, enabling mobile voice calls but falling short in data functionality and security.
- 2G transitioned to digital communication, facilitating SMS, improved call quality, and restricted data services.
- 3G introduced mobile internet connectivity and multimedia capabilities, facilitating video calls and browsing the mobile web.
- 4G provided fast internet, enabling high-definition (HD) video streaming, mobile gaming, and cloud computing.
- 5G, the existing standard, allows for extremely high data speeds, extensive device connectivity, and near-instant latency, rendering technologies such as smart cities, IoT, and remote surgery feasible.

Figure 2.1 shows the evolution of mobile networks.

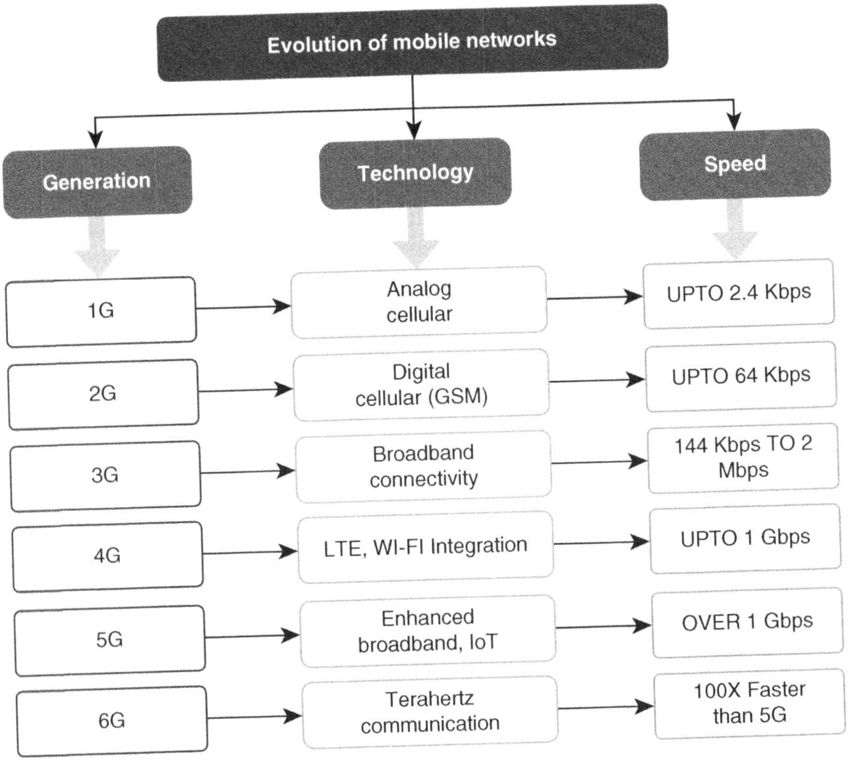

Figure 2.1 Evolution of mobile networks.

2.2.1 What Will 6G Bring?

Looking forward, 6G aims to deliver data rates up to 1 Tbps, with sub-millisecond latency. Beyond speed, however, its true transformation lies in its integration of AI, new spectrum (THz bands), and globally connected systems, including satellites, drones, and underwater networks. Some of the key innovations expected include:

- **AI-Driven Networks:** Intelligent, adaptive systems that optimize themselves in real time for speed, reliability, and energy efficiency.
- **Holographic and XR Communication:** Real-time 3D and holographic interactions for remote education, work, and entertainment.
- **Global and Seamless Coverage:** Connectivity extended to rural, remote, and even oceanic regions using satellite and aerial infrastructure.
- **Sustainability Focus:** Eco-friendly network designs to reduce power consumption and carbon emissions despite increased usage.

The vision for 6G includes much more than just technological advancements—it signifies a crucial transformation in our engagement with the digital and physical environments. By incorporating AI, enhancing communication, providing global connectivity, and ensuring sustainable practices, 6G will transform industries, communities, and everyday human interactions. The innovations described here are merely the starting point, with additional breakthroughs anticipated as research progresses and real-world trials commence. Like any generational transition, the advancement of 6G will necessitate international collaboration, ethical anticipation, and a dedication to inclusiveness and strength.

Figure 2.2 shows the 6G-based applications.

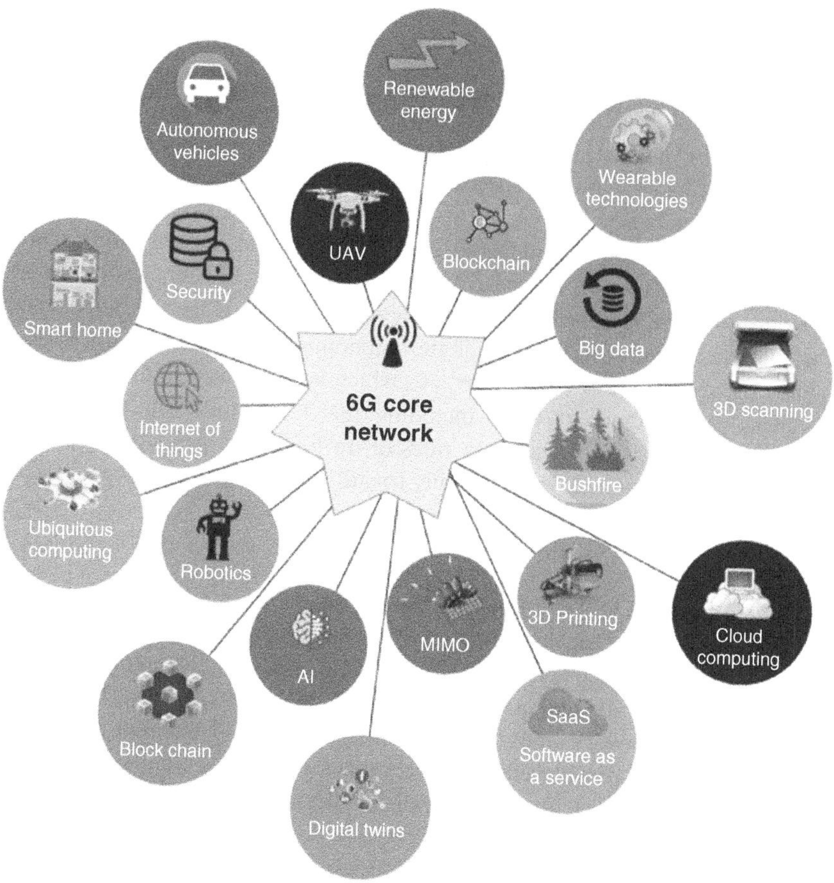

Figure 2.2 6G-based applications. *Source:* Zakria Qadir/with permission of Elsevier/ https://www.sciencedirect.com/science/article/pii/S2405959522000959#fig1 /CC BY-NC-ND 4.0.

2.3 6G Network Architecture for IoT

The sixth generation of the Internet of Things (6G-IoT) is anticipated to address critical issues related to highly reliable and ultralow-latency communication, efficient use of communication protocols, and extensive geographic coverage of IoT networks. Furthermore, as AI develops, gadgets get more complex and, when combined with their expanded computer power, may now actively engage with the network rather than just sending data. The proposed architecture combines cloud-native and AI-driven design principles with essential technologies. Ultradense tiny cells, reconfigurable intelligent surfaces (RIS)-enabled settings, satellite–terrestrial integration, and software-defined networking (SDN) enabling programmability and agility are some of the salient features.

2.3.1 6G-Enabled IoT: Vision, Techniques, and Objectives of 6G

A comprehensive survey discusses the transition from 5G to 6G, highlighting the requirement for faster internet speeds, reliability, and energy efficiency. It outlines enabling technologies such as THz communications, edge intelligence, and blockchain integration and addresses open issues like data security and network scalability. The 6G concept goes beyond small-scale gains in latency and throughput rates. It seeks to develop a completely immersive, intelligent, and pervasive digital infrastructure that can facilitate autonomous systems, human–machine symbiosis, and real-time holography. 6G, which is anticipated to function in frequencies higher than 100 GHz, including the THz band, will provide sub-millisecond latency and peak data rates of up to 1 Tbps. Among its goals are smooth AI integration, environmentally friendly energy use, high dependability, and support for worldwide, ubiquitous coverage via NTNs.

2.3.1.1 Toward 6G IoT: Advances and Challenges

This paper examines the challenges of 5G in supporting the IoE and the role of 6G in overcoming these challenges. It highlights emerging technologies like holographic beamforming and quantum communications and discusses the integration of AI for network optimization. Thus, it is anticipated that the 6G-IoT framework would provide effective techniques for remote applications, smart cities, and industrial IoT, all of which will significantly impact the next generation of smart agriculture. The key technologies that are needed to enable Agriculture 5.0 and meet 6G-IoT criteria. The computing power of digital network devices, including those at the network's edge, is getting more and more advanced. As a result, we may use their processing capacity to help with and, in certain cases, improve network performance. Furthermore, as AI advances, these gadgets will be able to detect and classify services using learning models by drawing knowledge and experience from the data they exchange with other networked entities.

2.3.1.2 Open Challenges and Future Directions

a) **Security and Privacy:** Ensuring robust protection mechanisms against cyber threats.
b) **Network Scalability:** Managing the growing number of connected devices and data traffic.
c) **Integration of Emerging Technologies:** Seamlessly incorporating AI, quantum computing, and blockchain into network architectures.
d) **Standardization and Regulatory Concerns:** Establishing global standards and policies for 6G deployment.
e) The integration of 6G and the IoT is set to reshape technological landscapes, driving advancements across various sectors. Here's an overview of their significance.

2.3.1.3 Key Advantages of 6G and IoT Integration

The integration of 6G technology with the IoT presents numerous transformative advantages across industries, services, and daily life. Some of the key benefits include:

a) **Ultralow-Latency and Real-Time Communication:** It is anticipated that 6G will enable nearly instantaneous communication between IoT devices by lowering latency to less than 1 ms. For time-sensitive applications like remote surgery, autonomous driving, industrial automation, and mission-critical systems, this is essential.
b) **Massive Device Connectivity:** With up to 10 million devices per square kilometer, 6G networks are expected to significantly outperform 5G. In smart cities, factories, and urban settings, this will enable dense IoT deployments, guaranteeing smooth communication for billions of devices.
c) **High Data Rates and Bandwidth:** HD video streaming, real-time data processing, and sophisticated augmented/virtual reality (AR/VR) experiences via IoT devices will all be made possible by 6G's much increased bandwidth, which is expected to reach up to 1 Tbps.
d) **Advanced AI and Edge Computing Support:** At the network edge, 6G is anticipated to natively support AI and machine learning. This will enable IoT devices to process data locally with less dependence on cloud infrastructure, which will increase data privacy, speed up decision-making, and use less bandwidth.

e) **Enhanced Energy Efficiency:** IoT devices frequently run on batteries. More energy efficiency is the goal of 6G networks, which will prolong battery life and allow IoT networks to run sustainably for extended periods of time, particularly in rural or difficult-to-reach areas.

f) **Improved Reliability and Resilience:** Even under dynamic or unfavorable circumstances, 6G is extremely dependable due to its improved network resilience and self-healing features. Critical applications like IoT systems for public safety, healthcare, and disaster management require this.

g) **Global Coverage with 3D Connectivity:** In order to provide seamless worldwide coverage, 6G will integrate satellite, aerial, and terrestrial components. Applications like agriculture, animal monitoring, and logistics will be supported by this 3D network design, which will allow IoT connectivity even in isolated, rural, or underdeveloped locations.

h) **Hyper-personalization and Context Awareness:** 6G networks will make it possible to provide hyper-personalized services that are adapted to user behavior and environmental context by utilizing AI and real-time data from IoT devices. This will improve the user experience in wearable technology, smart homes, healthcare, and other areas.

i) **Support for New Use Cases:** Advanced applications like brain–computer interfaces, tactile internet, holographic communication, and completely autonomous systems will become possible with 6G–IoT integration. Future interactions between humans and robots will be redefined by these new use cases.

2.3.2 Synergy of 6G and IoT Networks

Concentrating on transformative implementations, this research explores how 6G technologies are set to revolutionize crucial sectors such as healthcare, autonomous transportation, and the Industrial Internet of Things (IIoT). It emphasizes how 6G's fundamental features—like ultralow latency, exceptional reliability, and extensive device connectivity—can facilitate mission-critical applications including remote surgery, real-time vehicle coordination, smart manufacturing, and automation. In examining the facilitating technologies driving these advancements, the research evaluates the functions of RIS for adaptable signal management, edge computing for local processing, and AI-native networking for optimizing autonomous systems. Additionally, the conversation focuses on the real-world obstacles to execution, including spectrum limitations, infrastructure intricacies, energy effectiveness, and security issues. This research provides a comprehensive perspective on the challenges and innovations influencing 6G implementation in practical situations by examining both the opportunities and the constraints.

Figure 2.3 shows the emerging applications and use cases of IOT.

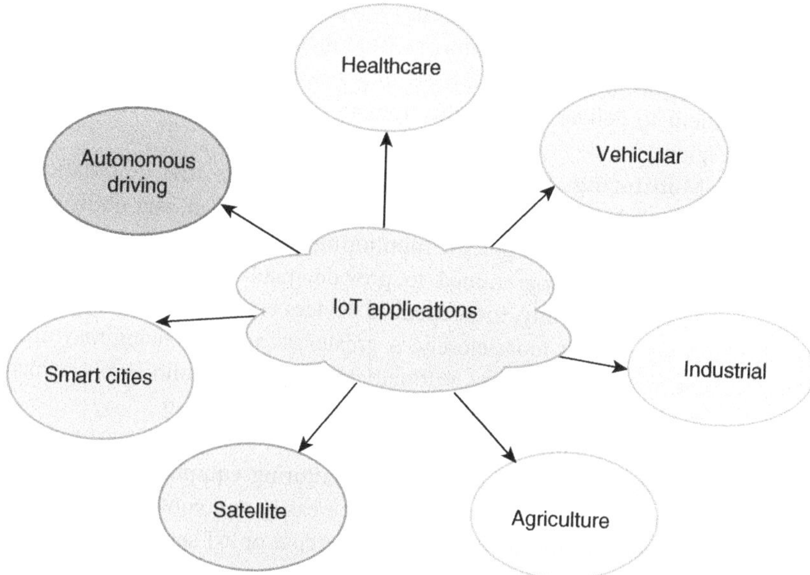

Figure 2.3 IoT applications.

a) **Healthcare IoT:** The collaboration of 6G and IoT will facilitate remote patient monitoring, telemedicine, and even remote surgeries, making healthcare more accessible and efficient. Enabling real-time monitoring and personalized treatment through ultrareliable communication. Wearables and even implantable medical gadgets that can assist with real-time health monitoring are examples of the IoT, which is basically a network of physical devices connected via the internet for medical purposes. Radio frequency identification, or RFID, is frequently used by these devices to enable automatic tracking and identification. Data is gathered, shared, and ultimately analyzed by the linked devices. As a result, they build a link between the digital and real worlds. Real-time monitoring, customized treatment plans, and enhanced medical decision-making have completely transformed patient care, which eventually improves health outcomes and streamlines healthcare delivery. With their incredibly varied applications, IoT devices have the potential to become the new norm in healthcare.
b) **Remote Patient Monitoring (RPM):** Health metrics are automatically gathered by wearable technology, including smartwatches, phones, and wearable sensors. These remote patient monitoring devices gather medical data, like blood pressure, body temperature, body fat percentage, and more, and use that data to prescribe medical interventions or send out alarms. A stay

at the medical facility would be necessary to collect data for many conditions that call for ongoing health monitoring. RPM mainly resolves this problem by giving healthcare professionals access to a steady flow of information that enables them to deliver high-quality remote patient care to even the most remote regions.

c) **Glucose Monitoring**: Wearable or embedded sensors that automatically check a patient's blood sugar levels, record the information, and notify medical experts are examples of glucose monitoring systems. Certain implantable devices can even be programmed to provide insulin automatically when required. More importantly, these medical devices enable patients to monitor their blood sugar levels more closely. A greater standard of living may arise from certain gadgets' ability to automatically control insulin levels, which together result in more accurate patient care and monitoring.

d) **Heart-Rate Monitoring:** IoT devices that monitor a patient's heart rate are already portable, in contrast to glucose monitoring equipment. A person's heart rate or other vital signs can already be easily and continuously monitored by a number of gadgets, such as smartwatches or IoT sensors. As a result, information can be gathered even when the patient is asleep. The patient's electronic medical records may then contain that information. Since heart-rate variability is influenced by a variety of activities throughout the day, it is more difficult to monitor than glucose. Patients can avoid hospital visits while still having their condition monitored, thanks to continuous cardiac monitoring. These gadgets can also act as emergency response systems, warning medical personnel if a patient's heart rate deviates from normal.

e) **Hand Hygiene Monitoring**: In order to maintain appropriate cleanliness standards, a lot of hospitals and other healthcare institutions have recently started utilizing hygiene monitoring IoT devices. These medical devices are used to remind people to wash their hands frequently or whenever they enter a certain room. Critically sick patients are frequently housed in numerous hospital rooms. Any pathogen introduction could make these patients' symptoms worse. Any hospital's operations require appropriate decontamination protocols, even when dealing with comparatively healthy patients. In addition to reminding individuals to wash their hands, IoT health gadgets can also provide guidance on how to sanitize objects or clean hands. Infection rates are thought to drop by 60% in areas where these devices are used.

f) **Depression or Mood Monitoring:** IoT medical equipment can determine a patient's present emotional state and mental health by tracking biometric data such as blood pressure, heart rate, and other parameters. Even a person's eye movements can be tracked by more sophisticated IoT devices. It has historically been difficult to continuously evaluate patients' mental health. To make a diagnosis, doctors must periodically inquire about the patient's emotional state. However, the experts are unable to forecast patients' mood swings, and individuals may choose to withhold certain information from their medical records or decline to provide the professional with their medical history. The main advantage of IoT devices is that they use objective biometrics instead of subjective testimonies. These health indicators provide a reliable picture of an individual's mental health and are not under the patient's control. Additionally, patients may still be monitored while they sleep.

g) **Parkinson's Disease Monitoring:** The treatment of Parkinson's disease patients is one unexpected application of IoT technology in healthcare. These medical gadgets are able to monitor Parkinson's disease symptoms and how their intensity changes throughout the day. Parkinson's disease symptoms are erratic; thus, they must be continuously monitored. Hospitals typically require patients to remain in their care for an extended period of time in order to properly monitor them. IoT devices greatly simplify this process by automatically gathering patient data. This remote monitoring device improves patient outcomes in a number of ways. In addition to lowering healthcare expenses significantly, it also gives patients greater freedom by reducing the length of time they must spend in hospital rooms.

h) **Connected Inhalers**: In essence, these IoT gadgets are smart inhalers that are linked to a mobile application. Their function is to periodically remind users to take their medications by means of visual or auditory cues. They can even provide guidance on how to use these inhalers correctly. The primary problem with diseases like asthma or COPD is that they cause unexpected attacks without warning. Better medicine, timely reviews of care, or improved asthma management in general can save two out of three asthma-related deaths. As the name implies, these medical gadgets are tiny sensors that are swallowed and made to disintegrate or pass through the digestive system. Many of these gadgets are shaped like pills that people have to consume. Their goal is to gather different biometric information, such as the stomach's pH level or whether internal bleeding is present. Patients can select a far less invasive technique by using edible sensors. Because ingestible sensors can disintegrate or transit through a patient's digestive tract on their own, they are far less dirty and disruptive than previous approaches. Because a medical practitioner does not have to physically place the sensor into the digestive system, the procedure runs more smoothly. Recent developments in medical technology have even raised the prospect of creating tiny, readily absorbed tablets.

i) **Ingestible Sensors**: As the name suggests, these medical gadgets are tiny sensors that are designed to dissolve or move through the digestive tract when eaten. Many of these gadgets are shaped like pills that people have to consume. Their goal is to gather different biometric information, such as the stomach's pH level or whether internal bleeding is present. Patients can select a far less invasive technique by using edible sensors. Because ingestible sensors can disintegrate or transit through a patient's digestive tract on their own, they are far less dirty and disruptive than previous approaches. Because a medical practitioner does not have to physically place the sensor into the digestive system, the procedure runs more smoothly. Recent developments in medical technology have even raised the prospect of creating tiny, readily absorbed tablets.

j) **Connected Contact Lenses:** These linked gadgets have a camera and a sensor, unlike regular contact lenses. In addition to improving eyesight, these IoT lenses can detect early signs of illness. The primary problem is that over time, the eyes gradually deteriorate or acquire certain illnesses. These illnesses can occasionally worsen for a long time without the patient seeking treatment. Early indicators of some eye conditions can be detected by connected contact lenses. However, IoT lenses can also identify some general medical issues, like diabetes, early on by measuring the glucose levels in the tears. Last but not least, the technology is being developed by businesses outside the healthcare industry because the cameras in the lenses allow patients to snap images with them.

k) **Robotic Surgery**: IoT-enabled medical devices are a novel method of surgery. Surgeons can perform surgery on humans by injecting tiny equipment, such as Internet-connected robots, into the body. When an operation calls for precision, professionals employ these to minimize human mistakes. The primary problem is that compared to their robotic equivalents, the incisions needed to perform procedures by hand are larger. Complex operations require greater room for human hands. Surgeons can do more accurate and efficient procedures when they use these IoT robotic devices. Because the IoT gadgets are easier to insert, the incisions are smaller. This results in a less invasive process for the patient and faster recuperation durations. The primary prerequisite of these devices is the fact they need to be sufficiently compact and dependable to carry out intricate processes. Additionally, they must be able to precisely assess the inside circumstances of a patient and make an informed surgical decision.

l) **Vehicular IoT**: It facilitates seamless connectivity for intelligent transportation systems. Internet access is quickly becoming a necessary and required function of contemporary automobiles. When it comes to offering internet services to car users, wireless access technologies are crucial. The two most viable options are Wi-Fi and cellular. Cellular networks like 3G and 4G-LTE are capable of offering dependable and ubiquitous access services. When driving, one can utilize the roadside Wi-Fi connection points to browse the Internet outdoors. There are currently two types of options for connecting cars to the Internet using extensively used cellular network infrastructure: built-in and brought-in.

m) **Autonomous Driving**: Unmanned aerial vehicles (UAVs) enhance communication for drone operations in various sectors. Imagination and reality collide in the field of autonomous automobiles. These self-driving wonders, powered by the IoT, use a complex network of interconnected sensors and gadgets to navigate, make judgments, and maintain safety. In addition to advancing the automotive sector into the future, the combination of IoT and autonomous vehicles also changes how we think about transportation.

The following are some of the advanced usages of 6G:

a) **Industrial IoT:** 6G is anticipated to be a key driver of Industry 5.0, the forthcoming stage of industrial evolution that highlights not only automation and efficiency but also human-centeredness, resilience, and sustainability. By integrating ultralow latency, exceptional reliability, extensive device connectivity, and AI-driven insights, 6G will enable highly interconnected, adaptable, and autonomous manufacturing environments. These abilities will propel innovations throughout all levels of the industrial value chain, from factory operations to worldwide supply networks.

   i) **Predictive Maintenance**: A key advantage of 6G in Industrial IoT is the improved predictive maintenance systems. By utilizing AI at the edge, manufacturing systems can handle high-frequency sensor data in real time without depending on centralized cloud computing. This allows for:

      a) Prompt identification of mechanical wear, misalignment, or vibration irregularities.
      b) Continuous health monitoring of machinery and infrastructure.
      c) Reduced downtime and repair expenses by implementing proactive measures.
      d) Extended equipment durability and enhanced operational effectiveness.

   With 6G's assured low-latency and high throughput, even sophisticated pattern recognition algorithms can operate with sub-second response times, providing almost immediate feedback and action.

   ii) **Digital Twins:** 6G will additionally facilitate the creation of high-fidelity digital twins—virtual representations of real-world systems that are perpetually refreshed with real-time data from the physical realm. These virtual settings allow for:

      a) Simulation and testing of manufacturing scenarios in real time.
      b) Improvement of processes, arrangements, and distribution of resources.
      c) Detecting anomalies and analyzing root causes without interrupting production.
      d) Improved collaboration in design and engineering among worldwide teams.

   Leveraging the high bandwidth and instantaneous synchronization features of 6G, digital twins will transform from static models to dynamic, self-refreshing systems that capture every subtle detail of physical processes in real time.

iii) **Human–Robot Collaboration:** A key characteristic of Industry 5.0 is the integration of human and machine intelligence. In factories powered by 6G, collaborative robots (cobots) will operate alongside human workers with unparalleled responsiveness, safety, and flexibility. Significant progress will encompass:
   a) Exact motion synchronization and responsive actions aligned with human intention.
   b) Immediate feedback via haptic interfaces and visualizations based on XR technology.
   c) Enhanced safety features utilizing combined sensing and AI for risk assessment.
   d) Context-sensitive task delegation in which robots provide help with strenuous, monotonous, or dangerous activities.

iv) **Supply Chain Automation**: With 6G, supply chain processes will undergo a significant change, improving visibility, traceability, and automation. The combination of AI, edge computing, and fast communication will enable manufacturers to oversee logistics in real time, monitoring the location of products and detecting environmental conditions such as temperature and humidity. Self-driving cars will enhance transportation efficiency, while intelligent inventory control with RFID tags and sensors will improve stock management and minimize waste.

   6G's uRLLC guarantees that machine actions and decisions are adjustable within microseconds, facilitating natural and secure interactions in changing environments

b) **Agriculture and Precision Farming:** 6G will revolutionize agriculture by enabling a fully connected and intelligent ecosystem of machines, sensors, and drones, fostering sustainable farming practices. Key applications include:
   i) **Soil and Crop Monitoring:** Real-time sensing of soil moisture, nutrient content, and crop health via IoT-enabled sensors for focused response.
   ii) **Weather Prediction and Climate Adaptation**: Improved microclimate data gathering for adaptive farming and hyper-local weather modeling.
   iii) **Autonomous Equipment**: For planting, spraying, and harvesting operations, self-driving tractors and drones communicate over 6G networks.
   iv) **Livestock Monitoring**: Employing wearable technology and implanted sensors to follow an animal's location, health, and behavioral tendencies in real time.

c) **Satellite IoT: Expanding Connectivity to Remote and Underserved Areas**

As a crucial part of 6G's NTNs architecture, satellite IoT is essential to delivering continuous worldwide connectivity. Terrestrial networks have transformed communication in urban and heavily populated areas, but they are unable to provide reliable, high-quality service in mountainous, remote, rural, and maritime locations. By providing dependable data transfer in situations where traditional infrastructure is impractical or financially unviable, satellite IoT helps close this digital divide.

The following are the key advantages of Satellite IoT in a 6G World:

i) **Global Coverage**: Low Earth Orbit (LEO) constellations, such as OneWeb or Starlink from SpaceX, will guarantee worldwide coverage, including seas and Polar Regions.

ii) **Always-On Connectivity:** Offers uninterrupted support for vital systems in situations where outages could be disastrous, including in remote industrial operations or disaster areas.

iii) **Low-Power Operation** Energy-efficient next-generation satellite IoT devices are being developed, making them perfect for usage in battery-powered settings and remote sensor nodes.

iv) **Scalable and Flexible Networks**: Depending on demand, the weather, and mission priorities, satellites can dynamically modify bandwidth and coverage.

2.3.3 Applications and Use Cases

a) **Disaster Response and Emergency Communication:** Terrestrial infrastructure is frequently overloaded or damaged during natural catastrophes. Emergency responders can coordinate relief efforts, stay in touch, and use satellite IoT to launch connected drones for search and rescue missions.

b) **Remote Asset Monitoring:** Real-time monitoring of pipelines, drilling platforms, and environmental sensors in remote areas is essential to sectors, including mining, forestry, and oil and gas. Continuous data flow is guaranteed by satellite IoT for operational efficiency and safety.

c) **Maritime and Aviation Tracking**: Satellite IoT is used for navigation, performance monitoring, weather updates, and emergency signaling by ships and airplanes that are operating distant from ground-based communication networks.

d) **Agricultural and Environmental Monitoring:** In order to assist sustainable practices and conservation efforts, satellite-connected sensors can collect data on soil quality, moisture levels, deforestation, and wildlife activity in remote farms, forests, and natural reserves.

e) **Infrastructure Development:** Without waiting for fiber-optic or cellular buildouts, satellite IoT provides a quick and affordable means to allow fundamental digital services (such as mobile banking, remote education, and telemedicine) in developing nations.

f) **Smart Cities:** 6G will enhance urban living by enabling real-time traffic management, efficient energy use, and improved public safety through interconnected IoT devices.

Figure 2.4 shows the application of IoT in smart cities.

The increase in urbanization is causing cities of all sizes to become densely populated. Municipalities are using cutting-edge city solutions, like the IoT, to effectively handle issues like traffic congestion, waste management,

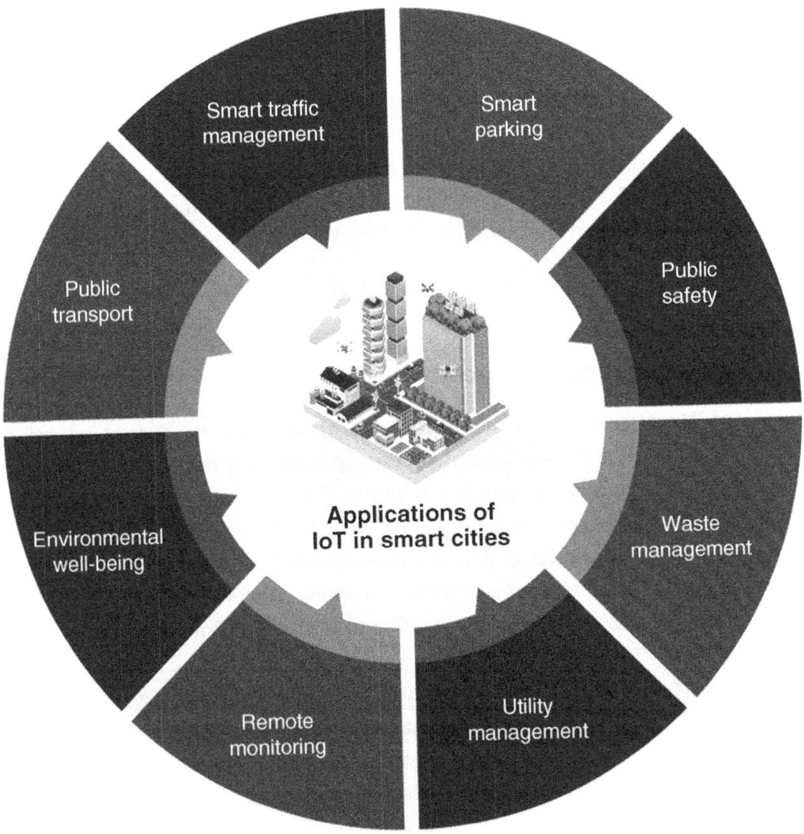

Figure 2.4 Applications of IoT in smart cities. *Source:* Rishabh Software/ https://www.rishabhsoft.com/blog/iot-in-smart-cities-applications-benefits#elementor-toc__heading-anchor-0/.

sanitation, environmental deterioration, and rising energy usage. Urban living is becoming safer and more convenient with the use of IoT-based smart city solutions. It is done while assisting cities in enhancing public utility services and infrastructure. With real-world examples of IoT smart cities, possible advantages, and more, this chapter will examine the possibilities of IoT technology.

g) **Smart Traffic Management**: To gather information on traffic patterns, congestion, and accidents, IoT sensors can be mounted on traffic lights, roads, and automobiles. By using this data, traffic flow may be optimized, congestion can be decreased, and road safety can be enhanced. These systems use sensors and GPS data from the rider's smartphone to report the location and speed of a vehicle. Additionally, historical data can be used to forecast preferred routes and prevent potential congestion problems.

h) **Smart Parking:** IoT sensors can be installed in parking spaces to detect occupancy and transmit that information to a central server. This information can save traffic and search time by guiding vehicles to available parking spots. When a neighboring parking space becomes available, the car is immediately alerted by the sensors buried in the ground, which send the data to the cloud.

i) **Public Safety:** Public areas can be equipped with IoT-enabled cameras and sensors to keep an eye out for possible security risks like unsecured luggage or suspicious activities. Real-time tracking, analytics, and decision-making are all integrated into IoT-enabled solutions. Predicting possible criminal situations is made easier by analyzing data from social media feeds, CCTV cameras, and audio sensors placed across the city. This can help law enforcement officials react to potential threats quickly and efficiently.

j) **Waste Management:** Using IoT-powered technologies, waste collection operators optimize collection schedules and routes while continuously analyzing waste levels, fuel consumption, and container usage. Trash cans and recycling bins can be equipped with IoT sensors to monitor fill levels and optimize waste collection routes, reducing costs and their impact on the environment. Each container has a built-in sensor to monitor the quantity of trash. When a container reaches the threshold level, the truck driver is instantly alerted using a mobile app, reminding him to dump the full container instead of the partially filled one.

k) **Utility Management:** Smart solutions with IoT capabilities allow citizens to reduce their household utility costs, deploying sensors connected to the IoT in homes and businesses, and energy use may be optimized and monitored, lowering costs and carbon emissions.

l) **Smart lighting:** Streetlights with IoT sensors can have their illumination settings changed in reaction to outside light, reducing energy consumption and pollution.

m) **Water management:** By installing IoT sensors in water distribution systems, water quality can be tracked, leaks can be found, and water use can be optimized to save money and resources.
n) **Remote Monitoring:** Effective utility management for residents is another benefit of IoT-based smart city technologies. They enable locals to monitor and manage resource usage using their meters. For example, a homeowner can use a smartphone app to turn off their HVAC system. Utility companies can also alert families to problems like leaks or possible failures and dispatch experts for prompt maintenance and repairs.
o) **Environmental Well-Being**: Municipalities can remotely monitor environmental conditions with the aid of IoT-enabled systems. Water grids, for example, have sensors installed to check the quality of the water and send out alerts in the case of leaks or fluctuations in the water's chemical makeup. The same technology is essential for suggesting remedies that improve air quality and is also used to measure the quality of the air in polluted places.
p) **Public Transport:** Traffic controllers can examine and spot trends in public transportation use by using data from sensors implanted in various sources. While maintaining a consistent level of safety and punctuality, this information shortens wait times and enhances the travel experience for citizens. Additionally, a smart city can install BLE beacons on bridges and roadways to track deterioration and make quick repairs if harm is imminent.
q) **Enhanced Energy Efficiency:** To make 6G networks sustainable, research is being done on AI-based power management systems, adaptive network protocols, and energy-efficient hardware. For instance, smart antennas and low-power IoT devices could help conserve energy across a 6G network, reducing both operational costs and environmental impact.

2.3.3.1 Seamless Integration of Physical and Digital Worlds

With the use of technologies like wearable technology, AR, and VR, 6G seeks to close the gap between the digital and physical worlds. Applications where users need both connectivity and a high level of involvement, such as virtual collaboration, immersive education, and remote surgeries, make this integration particularly pertinent.

2.3.4 Key Technologies Enabling 6G

To realize the ambitious vision of 6G, a host of revolutionary technologies are being researched and developed. These technologies aim to deliver ultrahigh speeds, extremely low latency, and intelligent, adaptive network behavior.

This 3D network architecture guarantees dependable, fast connectivity for users in remote locations, high-rise buildings, and airplanes. As a result, the network is genuinely ubiquitous and performs consistently wherever it is used.

2.3.5 IoT in the 6G Era

With formerly unprecedented levels of device connectivity, data interchange, and intelligent automation, the IoT is set to undergo a substantial change in the 6G era. In addition to increasing the size and reach of IoT deployments, 6G will solve the speed, dependability, latency, and security issues of earlier generations. Massive IoT, critical IoT, and tactile IoT are the broad categories under which 6G-enabled IoT is anticipated to cope with a large number of applications due to enhanced network capabilities.

Figure 2.5 shows 6G-enabled IoT.

a) **Massive IoT:** The deployment of billions of inexpensive, low-power devices that need sporadic or occasional connectivity is referred to as the "massive IoT." These devices, which include asset trackers, smart meters, and environmental sensors, usually send out brief data packets from a variety of dispersed places. To support these high-density IoT deployments, 6G will include energy-saving protocols, enhanced spectrum efficiency, and ultra-scalable network designs. Even in extremely congested situations, innovations like grant-free access, sophisticated modulation techniques, and device-to-device (D2D) communication will guarantee uninterrupted connectivity and lessen network congestion. Predictive maintenance and efficient resource allocation will also be made possible by AI-powered network management, which will improve device performance and prolong battery life.

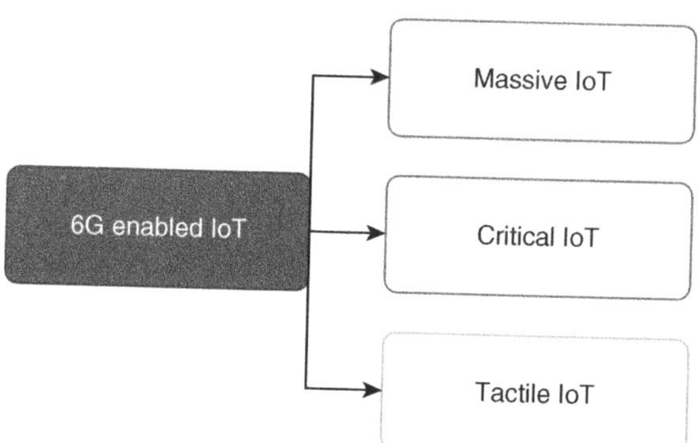

Figure 2.5 6G-enabled IoT.

b) **Critical IoT:** The critical IoT includes applications that need uRLLC and constant service quality. Autonomous cars, remote robotic surgery, smart grids, industrial automation, and emergency response systems are some examples of these use cases. Even a millisecond delay or a missed packet might have detrimental effects for these kinds of applications. 6G promises reliability of over 99.9999% and end-to-end latencies of less than 1 ms. In order to guarantee the deterministic performance required for mission-critical services, technologies such as edge computing, AI-driven prioritization, network slicing, and redundant routing will be essential. Additionally, these systems' integrated security and authentication features will protect them from online attacks and guarantee data integrity.

c) **Tactile IoT:** Systems that allow for interactive, real-time experiences with distant or virtual settings are referred to as tactile IoT. These include uses in industrial haptics, remote surgery, AR/VR, and telepresence. The requirement for incredibly low latency—typically less than 1 ms—to enable instant feedback and human-like reaction is what distinguishes tactile IoT. This will be made possible by 6G's integrated sensing and actuation technologies, excellent dependability, and ultralow-latency communication. Furthermore, in order to facilitate rich media transmissions—like 8K video or holographic content—which are frequently a component of tactile applications, high-speed uplink and downlink bandwidth will be necessary. These experiences will be seamless, immersive, and incredibly responsive, thanks to multi-access edge computing (MEC), intelligent edge nodes, and 3D connection.

2.3.6 Edge and Fog Computing in 6G IoT

As IoT and real-time applications grow, 6G will leverage edge and fog computing architectures to analyze data nearer to the source. This reduces latency and bandwidth usage to support mission-critical operations. It will be crucial to integrate with distributed computing and AI systems. Because of its success across a variety of agricultural applications, edge AI applications have become a prominent field of study. For example, deep learning computer vision techniques can help field robots derive guidance lines in complex farmland situations. Furthermore, real-time crop detection by harvesting robots is made possible by the combination of edge computing and lightweight neural networks. In the future of edge AI, the traditional approach of training models on powerful servers before deploying them to edge devices will also be altered.

With the advent of federated learning (FL), the training process is decentralized, and the training data is stored on the edge devices that generated it. By interacting with a central server to transmit locally computed model updates based on their own data, edge devices take part in the training process. After integrating updates from many devices and modifying the global model, the central server sends the revised version to the edge devices. This iterative approach is repeated until the global model performs at the desired level. As a result, a significant amount of data needs to be sent from the edge of the network to the cloud, increasing security and expediting the training process.

The benefits of processing data locally, which are often offered by edge computing and edge AI and reduce the need to move data to the cloud. As a result, the backbone network's traffic strain is lessened. It is crucial to remember that not all deployment scenarios may benefit from the ubiquitous application of edge computing and edge AI. Their main purpose is to respond instantly to real-time applications, like guiding, coordinating, and cooperating with autonomous vehicles and robots in agricultural activities. However, they do not perform well in data-intensive tasks such as augmented reality applications, high-resolution image processing, and large-scale analytics. As a result, sending data to the cloud is still required for complicated operations requiring a lot of processing power. The extent to which edge computing and edge AI may do away with the need for a cloud connection will depend on the specific application's requirements, the type of data, and the capabilities of the edge device.

2.3.6.1 Four-Dimensional Communication

Four dimensions will be used for communication: land, air, space, and undersea or beneath. Geographic coverage will surpass present bounds. As a result, 6G–IoT will create a unified communication environment where connectivity and services are continuously offered. This situation will be made possible by two key enabling technologies: UAVs and telescopes. Satellites provide wireless coverage in a range of Earth orbits, including low, medium, and geostationary Earth orbits, to enhance the current Earth-based communications networks and enable efficient network coverage of rural, remote, and geographically isolated areas. Additionally, new VLEO satellite systems are created, offering significant benefits over conventional satellite systems, including smaller satellites and higher energy efficiency. Because VLEO systems offer reduced latency and signal route losses, they are ideal for telecommunications applications. UAVs will serve as relays or even as mobile base stations in the case of aerial networking in the event that a portion of the terrestrial network is compromised or destroyed during an emergency. Additionally, they can be combined with satellite networks to produce a sophisticated combination NTN.

2.3.7 Issues with Safety and Confidentiality in the Internet of Things Ecosystems Powered by 6G

As the number of connected devices increases, maintaining strong security and privacy is crucial. In an effort to safeguard data and guarantee privacy, 6G networks are anticipated to include cutting-edge safety protocols and encryption techniques, hence offering a safe operating environment for IoT devices. With intelligent automation, integrated sensing, and omnipresent communication, 6G technology creates the framework for an ultra-connected future, but it also brings with it difficult security and privacy issues. Because of the size, diversity, and importance of 6G-enabled IoT applications—from remote healthcare to autonomous cars to tactile internet—security frameworks must be not only reliable but also flexible, intelligent, and scalable.

2.3.7.1 Expanding Threat Surface and Vulnerabilities

The threat surface is further increased by the enormous number of linked devices, many of which have little computational power. Endpoints in these distributed systems are vulnerable to distributed denial-of-service (DDoS) assaults, malware injection, identity spoofing, data eavesdropping, and physical tampering. Additionally, the system is more vulnerable to physical and cyber-attacks due to the incorporation of software-defined network components, new frequency bands (THz), and NTNs (satellites and UAVs). Mission-critical services and tactile IoT demand real-time connectivity, which leaves limited opportunity for latency-induced security processes and increases vulnerabilities.

2.3.7.2 Privacy Risks in Data-Driven Environments

Personal, behavioral, and environmental data will be continuously collected, processed, and shared across several domains in 6G environments. Due to user tracking, profiling, and unlawful data sharing, this poses serious privacy hazards. These worries are made worse by the combination of AI, big data analytics, and edge computing, since private information could be intercepted during transmission or made available to other service providers. One of the main research challenges is ensuring privacy preservation while preserving system responsiveness and performance.

2.3.8 Emerging Solutions and Technologies

The following are several emerging technologies that offer promising avenues to address these security and privacy concerns:

a) **Blockchain and Distributed Ledger Technologies (DLTs):** These offer immutable transaction records, secure data provenance, and decentralized authentication. Blockchain is especially helpful for transparent access control systems, safe firmware updates, and managing the identities of IoT devices.
b) **Post-quantum Cryptography (PQC):** In quantum computing advances, conventional encryption techniques are growing more susceptible. PQC aims to provide cryptographic solutions that prevent adversaries with quantum

capabilities from compromising the long-term privacy and reliability of 6G communications.

c) **AI-Based Intrusion Detection Systems (IDS):** These systems use deep learning and machine learning models to identify anomalies in real time and react to attacks on their own. Distributing these systems at the edge can speed up reaction times and help them adjust to changing attack patterns.
d) **Federated Learning and Privacy-Preserving AI:** These methods greatly lower privacy risks and data leakage by enabling AI models to be trained across decentralized edge devices without disclosing raw user data.

2.3.9 Key Research and Standardization Challenges

Despite these advances, 6G still faces substantial unresolved challenges:

a) **Global Standardization**: Interoperability is hampered as cross-domain IoT applications become vulnerable due to the lack of widely recognized security frameworks and communication protocols.
b) **Mobility and Session Management:** New models of authentication and session continuity that take into consideration shifting contexts and topologies are needed to secure extremely mobile and transitory connections over 3D networks (land, air, and space).
c) **Spectrum and Resource Management:** Trusted access control, interference reduction, and jamming resistance are among the major technological issues associated with secure and dynamic spectrum sharing, particularly in high-frequency THz bands.
d) **THz Hardware Vulnerabilities**: New hardware-level threats, such as side-channel assaults and physical layer breaches, may surface as THz technology develops, requiring hardware-assisted security design from the start.
e) **AI Trustworthiness and Explainability**: Despite their strength, AI systems are susceptible to manipulation by hostile attacks or tainted data. A crucial topic of continuing research is ensuring AI's dependability, interpretability, and ethical alignment in crucial security applications.

2.3.10 Collaborative Approach to Security Innovation

Addressing these complex security and privacy issues requires a multi-stakeholder approach. Governments must create supportive regulatory frameworks and invest in national cybersecurity infrastructure. Academia must drive foundational research in cryptography, AI robustness, and formal verification. Industry stakeholders must adopt secure-by-design principles, participate in global standardization bodies, and ensure the implementation of scalable and interoperable solutions. Cross-sector collaboration will be essential in building trust and resilience into the global 6G ecosystem.

2.3.10.1 Energy Efficiency and Sustainability in 6G Networks

Future network generations' effects on the natural world and energy sources must be carefully considered, as the need for pervasive, fast, and minimal latency wireless communication increases. The rollout of 6G presents a pivotal opportunity to align technological advancement with the global sustainability agenda. With its projected massive device connectivity, ultrahigh data rates, and intelligent applications, 6G must deliver on both performance and ecological responsibility.

2.3.10.2 Sustainability Challenges in Future Networks

Due to the deployment of dense infrastructure, intensive computing at the edge, and an increase in connected devices, previous mobile communication generations—especially 4G and 5G—introduced notable increases in power consumption. In the absence of intelligent energy-saving technologies and architectures, 6G networks—which are anticipated to facilitate widespread IoT, real-time AI inference, THz communications, and NTNs—run the danger of increasing energy needs.

Key challenges include the following:

a) Sustaining the long-term operation of billions of low-power IoT devices.
b) Controlling the energy consumption of AI-powered network operations and intelligent base stations.
c) Reducing the carbon emissions from edge computing nodes and data centres.
d) Ensuring that satellite and aerial components are integrated into NTNs in a sustainable manner.

2.3.10.3 Green Enabling Technologies

6G will employ several state-of-the-art technologies intended to improve environmental and energy efficiency as a way to deal with these problems:

a) **Energy Harvesting:** In order to reduce their reliance on batteries and external power sources, future 6G-enabled devices will depend more and more on ambient energy sources, including solar, thermal, kinetic, and radio frequency energy. For remote or inaccessible IoT deployments, such as environmental sensors, agricultural equipment, and structural health monitors, this is especially crucial.
b) **Reconfigurable Intelligent Surfaces (RIS):** RIS technology eliminates the need for energy-intensive active transmission by enabling electromagnetic waves to be intelligently controlled and passively reflected. By improving signal

propagation and reducing path loss, RIS can drastically reduce transmission power requirements, particularly in crowded urban or indoor settings.

c) **AI-Based Resource Optimization:** The dynamic and predictive optimization of energy use throughout the 6G network will be greatly aided by AI and ML. AI can minimize energy consumption while preserving Quality of Service (QoS) by optimizing beamforming techniques, load balancing, user scheduling, and base station sleep modes. More environmentally friendly AI applications will also benefit from federated learning and energy-conscious neural network pruning.
d) **Network Virtualization and Softwarization:** 6G can enable energy-efficient provisioning using technologies like network slicing, SDN, and network function virtualization (NFV) by dynamically scaling resources based on traffic demand and application priority.
e) **Network Virtualization and Softwarization**: By dynamically scaling resources according to traffic demand and application priority, 6G can facilitate energy-efficient provisioning using technologies such as network slicing, SDN, and NFV.
f) **Low-Power Communication Protocols:** New MAC and routing protocols designed specifically for energy-constrained IoT devices will reduce idle listening, retransmissions, and unnecessary signaling. Lightweight protocols such as MQTT, CoAP, and novel 6G-specific alternatives will play a key role.

2.3.10.4 Design Techniques and Architectural Aspects of 6G in IoT

Sustainability needs to be ingrained in the network's architecture and lifetime, going beyond component-level effectiveness, which includes;

a) **Green Data Centers and Edge Nodes:** Adoption of improved cooling systems, energy-conscious work scheduling, and renewable energy sources in data processing infrastructures are examples of green data centers and edge nodes.
b) **Circular Economy in Device Design:** In device design, the circular economy promotes modularity, extended hardware lifecycles, and recyclable materials to lessen the impact on the environment and electronic waste.
c) **Carbon-Neutral Network Operations**: By including lifecycle emissions analysis into network planning, deployment, and operation, operators and vendors are expected to demonstrate their commitment to carbon-neutral targets.

2.3.11 Toward a Sustainable 6G Environment

Sustainability is not a stand-alone goal in 6G; rather, it is a cross-cutting requirement that must impact all facets of network development, from business models and policy frameworks to technological design and spectrum regulation. There has never been a more pressing need for digital infrastructure to support global ecological goals as climate change, resource shortages, and environmental degradation worsen. Building wireless systems that are not just quicker and smarter but also resource-conscious, energy-efficient, and climate-resilient is made possible by 6G.

To realize this vision, international standardization bodies and national governments must collaborate to establish environmental performance standards that quantify and enforce sustainable practices across the telecommunications ecosystem. Energy-per-bit measurements, lifespan carbon footprint analyses, and specifications for low-impact and recyclable hardware components are a few examples of these standards. The advancement of green spectrum management, which minimizes energy consumption and prevents infrastructure redundancy through dynamic spectrum access, environmentally friendly licensing, and spectrum sharing, is equally crucial.

In order to dynamically optimize energy consumption, 6G networks will mainly rely on AI, edge computing, and RIS. For example, edge processing can lower the energy load on centralized data centers by managing processes locally, while AI can be utilized to route data through the most energy-efficient channels. By improving signal transmission without requiring high-power transmitters, RIS technology can further reduce energy consumption. Additionally, base stations and other network components will be built to go into deep sleep states when traffic is light, improving overall energy efficiency.

Furthermore, the future viability of 6G depends on the hardware layer. The usage of clean sources of energy like wind and solar power, along with modular, upgradeable architecture, will make 6G infrastructure long-lasting and low-emission. In an effort to lessen the environmental impact of electronic trash, research into recyclable or biodegradable materials for devices and antennas is also gaining traction.

In conclusion, 6G offers a singular and opportune moment to radically reinvent wireless networks with sustainability at the forefront of their architecture. 6G has the potential to revolutionize ecological well-being worldwide by combining energy-conscious technologies, intelligent network management, and green regulatory frameworks. The creation of 6G, if done carefully and cooperatively, has the potential to guarantee that the digital revolution is not only intelligent and interconnected but also sustainable and just for future generations.

2.3.11.1 Future Roadmap

The path to 6G is expected to move from theoretical study to real-world deployment between 2025 and 2030. As international stakeholders coordinate efforts across standardization, testing, and deployment, this era will be crucial in determining the direction of wireless communication in the future. Field tests, pre-commercial installations, harmonizing international standards, and maturing enabling technologies will be important initiatives.

The main elements of this developmental roadmap are as follows;

a) **Standardization and Framework Finalization:** Standardization ensures interoperability, security, and performance across devices and networks. By 2026–2027, major standardization bodies such as ITU-R, 3GPP, and IEEE are expected to converge on:
   - **6G Vision and Usage Scenarios:** Completing the worldwide 6G vision, which includes massive machine-type communications (mMTC), uRLLC, and extreme mobile broadband (xMBB).
   - **KPI Definitions**: Agreement on important performance metrics (like goals for spectral efficiency, energy efficiency, latency under 0.1 ms, and top transmission speeds over one Tbps).
   - **Architecture Guidelines**: Model architecture that incorporates network slicing, AI-native features, and support for both terrestrial and non-terrestrial multilayered networks.
   - **Standards for Security and Privacy:** Combining reliable AI systems with quantum-resilient encryption.

b) **THz Communications**
   i) Establishment of regional and global THz testbeds for evaluating short-range ultrahigh-speed data transmission.
   ii) Development of THz-compatible hardware (antennas, RF circuits, and power amplifiers).
   iii) Tackling obstacles including thermal noise, chemical absorption, and excessive route cost.

c) **Reconfigurable Intelligent Surfaces (RIS)**
   i) Testing RIS in smart cities and industrial zones to enhance coverage and capacity.
   ii) Investigating dynamic beamforming and programmable propagation environments.
   iii) Standardizing control protocols for RIS elements embedded in infrastructure.

d) **Non-terrestrial Networks (NTN)**
   i) Satellite connection, UAV, and HAPS in 6G structure.
   ii) Pilot projects focusing on rural and maritime connectivity.
   iii) Adaptive routing, latency management, and hybrid network slicing across NTNs and terrestrial segments.

e) **AI-Native Networks**
   i) Implementation of AI for predictive maintenance, network optimization, anomaly detection, and intent-based orchestration.
   ii) Experimentation with federated and edge learning models to preserve privacy while enabling real-time intelligence.
   iii) Validation of AI/ML model accuracy, interpretability, and ethical governance.

f) **Spectrum Allocation and Management**
   The vital component of wireless networks is spectrum.
   The following will be the main focus between 2025 and 2030:
   a) **Finalizing Frequency Bands:** THz and sub-THz allocations will be determined by the results of the ITU's World Radio-communication Conference (WRC-27).
   b) **Spectrum Sharing and Dynamic Access:** Coexistence frameworks with incumbent users and legacy technologies are being developed for spectrum sharing and dynamic access.
   c) **Harmonization Efforts:** Global device compatibility, economies of scale, and regional alignment to prevent fragmentation.
   d) **Green spectrum policies**: promote low-power technology design and energy-efficient spectrum use.

g) **Pilot Deployments and Pre-commercial Trials**
   By 2028–2029, pre-commercial 6G systems will begin limited deployment to test scalability, user experience, and operational challenges.
   a) **Urban Testbeds:** Deployments in smart cities integrating autonomous transport, public XR services, and Virtual twins.
   b) **Industry 5.0 Environments:** Intelligent production, teleoperation, predictive analytics in factories.
   c) **Healthcare Trials:** Use of 6G for real-time remote surgery, biometric monitoring, and immersive medical training.
   d) **Public Safety and Disaster Management:** NTN-based resilient communication systems for emergency scenarios.

2.4 Conclusion

The fundamentals of 6G and its intersection with the IoT have been thoroughly examined in this study. As we move beyond the capabilities of 5G, 6G emerges not just as a faster or more efficient communication network, but as a foundational technology for the next-generation digital ecosystem. At the heart of this transformation lies the convergence of several advanced technologies: AI, THz communications, RIS, and edge computing.

These technologies collectively redefine what is possible in terms of connectivity, latency, bandwidth, and autonomous decision-making. AI-driven automation, THz-enabled ultrahigh-speed data transfer, environment-aware RIS for signal optimization, and localized edge processing will power an ecosystem where devices and systems not only communicate but also intelligently collaborate. The World Wide Web of Contents transforms into the Intelligent Web of Everything in this context, which is a self-optimizing network of linked surroundings, services, and gadgets.

When combined, 6G and IoT offer the potential for unprecedented levels of sustainability, intelligence, and ubiquitous connectedness. From autonomous factories and smart cities to real-time healthcare and precision agriculture, these integrated technologies can dramatically improve human life and environmental sustainability. However, the road to this hyper-connected, intelligent future is not without challenges. Achieving the full potential of 6G-enabled IoT will require interdisciplinary collaboration across domains such as wireless engineering, computer science and AI, cybersecurity and cryptography, law and public policy, and environmental sciences.

Furthermore, robust international cooperation is essential to ensure harmonized standards, ethical guidelines for AI, and secure cross-border data flows. Regulatory frameworks must evolve alongside the technology to address issues such as spectrum allocation, AI governance, and digital rights.

In conclusion, 6G is more than a technological evolution—it is a transformational force that will redefine how we interact with the world. Its synergy with IoT promises to unlock a new era of intelligent connectivity that is not only faster and more efficient but also more inclusive, resilient, and sustainable. Realizing this vision will depend on our ability to navigate complex technical and policy landscapes with foresight, innovation, and collaboration.

Further Reading

1 Gupta, A. and Jha, R.K. (2015). A survey of 5G network: architecture and emerging technologies. *IEEE Access* 3: 1206–1232.

2 Saad, W., Bennis, M., and Chen, M. (2019). A vision of 6G wireless systems: applications, trends, technologies, and open research problems. *IEEE Network* 34 (3): 134–142.

3 Dang, S., Amin, O., Shihada, B., and Alouini, M. (2020). What should 6G be? *Nature Electronics* 3 (1): 20–29.

4 Zhang, Z., Xiao, Y., Ma, Z. et al. (2019). 6G wireless networks: vision, requirements, architecture, and key technologies. *IEEE Vehicular Technology Magazine* 14 (3): 28–41.
5 Rimal, B.P., Maier, M., and Van, D.P. (2017). Mobile-edge computing vs. centralized cloud computing: a comparative analysis. *IEEE Communications Magazine* 54 (1): 20–27.
6 Charu, Anshul Kumar, Anurag Upadhyay, Priya Chauhan, and Vaishnavi Sharma (2023). A 6G Network: Future of Nations? Challenges in 6G Communications. *Tuijin Jishu/Journal of Propulsion Technology*. ISSN: 1001-4055. Vol. 44 No. 04 (2023).
7 Ali, A., Hussain, S.M., and Begh, G.R. (2023). 6G and IOT use cases. In: *6G Wireless: The Communication Paradigm Beyond 2030* (ed. F. Tariq, M. Khandaker, and I.S. Ansari), 354. CRC Press.
8 Puspitasari, A.A., An, T.T., Alsharif, M.H., and Lee, B.M. (2023). Emerging technologies for 6G communication networks: machine learning approaches. *Sensors* 23 (18): 7709.
9 Hosseinzadeh, M., Hemmati, A., and Rahmani, A. (2022). 6G-enabled internet of things: vision, techniques, and open issues. *Computer Modeling in Engineering and Sciences* 133 (3): –509.
10 Qadir, Z., Le, K.N., Saeed, N., and Munawar, H.S. (2023). Towards 6G internet of things: Recent advances, use cases, and open challenges. *ICT Express* 9 (3): 296–312.
11 Chen, S., Liang, Y.-C., Sun, S. et al. (2020). Vision, requirements, and technology trend of 6G: how to tackle the challenges of system coverage, capacity, user data-rate and movement speed. *IEEE Wireless Communications* 27 (2): 218–228.
12 Alsharif, M.H., Kelechi, A.H., Albreem, M.A. et al. (2020). Sixth generation (6G) wireless networks: vision, research activities, challenges and potential solutions. *Symmetry* 12 (676): 4.
13 Khan, L.U., Yaqoob, I., Imran, M. et al. (2020). 6G wireless systems: a vision, architectural elements, and future directions. *IEEE Access* 8 (14): 14704–47029.
14 Kurt, G.K., Khoshkholgh, M.G., Alfattani, S. et al. (2021). A vision and framework for the high altitude platform station (HAPS) networks of the future. *IEEE Communications Surveys & Tutorials* 23 (2): 729–779.
15 Basharat, S., Ali, H.S., Pervaiz, H. et al. (2021). Reconfigurable intelligent surfaces: potentials, applications, and challenges for 6G wireless networks. *IEEE Wireless Communications* 28: 1–8.
16 Wang, H., Wang, W., Chen, X., and Zhang, Z. (2014). Wireless information and energy transfer in interference aware massive MIMO systems. *2014 IEEE Global Communications Conference*, pp. 2556–2561. IEEE.
17 You, X., Wang, C.-X., Huang, J. et al. (2021). Towards 6G wireless communication networks: vision, enabling technologies, and new paradigm shifts. *Science China Information Sciences* 64 (1): 1–74.

18 Nguyen, V.-L., Lin, P.-C., Cheng, B.-C. et al. (2021). Security and privacy for 6G: a survey on prospective technologies and challenges. *IEEE Communications Surveys and Tutorials* 23: 2384–2428.

19 Nguyen, D.C., Ding, M., Pathirana, P.N. et al. (2021). 6G internet of things: a comprehensive survey. *IEEE Internet of Things Journal* 9: 359–383.

20 Dao, N.-N., Pham, Q.-V., Tu, N.H. et al. (2021). Survey on aerial radio access networks: toward a comprehensive 6G access infrastructure. *IEEE Communications Surveys and Tutorials* 23 (2): 1193–1225.

21 Alwis, C.D., Kalla, A., Pham, Q.-V. et al. (2021). Survey on 6G frontiers: trends, applications, requirements, technologies and future research. *IEEE Open Journal of the Communications Society* 2: 836–886.

22 Tang, F., Mao, B., Kawamoto, Y., and Kato, N. (2021). Survey on machine learning for intelligent end-to-end communication toward 6G: from network access, routing to traffic control and streaming adaption. *IEEE Communications Surveys and Tutorials* 23 (3): 1578–1598.

23 Guo, F., Yu, F.R., Zhang, H. et al. (2021). Enabling massive IoT toward 6G: a comprehensive survey. *IEEE Internet of Things Journal* 8 (15): 11891–11915.

24 Fang, X., Feng, W., Wei, T. et al. (2021). 5G embraces satellites for 6G ubiquitous IoT: basic models for integrated satellite terrestrial networks. *IEEE Internet of Things Journal* 8 (18): 14399–14417.

25 Aggarwal, S., Kumar, N., and Tanwar, S. (2021). Blockchain-envisioned UAV communication using 6G networks: open issues, use cases, and future directions. *IEEE Internet of Things Journal* 8 (7): 5416–5441.

3

Green and Intelligent IoT in 6G Era: Machine Learning for Sustainable Smart Cities

Priyanka Datta[1], *Neha Garg*[2], *Amitesh Aggarwal*[2], *Amanpreet Kaur*[2], *and Yonis Gulzar*[3]

[1] *Department of Computer Science and Engineering, Laxmi Institute of Technology, Sarigam, Gujarat, India*
[2] *Chitkara University Institute of Engineering and Technology, Rajpura, Punjab, India*
[3] *Department of Management Information Systems, College of Business Administration, King Faisal University, Al-Ahsa, Saudi Arabia*

3.1 Introduction

6G technology integration with the Internet of Things (IoT) enables revolutionary advancements in machine learning (ML) data analytics capabilities. The combination of terahertz communication with smart reflecting surfaces and unified sensing technologies produces enhanced data collection and transmission accuracy in IoT systems, which leads to better ML application performance in intelligent environments. ML serves as a key factor to unlock IoT capabilities by analyzing behavioral data across connected devices for decision-making purposes. Using ML algorithms, IoT systems can recognize patterns, improve procedures, and optimize operations, enhancing the efficiency and adaptability of smart surroundings.

3.1.1 Overview of IoT and Data Analytics

IoT represents a network of interrelated physical devices, automobiles, and other things equipped with sensors, software, and other technologies. Such devices can gather and exchange data with minimal human interaction over the internet [1]. IoT seeks to offer continuous surveillance, automation, and expansion of various systems, from household automation to machines in industrial settings. IoT devices are commonly provided with communication protocols such as Bluetooth and Wi-Fi, as well as cellular networks, allowing them to connect easily with one another.

Synergies of 6G and IoT, First Edition. Edited by Himanshi Babbar.

Published 2026 by John Wiley & Sons, Inc.

Key features of IoT include:

- **Connectivity:** Devices are associated with the internet and may communicate with each other.
- **Sensors:** Devices extract information from the environment around them.
- **Automation:** Several IoT systems operate with limited human interaction.
- **Data Exchange:** Devices send acquired data to centralized networks as well as online platforms for processing.

3.1.1.1 Importance of Data Analytics in IoT

Data analytics has become essential in IoT because of the massive amount of data generated via linked devices. The value of IoT largely depends on the capability to evaluate this data and establish useful insights. Data analytics allows organizations to:

- **Monitor Systems**: The continuous processing of data through IoT devices enables real-time monitoring of surroundings and systems. For example, smart cities utilize IoT devices to track roadways and pollution levels [2].
- **Predictive Maintenance**: In industrial settings, data analytics may anticipate the breakdown of equipment before it happens, simply identifying irregularities in performance data, which helps minimize downtime as well as maintenance costs.
- **Decision-Making:** Data-driven findings allow enhanced decision-making. For instance, smart houses optimize energy consumption to optimize power usage.
- **Optimization**: Data analytics may assist in enhancing operations by discovering inefficiencies and proposing improvements, like management of supply chains or smart grid systems.
- **Personalization**: Consumer-facing IoT devices, such as fitness devices or intelligent appliances, use data analytics to offer customized services, like health suggestions or automated temperature control in homes [3].

3.1.2 Role of Machine Learning in IoT

ML plays a vital role in the IoT network by allowing cognitive processing and making choices based on the massive amount of data created by IoT devices. Although IoT emphasizes data collection and transmission, ML assists in data analysis and derives actionable insights [4]. There are various key roles ML plays in IoT.

3.1.2.1 Data Processing and Pattern Recognition

IoT systems acquire huge amounts of data from a number of sources, like cameras, sensors, and other linked devices [5]. ML methods are important for processing and recognizing patterns in this information. ML models can:

- Identify imperfections and trends in current data streams.
- Identify hidden relationships between variables.
- Handle and analyze data in real time for quick decision-making.
- For instance, in smart homes, ML can identify patterns in energy consumption and regulate cooling, heating, or lighting remotely based on human behavior.

3.1.2.2 Predictive Maintenance

ML is highly beneficial in maintenance predictions, particularly in industrial IoT (IIoT). By exploring both historical and current data from devices and machines, ML algorithms can anticipate feasible failures before they happen. This minimizes downtime, enhances efficiency, and lowers maintenance costs. The algorithms learn from the previous performance of equipment and constantly improve prediction accuracy over time [6].

3.1.2.3 Anomaly Detection

In many IoT-related applications, like healthcare, security, or industrial environments, identifying anomalies is important. ML models are great at learning what regular behavior looks like and immediately identifying variations from that behavior. For example:

- In healthcare IoT, ML can identify vital signs and unusual patterns, warning professionals of potential health risks [7].
- In cybersecurity, ML algorithms can identify unexpected network traffic patterns that could suggest a data breach or cyberattack.

3.1.2.4 Optimization and Automation

ML improves IoT systems by strengthening processes and automating the decision-making process. In smart settings (e.g., homes, factories, cities), ML algorithms identify data to determine optimal solutions, like:

- Minimizing energy consumption in smart buildings by improving usage patterns.
- Maximizing supply chains and logistics through forecasting demand and adjusting inventory levels.

- Through reinforcement learning, ML models can evolve and learn to continuously enhance the performance of automated devices in various IoT applications.

3.1.2.5 Personalization

ML algorithms allow IoT devices to offer customized experiences through learning from consumer behavior and choices. Devices such as intelligent thermostats, health trackers, and personal home assistants utilize ML to adapt to the activities of individual users, offering customized recommendations or modifications. For instance, intelligent speakers can use speech recognition to alter responses or song playlists to particular users' preferences.

3.1.2.6 Security Enhancements

ML plays an important role in improving IoT security. With numerous devices interlinked, the probability of security breaches rises. ML algorithms assist in detecting and responding to cyber threats more efficiently by learning from prior security incidents and evaluating current data to determine illicit activity. These models can recognize unusual behavior in systems or devices, warning of potential risks or attacks in real time.

3.1.2.7 Edge Computing and Real-Time Decision-Making

In IoT, data is frequently processed at the boundaries (nearer to the data source) to minimize latency and usage of bandwidth [8]. ML models can be deployed on edge devices to execute real-time decisions locally without forwarding data to a central server. This is extremely important in applications such as autonomous automobiles, where immediate responses are important for safety [2].

3.1.2.8 Improving User Experience

For household IoT devices, ML boosts customer experience by automating operations, learning natural language, and providing predictive insights. For example:

- Virtual assistants, like Amazon's Alexa as well as Google Assistant, use natural language processing (NLP), an aspect of ML, to comprehend and respond to individual commands.
- Wearable devices, such as fitness devices, use ML to analyze health data, deliver personalized recommendations, and encourage users to attain fitness goals.

3.2 IoT Data Characteristics

3.2.1 Types of IoT Data

There are three main types of IoT data as shown in Figure 3.1. Sensor data is one of the most prevalent forms of IoT data, produced by sensors embedded in gadgets. These sensors can track many physical properties like temperature, speed, humidity, motion, pressure, light, etc. For example, each room of a smart home might have temperature sensors, which help the thermostat keep the house at a reasonable temperature. In agriculture, soil moisture sensors offer information that helps farmers adjust irrigation so that farmers can make sure that crops are receiving the proper amount of water. Sensor data is usually measurable, easy to understand, and constantly updated in real time [9].

Image data consists of pictures, videos, or any visual data that is recorded by cameras or photographic sensors. This kind of data is complicated and demands a large amount of storage and processing power. Cameras in smart cities track traffic patterns, spot accidents, or check offenses like speeding. IoT devices in the medical field, such as attached medical imaging machines, generate X-rays, MRIs, or ultrasound images that assist with diagnosis. Processing picture data frequently includes modern technologies like computer vision and artificial intelligence to detect patterns, identify objects, and make judgments based on visual input [10].

Time-series data consists of data points that are being collected or recorded at predetermined intervals over time. It is frequently produced by sensors, meters, and other IoT devices that track changing circumstances. For example, a smart electricity meter can analyze the use of power every minute or hour and allow for the recording of daily, weekly, or seasonal usage patterns. In industrial environments, devices gather operational data such as vibration levels, temperature, or speed over time to track the equipment's health and anticipate malfunctions. Time-series data is useful in trend analysis, anomaly identification, forecasting, and preserving operational effectiveness [11].

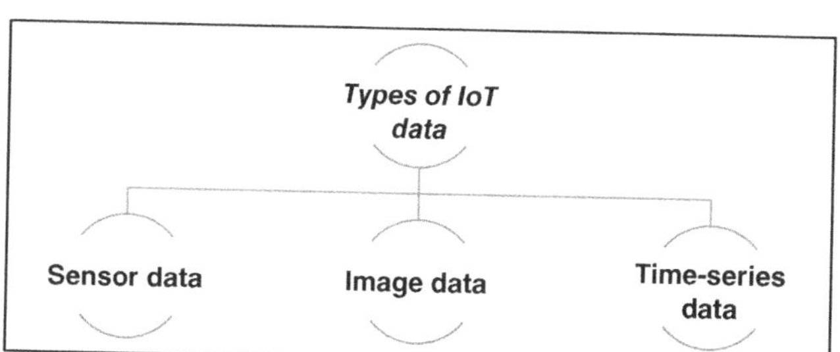

Figure 3.1 Types of IoT data.

3.3 Data Challenges

When data has an excessive number of variables or attributes, it becomes challenging to process and examine. In the world of IoT, devices gather a variety of data points such as temperature, humidity, speed, pressure, etc. For example, an advanced health-tracking gadget may simultaneously record heart rate, sleep habits, blood pressure, steps, and calorie intake. Keeping track of this "high-dimensional" data is like attempting to solve a challenging puzzle; the more pieces there are, the more challenging it becomes to find significant patterns. High dimensionality requires sophisticated algorithms and processing capacity to provide meaningful insights without becoming overwhelmed by the sheer amount of data [12] (Figure 3.2).

Noise is the unnecessary or unwanted information mingled with valuable data. For instance, a smart home security system motion sensor may be set off by pets or wind, causing "noisy" data that does not represent a security risk. Similarly, weather sensors can record false readings because of abrupt temperature spikes or equipment failures. Noise can affect the data analysis results, leading to erroneous conclusions [13]. To make meaning of data, sorting out or minimizing this noise is necessary, which can be difficult and time-consuming work [14].

Heterogeneity refers to the data that comes in a variety of forms and types. In IoT, devices produce a wide range of data, like numerical data from sensor readings, text inputs from users, image data from camera captures, and audio signals from microphone recordings. For example, a smart city traffic management system collects information from GPS devices, cameras, and speed sensors. Each of these

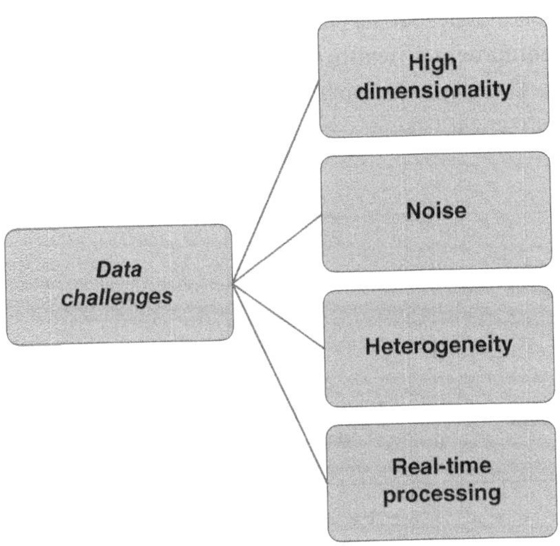

Figure 3.2 Types of data challenges.

data kinds has a unique format and structure, which makes it challenging to combine them for evaluation. This diversity necessitates flexible systems that are capable of handling and processing all types of data simultaneously, which can be difficult to set up and manage [15].

Real-time processing in some IoT applications, data must be processed and evaluated as soon as it is gathered. For instance, in self-driving cars, sensors always collect information about the situation on the road, nearby automobiles in the area, and pedestrians. The car's system should swiftly analyze this information to make split-second judgments, such as when to stop or turn. However, processing massive amounts of data in real time is difficult. It requires fast processing, powerful hardware, and intelligent algorithms to guarantee prompt decision-making without any delays. In case of failure in real-time processing, it might result in major problems like accidents or malfunctions in the system [16].

3.4 Data Preprocessing Techniques

Data preprocessing is an essential step in getting raw data for analysis. It entails a variety of methods to clean and transform data to prepare data for ML models. There are three main techniques of data preprocessing as shown in Figure 3.3.

Data cleaning is the process of finding and fixing mistakes or inconsistencies within a dataset to raise its quality. Real-world data is frequently untidy, with problems including missing values, duplicate entries, outliers, or errors. To show missing values, data cleaning includes the exchanging of them with meaningful replacements, such as the average (mean) or the most frequent value (mode) of the column. It also entails eliminating duplicate rows so that every piece of data is distinct, which also helps in diminishing redundancy. Furthermore, data cleaning fixes errors like misspellings, incorrect decimal points, or erroneous entries. By

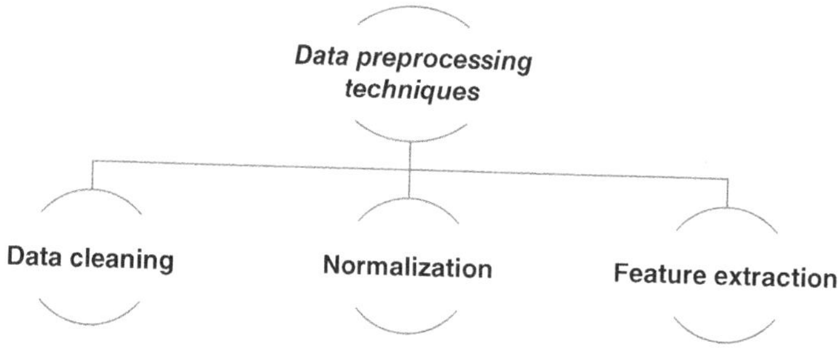

Figure 3.3 Techniques of data preprocessing.

creating data that is consistent and error-free, data cleaning can play a crucial role in increasing the accuracy and reliability of ML models [17].

Normalization is a process applied to modify the scale of data. In datasets, certain features, like income, may have a far wider range of values compared to other features, like age. This can cause problems for some ML algorithms. Normalization transforms the data to an acceptable range, usually ranging between 0 and 1, which makes it easy for algorithms to process the data. It helps to enhance the performance and accuracy of models, particularly those that are sensitive to variations in the scale of input features [18].

Feature extraction includes the selection of the most pertinent unit of information (features) from raw data, which makes it easy for ML models to recognize the patterns. Raw data may consist of many types of variables, but not all of them are equally significant. Feature extraction finds the data by recognizing the most valuable qualities [19]. For instance, if analyzing images, rather than using every pixel, you could extract elements like edges, textures, or colors. This process fewer the amount of data of the model that has to be exercised while preserving the critical information needed for generating predictions [20].

3.5 Machine Learning Techniques for IoT

ML methods have become important in IoT devices for making decisions and analyzing based on huge streams of data created by IoT devices [21]. These methods help predictive maintenance, intelligent automation, real-time decision-making, and anomaly detection in different IoT applications [22]. Some ML techniques are as follows:

3.5.1 Supervised Learning

It is widely used in IoT devices for tasks such as predictive maintenance, energy usage prediction, and health monitoring. The training of models in supervised learning on labeled datasets in which the input–output pairs are known in advance. For example, in smart homes, supervised learning algorithms like Support Vector Machines (SVMs) or decision trees can be used to forecast energy consumption patterns depending on past data from a variety of appliances [23].

3.5.1.1 Decision Trees

According to ML, a decision tree is a technique that classifies data into subsets depending on the values of features. This helps the technique to make decisions. At every node, the algorithm asks a question and then branches off in a diverse direction depending on the response, which ultimately leads to a choice being made at the terminal nodes, as shown in Figure 3.4. As a result of its ease of

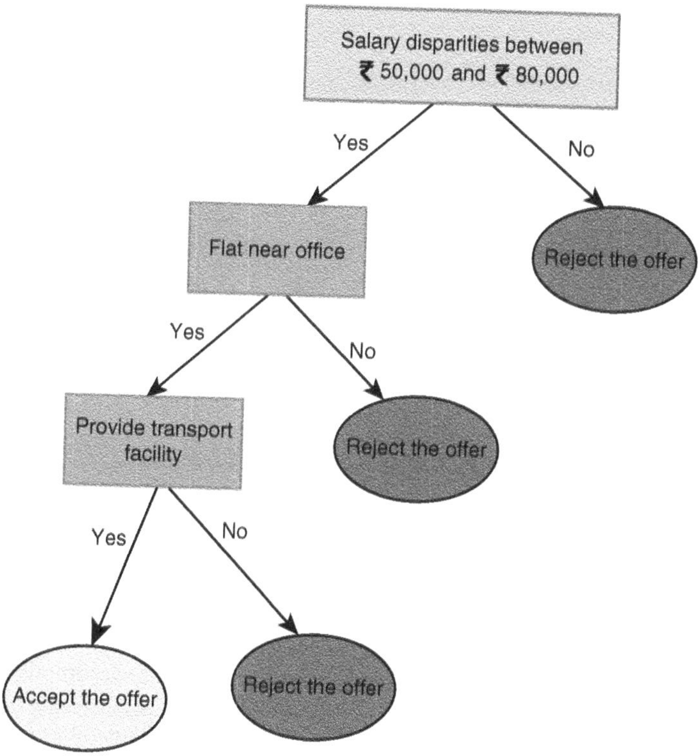

Figure 3.4 A decision tree.

use and interpretability, it is very frequently utilized for regression and classification applications [24].

3.5.1.2 Support Vector Machines (SVMs)

A supervised learning approach is SVMs; it is used for regression and classification problems. Finding the hyperplane in a feature space that most effectively separates data points belonging to other classes. The optimal hyperplane is the one that enhances the margin between the nearest data points (support vectors) of various classes, as illustrated in Figure 3.5. It is mainly effective for high-dimensional spaces and can also handle data of nonlinearity by using kernel functions [25].

3.5.2 Unsupervised Learning

Unsupervised learning is utilized in IoT devices for the identification of anomalies and clustering. The purpose of these methods is to reveal previously unknown groupings or patterns by working with unlabeled data. Particularly helpful in

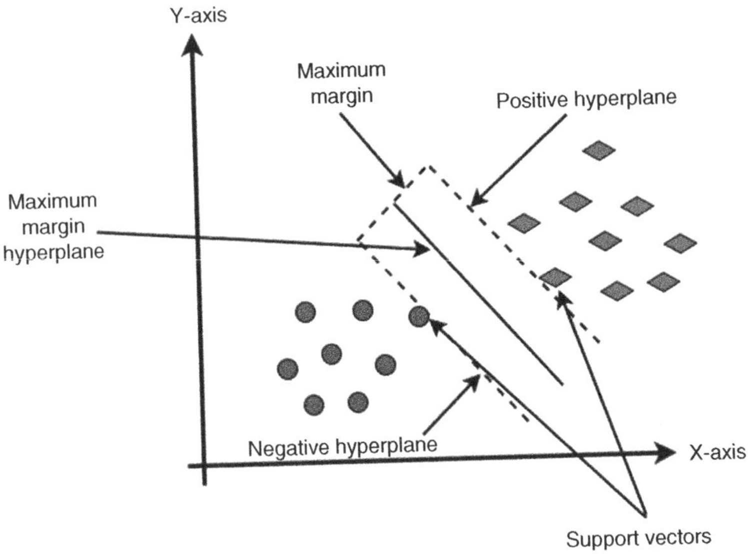

Figure 3.5 The support vector machine.

settings where the labeling of data is very expensive or impossible to accomplish [26]. For example, clustering algorithms such as density-based spatial clustering of applications with noise (DBSCAN) or K-means are utilized in smart cities to group data of sensors from environmental monitoring stations to identify pollutants at unusually high levels of traffic congestion.

3.5.2.1 K-means

In ML, K-means is a clustering method that depends on the similarity of features that classify data into K different clusters. Each data point is assigned to the nearest cluster centroid and then continuously updates the centroids till convergence. Due to its efficiency and simplicity, K-means is becoming popular for jobs including pattern identification, picture reduction, and segmentation of customers [27].

3.5.2.2 Density-Based Spatial Clustering of Applications with Noise (DBSCAN)

It is a clustering method based on density for data points in feature space. Noise is marked as outliers, and clusters are expanded from main points originating in dense areas. The technique is important in disciplines like image processing and spatial data analysis because it is very effective for noise handling and finding clusters of arbitrary forms [28].

3.5.3 Reinforcement Learning

To solve optimization issues in IoT devices, like network traffic management, resource allocation, and energy efficiency, reinforcement learning devices are utilized. A real-world agent is taught to behave in a given environment to maximize the accumulation of rewards over a certain amount of time. For example, it is possible to utilize RL algorithms in a smart grid to regulate energy distribution more effectively, thereby optimizing the utilization of renewable energy sources depending on real-time client demand [29].

3.5.3.1 Q-learning

A model-free RL method is Q-learning, which learns the best actions for an agent by maximizing total rewards. It updates a Q-value for every state-action pair applying the Bellman equation. Without utilizing a model of the environment, the agent explores the environment and over time improves its function policies depending on the reviews [30].

3.5.3.2 Deep Q-Networks (DQN)

DQN is a class of reinforcement learning techniques that merge Q-learning with deep neural networks. It seeks to optimize decision-making problems by teaching an agent to operate in an environment, hence maximizing cumulative challenges. The agent understands a policy using a Q-function, which calculates the expected future challenge of actions provided in the current state. Compared to preserving a table of Q-values for every state-action pair, DQN utilizes a neural network to resemble the Q-values, keeping it suitable for high-dimensional or continuous state spaces. To improve efficiency and training stability, it presented techniques consisting of a target network like a distinct network to provide stable q value and also experience replay, such as reusing and storing past events to disrupt the correlation between samples of successive data [31].

3.5.4 Deep Learning

Image identification, sensor fusion, and NLP are examples of activities that require high-level feature extraction. Deep learning (DL) techniques such as recurrent neural networks (RNNs) and convolutional neural networks (CNNs) are extremely effective in handling huge and complicated data received from IoT devices [32]. For example, RNNs can forecast patient trends of health depending on time-series information from devices that are monitoring health; on the contrary, CNNs are used in healthcare IoT devices to interpret medical images captured by wearable devices [33].

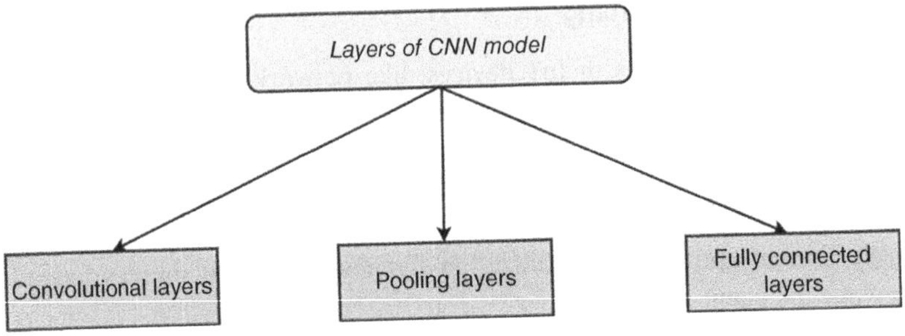

Figure 3.6 Key components of CNN models.

3.5.4.1 Convolutional Neural Networks (CNNs)

It is a type of DL technique commonly employed for processing tasks and image recognition. It has several layers, like convolutional layers, that apply filters to the input data to extract significant features such as patterns, textures, and edges. Since CNNs can capture local dependencies and spatial hierarchies, they are especially well-suited for object detection, image classification, and other tasks including visual input [34]. The layers of the CNN model are shown in Figure 3.6. CNN models are successful for applications including autonomous driving, facial recognition, and medical imaging because their architecture allows them to effectively and automatically learn and identify features from images.

3.5.4.2 Recurrent Neural Networks (RNNs)

It is a kind of neural network developed for sequential data that feeds the output from past steps as input to the present step. This design lets RNNs retain a memory of previous inputs, hence making them perfect for jobs like speech recognition, language modeling, time series, and other historical data. RNNs can, however, suffer from long-term dependencies because of problems such as vanishing gradients, so specially tailored variants like Long Short-Term Memory (LSTM) networks were designed to overcome the issue [35].

3.5.5 Federated Learning

The emerging ML technology is known as Federated Learning (FL), which helps train models on scattered IoT systems without having to centralize data [36]. This technology enhanced privacy and also diminished latency by keeping data on edge systems while sharing only the model changes. FL can be utilized in smart cities to train traffic forecasting algorithms across distributed edge systems (such as traffic cameras) without sending sensitive information to a central server. This is possible because smart city devices are becoming increasingly connected.

3.5.6 Transfer Learning

Transfer learning (TL) is used in IoT when models trained in a certain domain or system are adapted to another domain or system. This is especially helpful in IoT settings where labeled data are scarce since it enables the reuse of knowledge from a closely associated domain. A model that has been trained to identify patterns of human activity, such as running or walking, on a wearable device can be altered for use on another device with very little extra training. This is possible in the field of wearable technology [37].

These models provide a framework for tackling diverse issues concerning IoT, from real-time decision-making to predictive analysis. By using ML, IoT devices become more adaptive, responsive, and intelligent to dynamic environments.

3.6 Case Studies

3.6.1 Predictive Maintenance in Manufacturing

This section of the chapter illustrates how predictive maintenance is used by smart companies to avoid sudden breakdowns of their machinery, hence enhancing their profit.

3.6.1.1 Background

Predictive maintenance is a technique utilized in manufacturing to predict when the machinery is likely to fail to operate so that maintenance can be done on time, preventing a breakdown in companies. ML with approaches like anomaly detection, analysis, and time series using techniques like CNNs and RNNs plays a vital role in predictive maintenance [38].

3.6.1.2 Implementation

In companies with huge-scale manufacturing, different sensors are used along with machines to gather real-time data like pressure, machine operational status, temperature, and vibration. This data from sensors are used to train ML techniques, which understand the normal working conditions and can detect anomalies or variations that might indicate machinery failure [39]. The ML process is shown in Figure 3.7. By using this technology, industrialists can schedule maintenance of machinery before their failure, hence minimizing breakdown, enhancing machinery lifespan, and reducing maintenance cost.

3.6.1.3 Example

In a European automobile manufacturing company, ML models were utilized to predict failures of bearings in conveyor belts. The technology was able to forecast

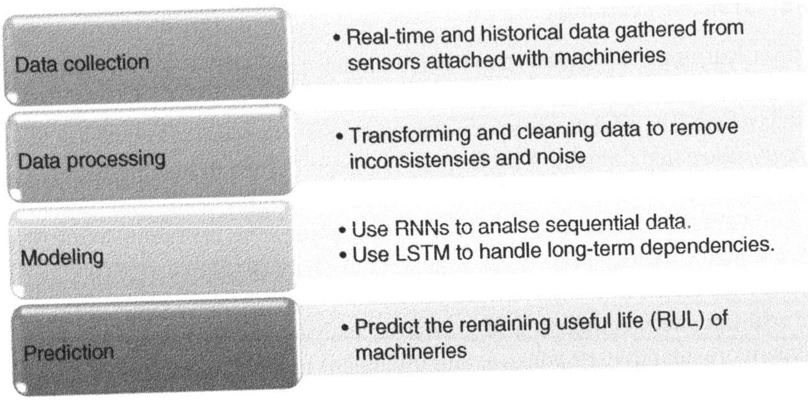

Figure 3.7 The process of machine learning.

failures with almost 90% accuracy by using past vibration information, hence reducing breakdown and enhancing manufacturing reliability [40].

3.6.1.4 Conclusion

Predictive maintenance applied in manufacturing companies has resulted in a notable enhancement of efficiency in productivity. 20% less machine breakdown since unforeseen machine issues are now predicted and avoided ahead of time. Since service is only done when required, preventing unneeded repairs, this proactive strategy has also resulted in a 15% drop in maintenance expenditures. Moreover, the general output throughput has grown since machinery is kept operational for longer periods and experiences less breakdown, thereby simplifying and improving the manufacturing process.

3.6.2 Smart Cities and Environmental Monitoring

Smart cities are metropolitan areas that employ technology and data to raise the standard of living for their citizens. One of the most crucial features of smart cities is environmental monitoring, which consists of traffic management and pollution control. With the swift growth of cities, keeping up with transportation congestion and cutting pollution have become serious concerns. Nowadays, ML is used to provide clever fixes for these issues [41].

Traffic congestion is a prevalent problem in most urban locations. The situation leads to longer travel times and increased fuel consumption and environmental pollution. Smart cities employ ML algorithms to track and analyze traffic patterns in real time for control purposes. Multiple sensors and cameras throughout the city gather information about vehicle speeds and traffic movement and road conditions. ML algorithms process data to identify traffic patterns and forecast congestion levels. The algorithm detects traffic jams through its sensors, which enables it to modify traffic light operations to reduce waiting times and decrease congestion while improving vehicle flow.

ML technology enables the creation of adaptive route planning systems for public transportation vehicles and emergency response vehicles. The system analyzes current traffic patterns to determine the quickest paths for buses and emergency vehicles, including ambulances and fire trucks. The system reduces travel duration while simultaneously decreasing fuel consumption and environmental carbon emissions. Smart cities use predictive models to forecast traffic patterns during peak hours and special events, which helps them optimize their traffic management systems [42].

The second major issue in cities is air pollution, which affects both the environment and public health. Smart cities employ ML for continuous pollution management and air quality monitoring. The city-wide sensors collect information about CO_2, NO_2, and $PM_{2.5}$ pollutants. The ML algorithms analyze the data to detect pollution hotspots and determine the pollution sources.

The ML model can notify local authorities about pollution hotspots so they can implement essential actions such as industrial restrictions or traffic management in specific areas. The ML system uses traffic volume, weather conditions, and industrial activities to predict pollution levels. The system provides early warnings through this information, which enables preventive measures such as indoor stay recommendations and public transit suggestions for days with poor air quality [43] (Figure 3.8).

The main benefit of implementing ML in smart cities involves uniting various systems. The combination of traffic management with pollution control systems produces an enhanced complete solution. ML algorithms can optimize traffic signals to reduce congestion while simultaneously decreasing vehicle idling time at junctions and lowering emissions. Cities can reduce their environmental impact by promoting electric and public transportation systems in areas with high pollution levels.

The implementation of ML in smart cities for traffic and pollution control offers a practical solution to create cleaner and more productive urban spaces. Real-time data and predictive analytics enable cities to make informed decisions that protect both environmental health and citizen quality of life [44].

3.6.3 Healthcare and Wearable Devices

The implementation of wearable gadgets in healthcare has revolutionized patient monitoring and health data acquisition processes. The electronic devices, which include health monitors and specialized medical equipment, generate real-time information that benefits both patients and healthcare professionals [45]. This case

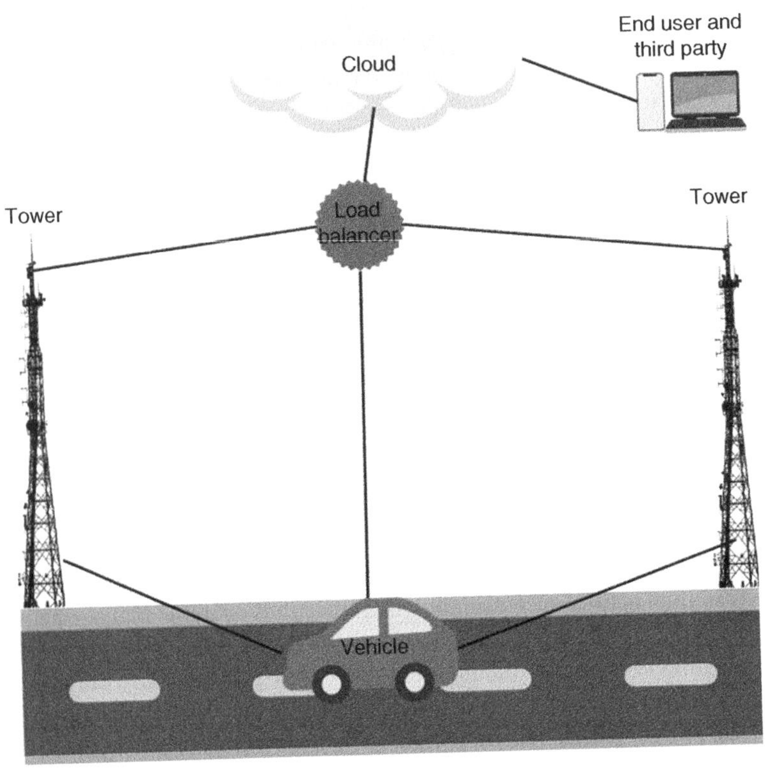

Figure 3.8 Smart city pollution.

study examines the applications of wearable gadgets for patient tracking and health data evaluation with a focus on enhancing healthcare outcomes.

3.6.3.1 Background

People use wearable electronic gadgets, which include smartwatches, fitness trackers, ECG monitors, and smart garments. The devices contain sensors that track multiple physiological indicators, including oxygen saturation (SpO_2), physical activity level, sleeping routines, and heart rate. The primary benefits of these devices include real-time data collection and remote monitoring and early warning systems for potential health risks.

Example Devices:

- Apple Watch Series 6: tracks heart rate, oxygen saturation, and ECG.
- Fitbit Sense—provides stress monitoring, ECG, and sleep tracking.

- AliveCor KardiaMobile—handy ECG monitor, particularly used for cardiac patients.
- Medtronic Guardian Sensor—constant glucose tracking for diabetic patients.

3.6.3.2 Applications in Patient Monitoring

- **Chronic Disease Management**: Individuals with chronic disorders such as cardiovascular diseases, diabetes, and hypertension benefit significantly from wearable electronic gadgets. For instance, glucose monitoring gadgets like Medtronic Guardian enable instant response on blood sugar levels, reducing the need for numerous hospital visits. Regular monitoring allows patients to maintain a suitable level of blood sugar while limiting complications like diabetic ketoacidosis. Likewise, wearable ECG monitoring aids in the early detection of irregular rhythms of the heart [46]. For patients who have atrial fibrillation, these electronic gadgets can alert them of irregular heartbeats, enabling them to go for appropriate medical intervention.
- **Postoperative Care:** Patients getting back from surgery usually need close monitoring of pulses. Wearable gadgets allow healthcare providers to track their healing progress virtually. For example, smartwatches are capable of tracking heart rate variability and level of oxygen, which constitute essential aspects of recovery. If any strange patterns are identified, the healthcare professional can take action before the condition escalates.
- **Elderly Care:** For the elders, wearable electronic devices can be a lifeline. The fall detection sensors fixed in wearables can instantly alert caregivers as well as emergency services. Furthermore, devices monitor health parameters such as heart rate and make sure that any inappropriate readings are immediately addressed, hence avoiding severe health crises [6].
- **Remote Patient Monitoring**: Remote patient monitoring (RPM) is one of the greatest transformative applications of wearable electronic devices. Patients with chronic illness, postsurgical recovery, or viral diseases can be observed from the convenience of their homes. This diminishes the pressure on healthcare facilities, reduces costs, and minimizes the chance of infection during hospital visits. For example, during the COVID-19 pandemic, external pulse oximeters were utilized extensively to follow patients' oxygen levels, warning them when their saturation headed below safe levels.

3.6.3.3 Applications in Health Data Analysis

- **Predictive Analytics**: Wearable devices gather huge amounts of data that can be examined to forecast health outcomes. For instance, by employing data from wearable ECG devices, predictive models can be developed to foresee cardiac events, such as heart attacks. ML algorithms employ patterns in heart rate, fluctuations, and rhythms to predict high-risk incidents, enabling preventive measures.

- **Personalized Healthcare**: Health data from devices enables healthcare professionals to provide personalized treatment suggestions. For example, data from health trackers can help modify diet and exercise routines for patients suffering from cardiovascular diseases or obesity. Patients suffering from sleep disorders can also get personalized recommendations based on sleep tracking data from devices such as Fitbit [47].
- **Improved Clinical Decision-Making**: Healthcare professionals benefit from the constant flow of data from devices, allowing them to make better clinical decisions. In situations such as ICU discharges, the device can track the vital signs of a patient for days or weeks, making sure no recurrence or undetected complications occur. This real-time data allows aggressive treatment adjustments depending on the patient's health patterns.
- **AI and Big Data Integration**: The huge amount of data produced by devices can be incorporated with artificial intelligence to find trends and connections. Big data analysis from devices helps in population health management, identifying health hazards among certain groups, and establishing public health strategies [48].

3.6.3.4 Challenges and Ethical Considerations

- **Data Privacy:** The gathering and transmission of personal health information causes significant privacy issues. Ensuring that this data is safe and only available to authorized persons is important to protect patient privacy.
- **Accuracy of Data:** As wearable devices are useful, the accuracy of the data they deliver can vary. False positives or negatives can lead to either anxiety or neglect of health issues. Devices must be thoroughly tested and validated to ensure reliability.
- **Integration with Healthcare Systems:** For wearable data to be beneficial, it must be incorporated into healthcare practitioners. Interoperability remains a difficulty, since different devices use different platforms and protocols, making it difficult for healthcare providers to access and understand the data efficiently [49].
- **Patient Adherence:** The efficiency of wearable technology is dependent on patient adherence. Users must consistently wear and keep up with the devices, as well as follow up on health alarms. Without proper adherence, the advantages of wearable monitoring may be compromised.

3.6.3.5 Conclusion

Wearable devices have altered the monitoring of patients and health data evaluation by giving continuous, real-time data that allows early detection of health hazards, personalized treatment methods, and enhanced clinical decision-making. While difficulties remain in terms of data quality, privacy, and interaction with healthcare systems, the potential advantages are important. As technology improves and becomes fully integrated into healthcare systems, wearable devices will go on to play an essential role in the future of healthcare.

3.7 Challenges and Future Directions

The integration of 6G and IoT opens new dimensions for addressing latency, security, and scalability challenges. With built-in AI support, distributed learning through federated and edge ML can succeed under 6G's high-speed, real-time architecture. Emerging trends comprise semantic communication, touchable IoT, and holographic engagements, all made viable by 6G-IoT synergies. IoT is quickly changing the world by linking commonplace objects, enabling them to communicate, process, and exchange data. This technology has many advantages, but it also faces several issues that must be resolved for successful and widespread implementation. It is necessary to tackle the full potential of IoT and determine its future course. There are four main challenges as shown in Figure 3.9.

One of the biggest problems in IoT is managing scalability and real-time processing. The expansion of IoT networks leads to a significant rise in the amount of data produced by millions of connected devices. The massive data stream needs robust systems to process continuous information flow without delays or mistakes. The massive data output from IoT devices, including sensors, cameras, and other

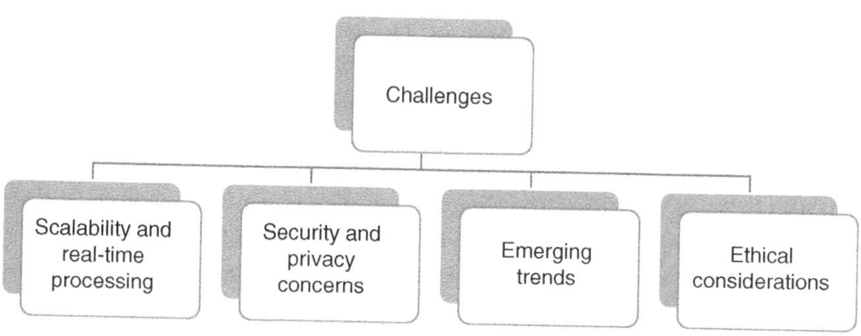

Figure 3.9 Types of challenges.

smart things, exceeds the capabilities of conventional data processing methods. Sensors in a smart city environment collect real-time data about traffic patterns and weather conditions and public safety. The processing of large data volumes needs to happen instantly to support decision-making activities such as traffic light control and emergency alert systems. The problem finds potential solutions through the implementation of cloud computing and edge computing systems. The processing of data through cloud computing occurs on remote servers, yet edge computing performs data processing near the source to achieve faster results with lower latency. A flexible and scalable infrastructure is required to support both options because it must handle increasing device numbers without compromising real-time processing capabilities. Additionally, maintaining various kinds of data (structured, unstructured, time series, etc.) poses another level of complexity. The analysis of different data formats at high speed is required for IoT systems to deliver meaningful data. The main challenge exists in creating systems that scale with device additions and adapt to handle diverse data outputs. As the IoT ecosystem expands, developing scalable, real-time processing frameworks becomes a priority area for researchers and practitioners [50].

Security and privacy are the biggest concerns in IoT. Since IoT devices are frequently used for gathering and transmitting sensitive data, they become easy targets for hackers. A single device with weak security measures can compromise the entire network by allowing data breaches and unauthorized access. The diverse range of devices in an IoT system makes it challenging to achieve robust security across all components, including industrial sensors and smart household appliances. The limited processing power and storage capacity of IoT devices prevents them from implementing standard security features, including encryption and multifactor authentication. The protection of personal information stands as a matter of equal importance to security concerns. The collection of personal data by IoT systems includes health measurements together with location data and behavioral patterns. The valuable data enables better insights and decision-making but creates privacy risks for users. The development of privacy-preserving ML serves as a solution to address this issue. The data remains encrypted throughout processing to prevent unauthorized parties from accessing it with decryption keys. The encryption process adds extra computational load to IoT devices, which have limited resources, thus requiring a compromise between system efficiency and data privacy protection. The General Data Protection Regulation (GDPR) of Europe, along with other regulations, require IoT systems to follow strict guidelines about personal data collection and processing and storage practices. The implementation of privacy-focused IoT system design at the beginning stage becomes essential to meet regulatory requirements while maintaining system performance. The IoT sector faces an ongoing battle to find the right equilibrium between delivering useful services and defending user privacy [51].

The solution to these problems comes from new trends in IoT data processing. The method of FL enhances data privacy because it allows multiple devices to train ML models without moving data to a central server. The model distribution to individual devices allows training on local data, while the aggregated updates enhance the global model. The local device maintains all sensitive data within its boundaries, thus reducing the risk of data breaches. The healthcare industry benefits from FL because it ensures essential patient data protection. The method introduces new security requirements to ensure safe model updates that protect both device and individual security. The emerging TinyML approach enables ML operations on small, low-power IoT devices. TinyML enables on-site data processing for devices, which reduces the need for external server data transfer and results in faster operations with lower latency and better privacy protection. A smart thermostat uses TinyML to analyze local temperature patterns for adjustments without cloud data transfer. The advantages of TinyML come with challenges to optimize ML models for efficient operation on limited resource devices, which demands continuous research and development. The field of IoT is increasingly adopting Explainable AI (XAI) as a significant technology. The decision-making process of conventional AI models remains unclear because they operate as "black boxes" without providing any information about their reasoning. The main goal of XAI is to enhance model transparency so users can understand the reasoning behind particular decisions. The need for transparency in AI-driven IoT systems becomes critical because it builds confidence in applications such as healthcare, finance, and law enforcement, where decisions produce significant real-world impacts. XAI provides explanations to ensure IoT technologies are used both properly and ethically [52].

The development of IoT technology has made fairness, transparency, and accountability issues more significant. The impact of IoT analytics on human life requires systems to operate with complete fairness. A smart healthcare system must provide equal benefits to all patients regardless of their socioeconomic status or background. The occurrence of unfair results from biased data collection and processing requires the use of impartial data and methods during IoT system development. Transparency stands as an essential ethical concern. Users need to understand IoT systems to make decisions, which is why XAI serves as a crucial component. The likelihood of users trusting AI model decisions leads to increased trust and acceptance of greater accountability. The use of IoT technologies in sensitive fields such as finance, law enforcement, and healthcare requires complete transparency. The system must have established protocols to identify and explain errors and correct them in IoT systems. The smart traffic management system should have a system to identify the cause of congestion failure and assign responsibility for fixing the issue. The development of accountability frameworks is necessary to establish dependable and credible IoT systems [53].

3.8 Conclusion

ML strengthens the IoT by transforming passive data collection groups into proactive executives. ML enhances IoT devices by streamlining duties, enhancing operations, recognizing abnormalities, and enabling real-time decisions. These breakthroughs allow regular resource maintenance and management and improvements in smart cities, industry, and healthcare. The most popular ML techniques were reinforcement, supervised, DL, and unsupervised. These simulations deal with IoT difficulties that involve high data volume, delay, and variance, allowing systems to investigate enormous amounts of data from various devices. In this case, ML is employed in scheduled servicing for IIoT, which helps prevent equipment breakdown, as well as in smart cities, wherein congestion and ecological data are analyzed in real time to improve the operation of cities. Among the most important issues highlighted is the requirement for scalable, rapid information processing. Solutions like edge computing and collaborative learning offer ways to analyze information nearer to its source statistically, minimizing latency and boosting system performance. Furthermore, ML plays a critical role in securing IoT networks, recognizing possible cyberattacks through anomaly detection, and enhancing user privacy using techniques such as FL. This study proposes persistent innovation in ML to deal with the growing complexity and data associated with IoT systems. To fully unlock the IoT's potential, future improvements must focus on advancing more efficient, flexible, and privacy-preserving ML models. By conquering these challenges, the incorporation of ML and IoT will result in smarter, more effective, and more adaptable systems, driving growth in various sectors and increasing the quality of life in associated circumstances.

CONSENT FOR PUBLICATION

Not Applicable

CONFLICT OF INTEREST

Not Applicable

ACKNOWLEDGEMENT

This work was supported by the Deanship of Scientific Research, the Vice Presidency for Graduate Studies and Scientific Research, King Faisal University, Saudi Arabia, under Grant No. KFU254757

References

1 Khare, S. and Totaro, M. (2019). Big data in IoT. *2019 10th International Conference on Computing, Communication and Networking Technologies (ICCCNT)*, pp. 1–7. IEEE. https://doi.org/10.1109/ICCCNT45670.2019.8944495.

2 Constante Nicolalde, F., Silva, F., Herrera, B., and Pereira, A. (2018). Big data analytics in IoT: challenges, open research issues and tools. *Advances in Intelligent Systems and Computing* 746 (May 2019): 775–788. https://doi.org/10.1007/978-3-319-77712-2_73.
3 Bzai, J., Alam, F., Dhafer, A. et al. (2022). Machine learning-enabled internet of things (IoT): data, applications, and industry perspective. *Electron* 11 (17): 1–33. https://doi.org/10.3390/electronics11172676.
4 Mahdavinejad, M.S., Rezvan, M., Barekatain, M. et al. (2019). Machine learning for internet of things data analysis: a survey. *Digital Communications and Networks* 4 (3): 161–175. https://doi.org/10.1016/j.dcan.2017.10.002.
5 Bishop, C.M. (2006). Pattern recognition and machine learning. In: *Information Science and Statistics*, 1e, 778. New York: Springer https://link.springer.com/book/9780387310732.
6 Umer, M., Aljrees, T., Karamti, H. et al. (2023). Heart failure patients monitoring using IoT-based remote monitoring system. *Scientific Reports* 13 (1): 1–19. https://doi.org/10.1038/s41598-023-46322-6.
7 Tiwari, D., Prasad, D., Guleria, K., and Ghosh, P. (2021). IoT-based smart healthcare monitoring systems: a review. *2021 6th International Conference on Signal Processing, Computing and Control (ISPCC)*, pp. 465–469. https://doi.org/10.1109/ISPCC53510.2021.9609393.
8 Hamdan, S., Ayyash, M., and Almajali, S. (2020). Edge-computing architectures for internet of things applications: a survey. *Sensors* 20 (22): 1–52. https://doi.org/10.3390/s20226441.
9 Teh, H.Y., Kempa-Liehr, A.W., and Wang, K.I.K. (2020). Sensor data quality: a systematic review. *Journal of Big Data* 7 (1): 1–49. https://doi.org/10.1186/s40537-020-0285-1.
10 Joshi, M.A., Raval, M.S., Dandawate, Y.H. et al. (2014). Introduction to image compression. *Image and Video Compression* 18–21. https://doi.org/10.1201/b17738-4.
11 Jie Li et al. (2017)"Networked Human Motion Capture System Based on Quaternion Navigation," Proceedings of the 11th International Conference on Body Area Networks, https://doi.org/10.4108/eai.15-12-2016.2267544.
12 Marjani, M., Nasaruddin, F., Gani, A. et al. (2017). Big IoT data analytics: architecture, opportunities, and open research challenges. *IEEE Access* 5: 5247–5261. https://doi.org/10.1109/ACCESS.2017.2689040.
13 Pandey, B.K., Veeramanickam, M.R.M., Ahmad, S. et al. (2023). ExpSSOA-deep maxout: exponential shuffled shepherd optimization based deep maxout network for intrusion detection using big data in cloud computing framework. *Computers & Security* 124: 102975. https://doi.org/10.1016/j.cose.2022.102975.
14 Barnaghi, P., Sheth, A., and Henson, C. (2013). From data to actionable knowledge: big data challenges in the web of things. *IEEE Intelligent Systems* 28 (6): 6–11. https://doi.org/10.1109/MIS.2013.142.

15 Abbasi, M.A., Memon, Z.A., Syed, T.Q. et al. (2017). Addressing the future data management challenges in IoT: a proposed framework. *International Journal of Advanced Computer Science and Applications* 8 (5): 197–207. https://doi.org/10.14569/ijacsa.2017.080525.

16 Ezema, E., Abdullah, A., and Mohd, N.F.B. (2018). Open issues and security challenges of data communication channels in distributed internet of things (IoT): a survey. *Circulation in Computer Science* 3 (1): 22–32. https://doi.org/10.22632/ccs-2017-252-63.

17 Kamiran, F. and Calders, T. (2012). Data preprocessing techniques for classification without discrimination. *Knowledge and Information Systems* 33 (1): https://doi.org/10.1007/s10115-011-0463-8.

18 García, S., Luengo, J., and Herrera, F. (2015). Instance selection. In: *Data Preprocessing in Data Mining*, vol. 72. https://doi.org/10.1007/978-3-319-10247-4_8.

19 Rodriguez, C.R., Angeles, D., Chafloque, R. et al. (2020). Deep learning audio spectrograms processing to the early COVID-19 detection. *2020 12th International Conference on Computational Intelligence and Communication Networks (CICN)*, pp. 429–434. https://doi.org/10.1109/CICN49253.2020.9242583.

20 Çetin, V. and Yıldız, O. (2022). A comprehensive review on data preprocessing techniques in data analysis. *Pamukkale University Journal of Engineering Sciences* 28 (2): 299–312. https://doi.org/10.5505/pajes.2021.62687.

21 Sardar, T.H. and Pandey, B.K. (2023). *Big Data Computing*. Boca Raton: CRC Press https://doi.org/10.1201/9781032634050.

22 Kumar, D. and Kukreja, V. (2022). Deep learning in wheat diseases classification: a systematic review. *Multimedia Tools and Applications* 81 (7): https://doi.org/10.1007/s11042-022-12160-3.

23 Khan, M.A. and Salah, K. (2018). IoT security: review, blockchain solutions, and open challenges. *Future Generation Computer Systems* 82: 395–411. https://doi.org/10.1016/j.future.2017.11.022.

24 Lu, Y., Ye, T., and Zheng, J. (2022). Decision tree algorithm in machine learning. *2022 IEEE International Conference on Advances in Electrical Engineering and Computer Applications (AEECA)*, pp. 1014–1017. IEEE. https://doi.org/10.1109/AEECA55500.2022.9918857.

25 Avendaño-Valencia, L.D. and Fassois, S.D. (2015). Natural vibration response based damage detection for an operating wind turbine via random coefficient linear parameter varying AR modelling. *Journal of Physics Conference Series* 628 (1): 273–297. https://doi.org/10.1088/1742-6596/628/1/012073.

26 Risteska Stojkoska, B.L. and Trivodaliev, K.V. (2017). A review of internet of things for smart home: challenges and solutions. *Journal of Cleaner Production* 140: 1454–1464. https://doi.org/10.1016/j.jclepro.2016.10.006.

27 Vankayalapati, R., Ghutugade, K.B., Vannapuram, R., and Prasanna, B.P.S. (2021). K-means algorithm for clustering of learners performance levels using machine learning techniques. *Revue d'Intelligence Artificielle* 35 (1): 99–104. https://doi.org/10.18280/ria.350112.

28 Daszykowski, M. and Walczak, B. (2020). A density-based algorithm for discovering clusters in large spatial databases with noise. In: *Comprehensive Chemometrics: Chemical and Biochemical data Analysis*, 2e, vol. 2, 565–580. https://doi.org/10.1016/B978-0-444-64165-6.03005-6.

29 Kato, N., Fadlullah, Z.M., Mao, B. et al. (2017). The deep learning vision for heterogeneous network traffic control: proposal, challenges, and future perspective. *IEEE Wireless Communications* 24 (3): 146–153. https://doi.org/10.1109/MWC.2016.1600317WC.

30 Khenak, F. (2010). V-learning. *2010 International Conference on Computer Information Systems and Industrial Management Applications (CISIM)* (Vol. 292, pp. 228–232). https://doi.org/10.1109/CISIM.2010.5643660.

31 Mnih, V., Kavukcuoglu, K., Silver, D. et al. (2015). Human-level control through deep reinforcement learning. *Nature* 518 (7540): 529–533. https://doi.org/10.1038/nature14236.

32 Kukreja, V. and Dhiman, P. (2020). A deep neural network based disease detection scheme for citrus fruits. *Proceedings of the 2020 International Conference on Smart Electronics and Communication (ICOSEC 2020)*, pp. 97–101. https://doi.org/10.1109/ICOSEC49089.2020.9215359.

33 Mohammadi, M., Al-Fuqaha, A., Sorour, S., and Guizani, M. (2018). Deep learning for IoT big data and streaming analytics: a survey. *IEEE Communication Surveys and Tutorials* 20 (4): 2923–2960. https://doi.org/10.1109/COMST.2018.2844341.

34 Hossain, M.A. and Alam Sajib, M.S. (2019). Classification of image using convolutional neural network (CNN). *Global Journal of Computer Science and Technology* 19 (May): 13–18. https://doi.org/10.34257/gjcstdvol19is2pg13.

35 Hochreiter, S. and Schmidhuber, J. (1997). Long short-term memory. *Neural Computation* 9 (8): 1735–1780. https://doi.org/10.1162/neco.1997.9.8.1735.

36 Li, Q., Wen, Z., Wu, Z. et al. (2023). A survey on federated learning systems: vision, hype and reality for data privacy and protection. *IEEE Transactions on Knowledge and Data Engineering* 35 (4): 3347–3366. https://doi.org/10.1109/TKDE.2021.3124599.

37 Jain, M., Jain, N., Lee, Y.H. et al. (2023). Transfer learning for cloud image classification. *The International Geoscience and Remote Sensing Symposium* 2023: 6005–6008. https://doi.org/10.1109/IGARSS52108.2023.10281791.

38 Karuppusamy, P. (2021). Machine learning approach to predictive maintenance in manufacturing industry - a comparative study. *Journal of Soft Computing Paradigm* 2 (4): 246–255. https://doi.org/10.36548/jscp.2020.4.006.

39 Çınar, Z.M., Nuhu, A.A., Zeeshan, Q., and Korhan, O. (2020). Sustainability-12-08211-V3.Pdf. *Sustainability* (Ml).

40 Lee, J., Bagheri, B., and Kao, H.A. (2015). A cyber-physical systems architecture for industry 4.0-based manufacturing systems. *Manufacturing Letters* 3 (December): 18–23. https://doi.org/10.1016/j.mfglet.2014.12.001.

41 Patti, E. and Acquaviva, A. (2016). IoT platform for smart cities: requirements and implementation case studies. *2016 IEEE 2nd International Forum on Research and Technologies for Society and Industry Leveraging a better tomorrow (RTSI)*, pp. 1–6. IEEE. https://doi.org/10.1109/RTSI.2016.7740618.

42 Shah, S.A., Ren, A., Fan, D. et al. (2018). Internet of things for sensing: a case study in the healthcare system. *Applied Sciences* 8 (4): 508. https://doi.org/10.3390/app8040508.

43 Zhang, N., Chen, H., Chen, X., and Chen, J. (2016). Semantic framework of internet of things for smart cities: case studies. *Sensors* 16 (9): https://doi.org/10.3390/s16091501.

44 Mehmood, Y., Ahmad, F., Yaqoob, I. et al. (2017). Internet-of-things-based smart cities: recent advances and challenges. *IEEE Communications Magazine* 55 (9): 16–24. https://doi.org/10.1109/MCOM.2017.1600514.

45 Alkhateeb, A., Catal, C., Kar, G., and Mishra, A. (2022). Hybrid blockchain platforms for the internet of things (IoT): a systematic literature review. *Sensors* 22 (4): https://doi.org/10.3390/s22041304.

46 Huang, Q. and Kieffer, K. (2019). An intelligent internet of things (IoT) sensor system for building environmental monitoring. *Journal of Mobile Multimedia* https://doi.org/10.13052/1550-4646.15122.

47 Wan, J., Al-awlaqi, M.A.A.H., Li, M. et al. (2018). Wearable IoT enabled real-time health monitoring system. *EURASIP Journal on Wireless Communications and Networking* 2018 (1): 1–10.

48 Kamruzzaman, M.M., Alanazi, S., Alruwaili, M. et al. (2023). AI- and IoT-assisted sustainable education systems during pandemics, such as COVID-19, for smart cities. *Sustainability* 15 (10): 1–18. https://doi.org/10.3390/su15108354.

49 Tawalbeh, L., Muheidat, F., Tawalbeh, M., and Quwaider, M. (2020). IoT privacy and security: challenges and solutions. *Applied Sciences* 10 (12): 4102. https://doi.org/10.3390/app10124102.

50 Kobyzev, I., Prince, S.J.D., and Brubaker, M.A. (2021). Normalizing flows: an introduction and review of current methods. *IEEE Transactions on Pattern Analysis and Machine Intelligence* 43 (11): 3964–3979. https://doi.org/10.1109/TPAMI.2020.2992934.

51 Chanal, P.M. and Kakkasageri, M.S. (2020). Security and privacy in IoT: a survey. *Wireless Personal Communications* 115 (2): 1667–1693. https://doi.org/10.1007/s11277-020-07649-9.

52 Irmak, E. and Bozdal, M. (2018). Internet of things (IoT): the most up-to-date challenges, architectures, emerging trends and potential opportunities. *International Journal of Computers and Applications* 179 (40): 20–27. https://doi.org/10.5120/ijca2018916946.

53 Ebersold, K. and Hartford, T. (2016). The internet of things: a cause for ethical concern. *Issues Information Systems* 17 (IV): 145–151. https://doi.org/10.48009/4_iis_2016_145-151.

4

Synergies of 6G and IoT

Harpreet Kaur, Kritika, Jasmine, and Vanshika Gupta

Chitkara School of Engineering and Technology, Chitkara University, Baddi, Himachal Pradesh, India

4.1 6G Technology Foundations

Sixth generation (6G) is the upcoming wireless standard that is expected to succeed fifth generation (5G) around 2030, with advanced features such as very high data rates, ultralow latency, integrated sensing and communication, artificial intelligence (AI)-native networking, and high energy efficiency and reliability.

6G aims to support around 10 million connected devices per square kilometer, deliver data rates exceeding 1 terabit per second (Tbps), and achieve ultralow latencies as low as 10–20 microseconds. Its extremely high data rate capacity makes it ideal for bandwidth-intensive applications such as virtual reality (VR), augmented reality (AR), and holographic communications, where massive data must be transmitted in real time.

4.1.1 Terahertz Communication

Terahertz (THz) communication plays a crucial role in 6G wireless networks; it is within the range of 100 GHz to 10 THz, where its range falls between microwave and infrared. This spectrum provides an expansive bandwidth, which is used to transfer data to reach remarkable levels, ranging from multi-Gigabits per second (Gbps) to Terabits per second (Tbps), significantly surpassing those of 5G.

THz waves have very short wavelengths (0.003–3 mm), facilitating the development of compact and high-density antenna arrays suitable for devices like smartphones or IoT sensors. This improves signal directionality, power, and spatial resolution, which are crucial for uses in real-time AR/VR, self-driving vehicles,

Synergies of 6G and IoT, First Edition. Edited by Himanshi Babbar.

Published 2026 by John Wiley & Sons, Inc.

and extremely fast wireless communication [1]. Moreover, its focused nature can give benefits to enhance the security of communication, as no signal loss or leakage should be encountered.

Still, some challenges THz have to face, like atmospheric attenuation, which is mainly due to water vapor, can easily be obstructed by obstacles such as walls or even the human body. Due to this reason only, it limits the short-range, line-of-sight scenarios such as indoor networks or data centers [2]. To improve this limitation, advanced beamforming, intelligent reflecting surfaces (IRS), and AI-driven routing.

The THz communication technology is in continuous progress. The improved transceiver designs and antennas based on new materials are beginning to tackle the key obstacles in bridging the THz gap. THz communication has the potential for delivering the ultrahigh capacity and extremely low latency demanded by 6G, but achieving this potential depends on overcoming significant challenges in signal transmission and hardware development.

4.1.2 Reconfigurable Intelligent Surfaces (RISs)

RISs are considered one of the most revolutionary techniques for improving the energy efficiency and spectrum of wireless communication systems, offering a new dimension of control and adaptability in communication systems [3]. These surfaces are the smart surfaces used to direct and enhance wireless signal communication for improved signal quality, interference control, and lower energy consumption.

RISs can be installed on large flat surfaces such as ceilings, walls, or buildings and are capable of independently modifying the phase, amplitude, or polarization of incident electromagnetic (EM) waves. This fine-grained control allows for the dynamic shaping of wireless signals to reinforce desired propagation paths and suppress unwanted ones, thereby enhancing signal quality, extending coverage, and optimizing resource utilization. Unlike conventional static wireless channels, RIS introduces the concept of a smart, programmable communication landscape, enabling unprecedented control over how signals traverse space. This is particularly significant in the context of the IoT, where reliable, efficient, and adaptive connectivity is essential for billions of interconnected devices [4, 5].

The architecture of RIS continues to evolve rapidly, with designs ranging from passive surfaces, which simply reflect signals in customized directions, to active surfaces incorporating signal processing, amplification, and even sensing capabilities. Emerging RIS structures include transmissive configurations, 3D stacked meta surfaces for volumetric coverage, and dynamic meta surface antennas that blend conventional beamforming with programmable surface control. These advanced designs support various operating modes, such as reflective, refractive, hybrid, and full-duplex operations, each suited for different deployment scenarios and performance objectives [3–5].

The integration of RIS into wireless networks allows for real-time adaptability of the communication environment. In practical terms, this means that RIS-equipped networks can dynamically reconfigure their radio environment in response to changing user locations, device densities, or environmental obstructions. This adaptability is crucial in fulfilling the demanding requirements of 6G networks and IoT ecosystems, where network performance must remain robust under dynamic and heterogeneous conditions. For example, in urban environments with dense infrastructure, RIS panels mounted on building surfaces or street furniture can mitigate signal blockages and dead zones, enabling continuous connectivity for mobile IoT devices. In industrial settings, RISs can reduce interference and latency in smart factories, while in healthcare, they can ensure reliable connections for wearable or implantable medical devices within complex indoor spaces.

Real-world validation of RIS is already underway. Prototypes integrated into commercial 5G networks have demonstrated substantial performance gains, particularly in improving coverage and throughput in challenging environments. Field trials involving RIS panels mounted on buildings or vehicles show that the technology can effectively compensate for obstructions and extend network reach. These practical deployments not only underscore the viability of RIS in future networks but also highlight the importance of addressing implementation challenges such as control signaling overhead, hardware imperfections, real-time channel estimation, and system scalability.

Despite its promising capabilities, RIS technology faces several technical and practical challenges. Managing large arrays of programmable elements requires efficient coordination and low-latency control mechanisms. Accurate and timely channel state information (CSI) is essential for effective RIS configuration, yet remains difficult to obtain, particularly in fast-changing or high-mobility scenarios. Furthermore, ensuring the security and privacy of RIS-assisted communications is an emerging concern, as programmable surfaces could be exploited for eavesdropping or signal manipulation if not properly secured [3, 4].

As the vision of 6G becomes a reality, the role of RIS in achieving intelligent, ubiquitous communication will be both fundamental and transformative, making it one of the most exciting frontiers in the evolution of wireless technology.

4.1.3 Massive MIMO and Beamforming in 6G

Massive MIMO is an essential wireless communication technique, especially effective for configurations with a high density of antennas. It employs numerous antennas at both the sending and receiving ends to significantly enhance data transfer speeds, spectral efficiency, and overall reliability. Building on traditional

MIMO, Massive MIMO increases the number of antennas, enabling efficient communication with multiple users simultaneously. Massive MIMO is set to be a foundational element of 6G networks for providing unmatched capacity, spectral efficiency, and user connectivity.

The implementation of dense antenna arrays in Massive MIMO improves beamforming and spatial multiplexing, which are critical for addressing the considerable path loss at THz frequencies. Transmission of multiple data streams over the same frequency is called spatial multiplexing, which significantly boosts the network capacity. Additionally, beamforming enhances signal quality, range, and efficiency by precisely directing signals to users or specific areas. This targeted transmission reduces interference, thereby enhancing user experience, especially in busy urban and indoor environments. With the rapid increase of technological innovation and rising commercial demands, massive MIMO is expected to evolve further in the future and reshape future telecommunications and related areas more broadly and deeply, especially as the industry advances toward 6G [6].

Extremely large-scale MIMO (XL-MIMO) is an upgraded version of Massive MIMO, characterized by narrower antenna spacing compared to typical systems and necessitating novel design approaches due to its distinct EM properties. The transition to ultradense antenna arrangements supports more accurate and intelligent beam management, crucial for achieving the low-latency and high-capacity targets of 6G. Furthermore, the installation of numerous antennas at the base station focuses energy in targeted regions, enhancing energy efficiency while reducing interference and fading, leading to better overall performance [7]. XL-MIMO also serves as a convergence point for various 6G enablers, including RIS, cell-free architectures, and AI-based management systems. Collectively, these advancements elevate massive and XL-MIMO from mere antenna technologies to foundational platforms enabling intelligent, energy-efficient, and space-integrated wireless communication in 6G. Figure 4.1 shows traditional, massive, and XL-MIMO systems for comparison.

These advancements contribute to improved spectral efficiency and increased spatial degrees of freedom. XL-MIMO is anticipated to provide ultrahigh capacity,

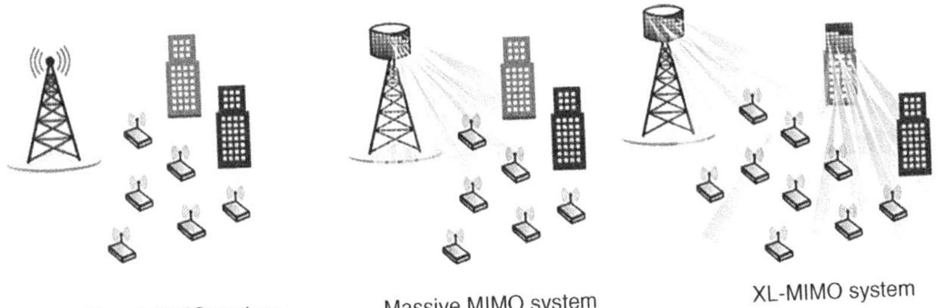

Figure 4.1 Evolution of MIMO systems from traditional MIMO to massive MIMO and XL-MIMO. *Source:* Adapted from Pan et al. [3].

near-field beamforming, and uninterrupted connectivity, especially in densely populated IoT environments [8].

Both Massive and XL-MIMO technologies unlock advanced applications such as wireless localization, physical layer security, vehicular communication, and non-terrestrial networks (NTNs), including satellites and unmanned aerial vehicles (UAVs) [6, 8]. They also support futuristic use cases like large-scale digital twins, remote multiplexing, and allowing spatial multiplexing [6].

However, alongside its growing importance, complex challenges are arising [8] in the context of large-scale, real-time, and energy-efficient communication for future IoT ecosystems. Massive MIMO leverages large-scale antenna arrays, but hardware costs are a limitation if antenna count scales up [6]. In XL-MIMO, challenges worsen with the extremely large number of antennas and near-field effects, which render conventional far-field models ineffective [8]. Channel estimation and CSI feedback, the information sent from the receiver to help improve signal quality, become difficult as antenna scale increases [7]. Additionally, hardware impairments like I/Q imbalance and phase noise affect system performance, particularly when low-cost radio frequency chain components are responsible for up/down-converting and processing radio signals, which may introduce performance limitations. Mutual coupling and space limitations in compact antenna arrays further impact system capacity [7].

To meet the complexity of future heterogeneous networks, AI-driven optimization techniques are increasingly essential for tasks such as channel prediction, dynamic beamforming, and hardware impairment compensation [7].

4.1.4 Integrated Space-Air-Ground-Sea Communication

It is a well-known fact that 71% of the earth's surface is covered by water. Most of the cargo transportation is done by sea. Conventional offshore communication cannot cover sea-surface connectivity between devices and users that are farther from the coastline. Thus, Space-Air-Ground networks have become a progressive solution for territorial communication in the 6G wireless communication network, owing to their capability of enabling global connectivity in remote locations such as islands and rural areas. The novel structure of Space-Air-Ground-Sea Integrated Network (SAGSIN) is represented in [9], which incorporates different types of relay stations for making communication links.

4.2 IoT Evolution and its Integration with 6G

The architecture of IoT has evolved significantly, transitioning from basic sensor-driven networks to sophisticated, intelligent ecosystems. 6G-integrated IoT acts as the nervous system of the future IoT landscape, enabling real-time intelligence, hyper-connectivity, and global scalability. IoT is poised to become more pervasive, autonomous, and seamlessly embedded into everyday life [10].

4.2.1 IoT Architecture Evolution

The evolution of IoT architecture plays a significant role in becoming more autonomous, pervasive, and integral to everyday life. Table 4.1 shows the evolution of IoT architecture from traditional IoT to intelligent 6G-powered IoT, reflecting a paradigm shift [11].

In Table 4.1, traditional IoT architecture has been compared with 6G-integrated IoT for different types of layers. 6G's advanced architecture provides a robust foundation for supporting autonomous and secure applications.

4.2.2 IoT Requirements for 6G

IoT is set to experience a major leap forward with the arrival of 6G technology. This next-generation network promises to reduce latency, extend coverage, and enhance energy efficiency. With 6G, IoT will power innovations like smart cities, autonomous vehicles, and immersive experiences, creating an interconnected world that operates more efficiently and seamlessly than ever before.

4.2.2.1 Latency

Latency refers to the time delay between initiating a request and receiving a response in a network or the time taken to travel from the source to the destination. As said earlier, low latency is one of the most important requirements that

Table 4.1 Evolving IoT Architecture Layers (6G Perspective)

Type of Layer	Traditional IoT	6G-Integrated IoT
Perception	Sensors and radio frequency identification (RFID)	AI-enhanced smart sensors
Network	IP based (3G/4G/5G mobile connectivity)	6G (THz, VLC, and NTNs) VLC—Visible Light Communication
Edge/Processing	Basic edge/fog computing	Edge AI, decentralized intelligence
Application	Alerts/monitoring	Predictive analysis, real-time control
Security	Basic encryption	Blockchain-based IoT, quantum security

helps in developing IoT. 6G aims to achieve latency as low as 10–100 microseconds, significantly reducing the 1 ms latency target of 5G. In IoT applications involving real-time data processing, such as autonomous vehicles, remote surgery requires ultralow latency to function effectively and safely. It ensures devices can communicate and respond in real time, enhancing the efficiency and safety of operations. 6G networks aim for a latency of less than 1 ms to support applications that demand instant feedback. But on the same hand, achieving such low latency necessitates advancements in network architecture, including edge computing, IRS, and AI-driven network management [10].

4.2.2.2 Coverage

Coverage is another important requirement for IoT in 6G. Coverage refers to the extent of the geographical area over which a network provides service. As the number of connected devices continues to grow, coverage becomes very crucial. 6G aims to provide coverage to over 99% of the Earth's surface by providing seamless connectivity in urban, rural, and remote areas, enabling IoT devices to function effectively regardless of their location. NTNs play a crucial role in extending coverage beyond traditional terrestrial infrastructures by utilizing satellites, drones, and high-altitude platforms. These technologies enable connectivity in remote and hard-to-reach areas where conventional networks may be limited or unavailable. Complementing this, seamless connectivity ensures uninterrupted service for users as they transition between different coverage areas, such as moving from urban environments to rural zones. Together, these advancements contribute to a more comprehensive and reliable network experience, supporting a wide range of applications and enhancing overall user satisfaction [10, 12].

4.2.2.3 Energy

Energy efficiency refers to the ability of a network to provide services while minimizing energy consumption. As IoT devices multiply, energy-efficient solutions are more pressing than ever. 6G technology is expected to support energy-neutral devices that can operate without traditional power sources, relying instead on energy harvesting techniques. This advancement will not only reduce operational costs but also contribute to sustainability efforts by minimizing the carbon footprint of IoT networks. IoT devices are being developed to harvest energy from sources like solar or vibrations, allowing them to run without batteries. Additionally, AI and IRS help optimize energy use in communication systems by improving signal efficiency, making 6G networks more energy-efficient and sustainable. IRS consists of smart surfaces that can dynamically reflect radio signals in targeted directions, enhancing coverage and reducing energy consumption. Controlled by AI, these surfaces adapt in real time to environmental conditions, making 6G networks more energy-efficient and sustainable [12].

Table 4.2 Target of 5G and 6G as per IoT Requirements

Requirement	Target of 5G	Target of 6G	Importance
Latency	1 ms	10–100 microseconds	Critical for real-time applications
Coverage	Approx. 70%	>99%	Ensures global connectivity
Energy	Standard	Ultralow	Supports sustainable IoT networks

Table 4.2 shows the targets of 5G and 6G technologies for fulfilling the key requirements of IoT. The lower values of latency and energy consumption parameters in 6G targets play an important role in real-time and sustainable IoT applications. In addition, extended coverage ensures global connectivity in 6G.

4.2.3 AI, Blockchain and Edge Computing for Secure IoT

Security remains a major challenge in expanding IoT ecosystems due to the increased surface area for cyberattacks. The integration of AI, machine learning (ML), edge computing, and blockchain is transforming the domain of IoT networks. AI and ML enhance data management by processing the huge amount of information provided by IoT devices. These technologies help in maintenance, flexible resource distribution, and self-operating system functions, which lessen the necessity for manual intervention. By identifying patterns and irregularities, AI anticipates failures and reacts to changing conditions in real time, thereby boosting both reliability and operational productivity [11].

Edge AI, a significant progress in this arena, brings computing capabilities nearer to where the data is generated. Rather than sending data to centralized cloud servers, edge devices handle it on-site. This approach greatly reduces latency and bandwidth use, making edge AI particularly suitable for applications that require quick responses, such as industrial automation, self-driving vehicles, and healthcare monitoring [13]. Additionally, decreased reliance on the cloud enhances energy efficiency and allows for more private and secure data management [14]. Edge computing addresses part of this concern by minimizing data transmission, thus reducing exposure to external threats. Furthermore, integrating blockchain technology (BCT) into IoT systems offers a decentralized and tamper-proof mechanism for securing data. Blockchain's cryptographic principles and smart contract functionality ensure secure, autonomous, and verifiable transactions between devices. This is particularly relevant in sectors where trust and data integrity are paramount, such as smart healthcare and supply chain management [14, 15].

Cognitive networks, powered by AI, extend these capabilities further by enabling IoT systems to learn from past experiences and adapt to new environments autonomously. These self-optimizing networks enhance communication efficiency, automate maintenance, and refine performance over time without manual reconfiguration. Together, these technologies establish a scalable, secure, and intelligent foundation for next-generation IoT systems, capable of meeting the demands of increasingly complex and data-intensive applications.

4.2.4 Sustainable IoT Design and Green Networking

The rapid expansion of the IoT has brought about tremendous benefits by connecting billions of devices to improve efficiency and convenience in various sectors such as healthcare, agriculture, and smart cities. However, this proliferation of IoT devices also raises significant environmental concerns, primarily due to the increasing energy consumption and electronic waste generated by these devices. This has led researchers and engineers to focus on sustainable IoT design, which aims to reduce the environmental footprint of IoT systems. Sustainable IoT emphasizes the use of energy-efficient hardware components, renewable energy sources, and designs that enhance device longevity and recyclability to minimize electronic waste and carbon emissions. Energy-efficient IoT systems not only reduce power consumption but also extend battery life, which is crucial for IoT devices deployed in remote or difficult-to-access locations [16].

A critical component of sustainable IoT is ensuring that communication networks between devices are also environmentally friendly. This has given rise to the concept of green networking, which focuses on reducing the energy consumption of network infrastructure such as routers, switches, and data centers. Green networking employs techniques like dynamic resource management, where network components scale their activity based on traffic demand, effectively reducing energy usage during periods of low network activity. Energy-efficient routing algorithms and virtualization of network resources can substantially lower the carbon footprint of communication networks while maintaining performance [17]. Moreover, integrating renewable energy sources into power network infrastructure further enhances the sustainability of the entire IoT ecosystem.

The integration of sustainable IoT design with green networking is essential for addressing the environmental challenges posed by the massive scale of IoT deployment. IoT devices that consume minimal energy but operate over energy-inefficient networks would still contribute significantly to carbon emissions. Therefore, holistic approaches optimize both the device-level energy consumption and the energy efficiency of networks. Eco-friendly IoT frameworks combine low-power sensor designs with energy-aware network management, enabling smart applications like precision agriculture and smart grid systems that are both effective and sustainable [18]. These frameworks are designed to reduce greenhouse gas emissions, lower operational costs, and support global sustainability goals.

Furthermore, advances in energy-efficient communication protocols for IoT play a pivotal role in sustainable design. Protocols that minimize data transmission, employ sleep modes, or use energy harvesting techniques are gaining attention. Several protocols tailored for IoT devices can optimize communication to extend device lifetime and reduce power consumption, contributing to sustainable IoT ecosystems. These innovations are critical for enabling scalable IoT applications in sectors that demand continuous monitoring without frequent battery replacements, such as environmental sensing and healthcare.

Therefore, sustainable IoT design and green networking are interconnected strategies vital to mitigating the environmental impact of the growing IoT infrastructure. By focusing on energy-efficient device design, eco-friendly materials, renewable energy integration, and energy-conscious network management, it is possible to harness the benefits of IoT technology while minimizing harm to the planet. Ongoing research and development in this field, supported by real-world implementations, are key to achieving a sustainable and green future for IoT and communication networks.

4.3 Network Architecture and Management

Network architecture and management are foundational in the 6G-IoT ecosystem. A scalable and flexible network infrastructure allows dynamic allocation of resources for numerous IoT services. AI-driven intelligent network management enables self-optimizing capabilities for improved performance and reduced downtime of large-scale 6G networks.

4.3.1 Network Slicing and QoS in 6G IoT

6G communication networks, expected by 2030, will transform the IoT, powering smart homes, hospitals, factories, and self-driving cars. Unlike 5G, 6G offers superfast data speeds, ultralow delays, and the ability to connect billions of devices, from simple sensors to advanced VR gadgets [19]. At the heart of 6G IoT is network slicing, a technology that splits a single physical network into multiple virtual networks, or "slices," each customized for specific tasks, like streaming videos or controlling robots. Network slicing is considered a fundamental pillar for meeting the varied needs of IoT devices [20]. Paired with QoS, which ensures 6G IoT efficiency and adaptability, creating what experts call the "Intelligent Network of Everything." [20]

Network slicing in 6G IoT is like dividing a highway into dedicated lanes for different vehicles—each lane (or slice) has its own rules and resources. This ensures that critical tasks get priority over less urgent ones [21]. Slices are tailored for specific needs such as enhanced mobile broadband (eMBB) for high-speed tasks like 4K video streaming; ultrareliable low-latency communications (URLLC) for time-sensitive tasks like self-driving cars; massive machine-type communications (mMTC) for connecting millions of devices like smart meters; and hybrid slices for complex setups like smart cities [19]. By using 6G's advanced tools, such as super-high-frequency THz waves and smart management systems, slicing saves resources, prevents interference, and connects devices in remote areas via satellites or drones [22].

QoS in 6G IoT ensures that every device gets the performance it needs, measured by factors like speed, reliability, and delay. For example, self-driving cars need delays as low as 2 ms and 99.999% reliability to avoid accidents, while video streaming requires high speeds up to 100 Mbps [23, 24]. Smart homes with thousands of sensors need networks that handle many connections without dropping data (less than 0.5% packet loss) [23]. However, achieving QoS is tricky because different devices have different needs, and managing millions of them in busy cities while keeping data secure is a big challenge [19].

Several technologies make network slicing and QoS possible in 6G IoT. Software-defined networking (SDN) acts like a traffic controller, managing slices to reduce delays and improve efficiency. Network function virtualization (NFV) turns network tasks, like security checks, into software, making it easy to set up new services. AI predicts how many devices will be active and adjusts slices automatically while also spotting security threats [21]. Edge computing processes data close to devices, speeding up tasks like robot control [19]. THz waves provide massive data speeds for video-heavy apps, and satellites connect IoT devices in far-off places like farms or oceans [22, 24]. These tools work together to make 6G IoT fast, reliable, and global.

Figure 4.2 shows different components in a layered architecture for modern IoT and network services. The first layer is the control and intelligence layer that provides centralized control using SDN. AI enables prediction and automation. The second layer allows several virtual networks to run on a shared infrastructure. It supports different use cases. The third layer replaces hardware with software functions like a firewall. The fourth layer is the connectivity layer that provides a network interface to devices. The fifth layer is the IoT device layer, which includes sensors and actuators for smart homes and self-driving cars, etc.

Despite its promise, 6G IoT faces hurdles. Connecting different devices, like sensors and cars, is complex, and keeping data safe in shared slices is critical, especially for medical or financial apps [14, 16]. Allocating too many resources can waste energy, and there is no global agreement on how slicing should work, which can cause compatibility issues [20, 21]. Furthermore, technologies like THz waves use a lot of power, which is not eco-friendly [24]. Solving these problems is essential for 6G IoT to succeed on a large scale.

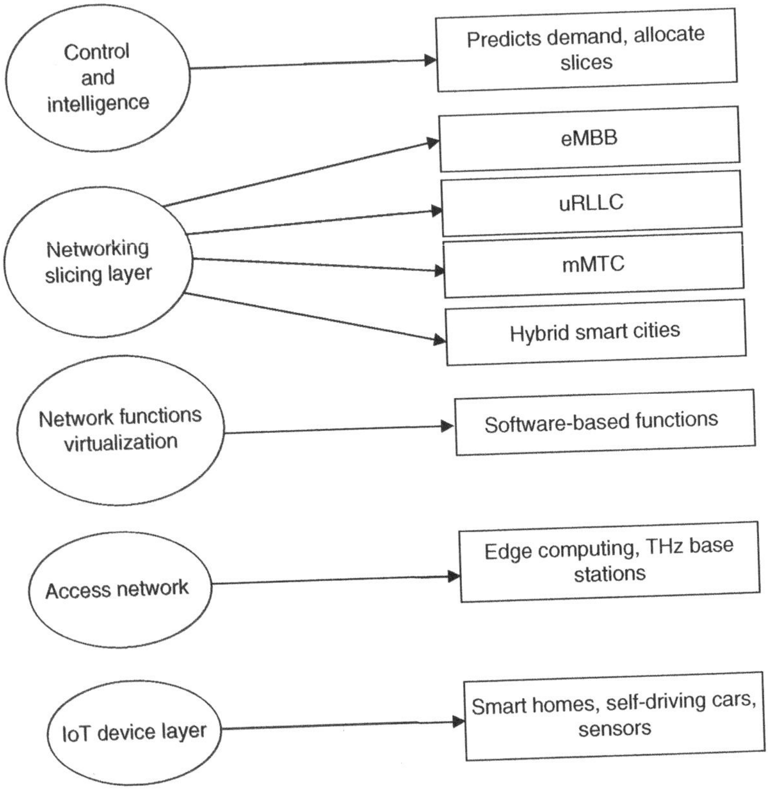

Figure 4.2 Hierarchical framework for smart IoT and network services.

Looking ahead, researchers are exploring exciting ideas for 6G IoT. Smarter AI could make slices adapt instantly to changing needs [21]. Quantum networking might create super-secure, ultrafast slices [24]. Combining ground networks with satellites could connect every corner of the world [22]. Eco-friendly slicing could reduce energy use, and privacy-focused AI could keep user data safe while ensuring QoS [23]. These advancements will make 6G IoT more powerful, sustainable, and inclusive, shaping a future where everything is smartly connected.

4.3.2 Edge Intelligence and Federated Learning

The advent of 6G networks introduces a paradigm shift in how data is generated, processed, and utilized. The anticipated surge in connected devices, ultralow-latency requirements, and the need for enhanced privacy protections render traditional centralized architectures inefficient. In response, EI and FL are emerging as foundational technologies that reshape network architecture and management strategies for next-generation intelligent systems.

EI refers to deploying AI capabilities directly on edge nodes, such as sensors, mobile devices, or gateways, thereby minimizing dependence on centralized cloud infrastructure. By enabling local data processing and real-time decision-making, EI reduces latency, alleviates core network traffic, and supports time-critical applications such as autonomous vehicles and industrial robotics. This architectural decentralization enhances responsiveness and empowers edge devices to act autonomously, aligning with the core objectives of 6G networks [25].

FL, a distributed ML paradigm, further strengthens EI by allowing multiple edge devices to collaboratively train shared AI models without transmitting raw data to a central server. Only model updates are exchanged, preserving data privacy and reducing communication overhead. This aligns with modern network management goals such as bandwidth efficiency, secure data governance, and regulatory compliance [26].

The combined architectural design of EI and FL redefines how intelligence is distributed and managed in 6G networks. By supporting decentralized computation, protecting user data, and enabling scalable, efficient model training, this integration lays the groundwork for intelligent, privacy-preserving, and resource-aware network ecosystems. As lightweight AI models and 6G connectivity advance, EI and FL—fortified with intelligent management strategies—will become central pillars of next-generation distributed systems. The EI-FL integrated architecture has transformed management frameworks in various sectors. For instance, in healthcare, wearable devices equipped with edge AI analyze patient data locally, while FL aggregates model insights across hospitals, enhancing diagnostic accuracy without compromising patient confidentiality [25]. In smart cities, edge devices control traffic systems in real time, and FL coordinates city-wide model training across districts for efficient urban planning [26]. Similarly, in Industrial IoT, edge computing supports predictive maintenance at the factory level, while federated collaboration enables global performance optimization across manufacturing units without sharing proprietary information [27].

However, managing such a distributed system poses significant challenges. A key concern is heterogeneity—the variability in hardware capabilities, network conditions, and data characteristics across edge nodes. Devices may differ in CPU power, memory, and energy constraints, making uniform training and model updates inefficient. Additionally, edge-generated data is often nonindependent and identically distributed (non-IID), which can hinder model convergence and reduce overall performance.

To overcome these management hurdles, heterogeneity-aware solutions are being actively explored. Ref. [28] introduced an asynchronous federated optimization strategy featuring temporally weighted aggregation, allowing devices with varying speeds and availability to contribute proportionally to the model without delay. This framework enhances adaptability, ensures timely convergence, and reduces idle times in real-world edge networks.

4.3.3 AI-Native Network Automation

The development of 6G networks is identified through its participation in AI-driven network automation, where AI transitions from being a supplementary feature to a core element within the network framework. In 5G systems, AI is optimized to do specific tasks, whereas in 6G, AI is deeply involved at all levels of the network; it can facilitate its decision-making, self-optimization, and quality to adapt in real time to changing circumstances [2].

The modification is very essential to handle the growing complexity and data requirements of applications like the IoT, autonomous systems, AR, and immersive media. By using the algorithms of advanced ML and deep learning (DL), 6G networks can analyze a vast amount of data, predict the behavior of the network, and, without human involvement, address problems proactively [29].

AI-driven dynamic network slicing is a crucial factor in enhancing resource efficiency, allowing several virtual networks to operate simultaneously on a shared physical framework. AI constantly tracks traffic patterns and reallocates resources to align with varying service-level demands, ensuring peak performance and scalability [30].

Another fundamental element of AI-centric 6G is the use of digital twin technology—virtual representations that reflect the real-time condition and functioning of network systems. These models continuously monitor live data support predictive maintenance, fault prevention, and strategic capacity planning. This capability is particularly essential for critical applications like remote healthcare, smart grid management, and autonomous transportation [31].

To facilitate these smart operations, a redesigned data plane is essential, one that is tailored for ultralow latency and immediate responsiveness. Edge computing plays a pivotal role by processing data closer to its origin, thereby reducing latency and allowing for quicker AI inference at the network's edge. Realizing this intelligent networking approach requires collaboration across various sectors. The strategic partnerships between telecom providers and AI technology firms. Hardware manufacturers and academic institutions are crucial. The Company NVIDIA is in partnership with a major telecom player to develop AI-optimized network systems. They mainly focus on improving spectral efficiency, automating end-to-end services, and ensuring smooth service integration.

4.3.4 Protocols and Standardization Roadmaps

The integration of IoT ecosystems with 6G networks is not simply a matter of speed and bandwidth—it demands a complete rethinking of how protocols function at every level. According to [32], 6G aims to achieve ultralow latency (below 1 ms), ultrahigh reliability (up to 99.99999%), and context-aware intelligence, which cannot be realized with the current TCP/IP model. These limitations become most evident in mission-critical IoT use cases like autonomous driving, telemedicine, and smart manufacturing, where minimal delays or unpredictable behaviors can lead to life-threatening consequences.

To overcome this, researchers propose a fundamental shift toward AI-native, adaptive network protocols. As outlined in [33], these protocols are built to be self-optimizing and context-aware by design. MLs are embedded within each layer of the stack, enabling predictive routing, intelligent scheduling, and real-time reconfiguration based on dynamic network and application states. Such protocol behavior marks a departure from static configurations and manual tuning, making networks capable of semantic and intent-based communication—where data is transmitted based on its meaning and relevance, not merely payload.

Moreover, in [34], the authors highlighted that the rise of EI plays a critical role in this evolution. Processing tasks at the network edge drastically reduces latency and allows immediate decision-making without dependence on centralized cloud systems. In 6G-IoT ecosystems, edge nodes equipped with lightweight AI models will interact directly with IoT devices, predicting behavior, optimizing bandwidth, and prioritizing mission-critical data. This localized intelligence is especially crucial in sectors like smart grids and autonomous mobility, where real-time responsiveness is paramount.

The road to 6G is not just about faster connectivity; it is about smarter, context-aware, and self-managing networks. Protocol architectures must evolve from rigid, general-purpose systems into flexible, AI-native stacks capable of powering an intelligent IoT ecosystem. The fusion of AI, edge computing, and semantic communication forms the backbone of this next-generation paradigm.

4.4 Key Use Cases of 6G-IoT Synergy

The convergence of 6G wireless technology and the IoT represents a transformative leap in the evolution of digital ecosystems. IoT has already revolutionized numerous sectors through massive interconnectivity and data-driven insights. 6G, with its promise of ultralow latency, terabit-level data rates, AI-native architecture, and integrated terrestrial and NTNs, is poised to overcome these barriers. By enhancing the scalability, intelligence, and responsiveness of IoT systems, 6G enables a new generation of applications that are not only faster and more reliable but also context-aware and energy-efficient.

This section explores the key use cases where the synergy between 6G and IoT can improve quality of life, redefine industry practices, and contribute to achieving global sustainability goals.

4.4.1 Smart Healthcare and Internet of Medical Things

Upcoming health systems are quickly evolving from a centralized, hospital-focused model to a distributed, patient-centric paradigm where continuous, personalized care is provided at a distance through smart technologies. This evolution is powered by the integration of 6G networks and IoMT, enabling smooth interaction among a wide range of medical equipment, sensors, and platforms. As healthcare applications become increasingly data-intensive—requiring ultrareliable, low-latency, and high-speed transmission—the shortcomings of existing systems become apparent. 6G, with its promise of ultrahigh data rates (up to 1 Tbps), microsecond-level latency, and ubiquitous connectivity, is poised to be the backbone of next-generation healthcare systems. The integration of AI with 6G-enabled IoMT devices empowers real-time monitoring of vital signs, early diagnosis through predictive analytics, and immediate medical intervention, especially for managing chronic conditions and supporting elderly care [35].

Smart hospitals exemplify this digital transformation, leveraging 6G-IoT synergy to facilitate remote patient monitoring, telemedicine, and automated decision-making. Through wearables and smart medical devices, patients can transmit health data continuously, while healthcare professionals can offer remote consultations, perform diagnostics, and adjust treatments with minimal delay. These advancements reduce the need for hospital visits, lower healthcare costs, and improve access to quality medical care, particularly in underserved and rural areas. Furthermore, 6G's support for immersive technologies such as holographic communication and tactile internet enables advanced surgical simulations and remote robotic surgeries. Patient empowerment grows through digital tools like mobile health apps and interactive portals. 6G-enabled healthcare systems will not only enhance clinical outcomes but also promote a collaborative, proactive, and data-driven healthcare ecosystem [31, 35].

4.4.2 Autonomous Vehicles and V2X

The integration of 6G and the IoT is set to transform vehicle-based networks by enabling highly interconnected, intelligent, and secure mobility environments. Utilizing the capabilities of 6G, such as very low latency, terabit-per-second data rates, and widespread connectivity, vehicular networks go beyond traditional vehicle-to-everything (V2X) frameworks to support massive IoT deployments and real-time decision-making in self-driving scenarios. The combination of IoT facilitates smooth interaction among vehicles, infrastructure, pedestrians, and cloud services, enabling advanced applications like collective sensing, sensor sharing, high-definition map updates, and remote vehicle diagnostics. Enhanced by edge AI implementation, these systems can make predictive decisions for collision avoidance, traffic optimization, and energy-efficient routing.

Moreover, 6G-enabled vehicular IoT networks will provide the scalability and reliability needed to manage millions of diverse smart devices, ensuring strong functionality even in highly populated urban areas. Together, 6G and IoT form the technological foundation for next-generation intelligent transportation systems, significantly enhancing road safety, user experience, and operational efficiency [36].

4.4.3 Industry 4.0 and Robotics

Wireless communication, particularly through the deployment of 6G technology, plays a pivotal role in establishing resilient and digitally interconnected infrastructures. With its unparalleled speed, low latency, and vast network capacity, 6G promises to advance industries such as healthcare, transportation, and manufacturing by providing high-quality digital services and enabling sustainable development. 6G is expected to not only address the increased data demands of the IoT but also integrate emerging technologies like AI, blockchain, edge computing, and digital twins to drive the Industry 4.0 revolution. By facilitating real-time data sharing, high-bandwidth applications, and seamless connectivity, 6G will support smart manufacturing, improve energy efficiency, and contribute to environmental sustainability. The synergy between 6G and IoT will enable the development of intelligent and adaptive systems that can optimize energy use, minimize waste, and ensure operational efficiency. This convergence will be instrumental in achieving the United Nations' Sustainable Development Goals (SDGs), particularly in fostering innovation, economic growth, and the advancement of sustainable industry practices. Ultimately, 6G will provide the infrastructure necessary for a fully digitalized, sustainable future, aligning with global efforts to achieve sustainability and resilience by 2030 [37].

4.4.4 Smart Cities and Environmental Sustainability

Nowadays, many cities are getting smarter with the use of information and communication technologies (ICTs), making traditional services and infrastructure more efficient. Despite many technological promises, there are still many challenges in realizing an envisioned future smart sustainable city (SSC) where humans and machines experience a hyper-connected society. In the hyper-connected society, it is expected that intelligent things such as urban air mobility (UAM), IoT sensors, and robots will be seamlessly connected in future wireless networks without limitations on the data rate, latency, and coverage. An SSC refers to an innovative city where ICTs are used primarily to improve the quality of life and efficiency of city operations and services while ensuring the needs of current and future generations in economic, social, environmental, and cultural terms. Moving to the post-pandemic world, the way we live and communicate with each other has changed rapidly. It is noted that people all over the world use various form factor devices such as smartphones, smartwatches, AR glasses, and VR headsets to join online meetings and connect with their family and friends. Consequently, it is necessary for current cities to get smarter by exploiting innovative technologies. The emergence of smart cities is growing and has attracted increasing interest from industry [38].

The rise of Industry 4.0 has redefined mobile telecommunications as a key enabler of sustainability, aligning closely with the United Nations' 2030 SDGs. While 5G networks have transformed digital connectivity by supporting massive device integration and smart applications, the shift toward advanced technologies like Beyond 5G (B5G) and 6G promises even greater innovation through AI integration, enhanced spectrum usage, and ultrareliable, low-latency communication. Sustainability in mobile networks hinges on core factors such as energy efficiency, scalable and adaptive infrastructure, broad network coverage, and robust privacy and security measures. These elements are crucial in ensuring that mobile communication technologies can support inclusive digital growth and environmental responsibility. As mobile networks evolve, they play an increasingly vital role in advancing all 17 SDGs by enabling access to digital services, fostering innovation, and minimizing ecological impact [39].

4.4.5 Remote Connectivity via Satellites and UAVs

The next-generation wireless network toward 6G is envisioned to enable intelligent, secure, reliable, and limitless connectivity. It is expected that 6G will bring a full-fledged framework for connected things and automation systems, from autonomous cars to UAVs, with stringent and diverse requirements in terms of reliability, latency, data rate, and energy efficiency. UAVs, commonly known as drones, play key roles in a wide range of use cases and scenarios that can go B5G and 6G. The examples of UAV applications include package delivery, media production, real-time surveillance, and remote construction. The deployment of UAVs is ever-increasing, and, as predicted by the Federal Aviation Administration (FAA), the number of commercial UAV fleets can reach up to 1.6 million by 2024. Meanwhile, UAS traffic management (UTM) systems are being developed to support safe and secure operation of low-altitude UAVs with their potentially high traffic density [40].

The integration of satellites with 6G is increasingly important, not only to leverage the successes of 5G integration, but also to increase the contribution that satellites can play in the ever-evolving global connectivity requirements. With 6G, the end user can be anything from a person with a device to an autonomous drone, a robot, a heart monitor, an AI security application, and more. 6G will add connective intelligence to the hierarchy of connected people and things. It will offer unrivaled performance characteristics through the creation of the world's largest neural network with integrated sensing, learning, and processing capabilities [41].

4.5 Threat Landscape in 6G-IoT

The threat landscape in 6G-IoT networks is increasingly complex and multifaceted, driven by the massive scale and interconnected nature of devices. As outlined in the paper "*Security and privacy in 6G networks: New areas and new challenges,*" the integration of billions of devices significantly broadens the attack surface, making endpoints more vulnerable to cyber intrusions. Advanced Persistent Threats (APTs) are expected to rise, enabling stealthy, prolonged attacks that threaten network integrity. Additionally, the exponential growth in data generation raises serious data privacy concerns, requiring robust protection mechanisms. Denial of Service (DoS) attacks pose heightened risks to critical, real-time applications by disrupting network availability. Insider threats remain a critical concern, especially in complex, multiuser environments. Altogether, securing 6G-IoT networks demands an autonomous, multilayered, proactive security strategy involving continuous monitoring, secure system design, and collaborative risk management [42].

4.5.1 AI-Powered Intrusion Detection

AI/ML algorithms can be used to detect intrusions, anomalies, and zero-day attacks in real time. AI and ML algorithms can automatically detect and respond to malicious or unauthorized activities in a network system. An AI-powered intrusion detection system can analyze a large amount of real-time data, distinguish complex attack patterns, and adapt to evolving threats. Self-healing networks and adaptive firewalls can also automate threat mitigation. Thus, accuracy and speed can be enhanced in cybersecurity defense.

4.5.2 Blockchain for Trust and Data Integrity

BCT plays as an immutable distributed ledger for performing secure financial transactions associated with cryptographic symbols. In other words, BCT establishes trusted secure communications among unknown entities in untrusted networks. BCT operated as a common, mutual, and distributed ledger used as a large-scale secure index measuring all communications architecture, particularly in the cybersecurity domain, to enhance the security of authentication. In fact, the blockchain may turn out to be one of the most disruptive Internet of Everything technologies. Moreover, by deploying the BCT in a spectrum sharing system, the problems of low spectrum utilization and spectrum monopoly could be overcome while securing spectrum utilization at the same time [15]. BCT has gained momentum for distributed applications and is recognized as the key enabling trusted technologies that have received deep attention in both research and industry communities in the context of IoT and 6G mobile communication.

Blockchain offers a scalable and distributed platform to record data permanently and validate transactions among different entities without involving central participants. The integration of blockchain with 6G wireless networks potentially allows MNOs to monitor and manage resource and spectrum utilization with some added advantages, like improving spectrum auctions, reducing administration expenditures, infrastructure sharing, and so on. Due to the inherited benefit of transparency, BCT can record real-time resource utilization and significantly improve spectrum efficiency through the dynamic allocation of spectrum. With the wireless networks paradigm shifting toward distributed solutions, plenty of base stations are deployed by MNOs where billions of multimedia devices communicate among themselves. Implementation of blockchain with the underlying infrastructure not only provides efficient spectrum sharing and resource management but also indulges incentives due to service sharing [43, 44].

4.5.3 Privacy by Design in 6G-enabled IoT

Some key technologies such as Zero Trust architecture (ZTA) and end-to-end encryption have already proven useful in crucial areas of the 6G network. They are giving 6G networks high reliability, low latency, and secure and efficient transmission services. However, most of these technologies come at the cost of new security and privacy concerns. Thus, FL should be implemented to train AI models without exposing user data. Moreover, differential privacy can be applied to anonymize data collected by IoT devices. Smart IoT devices come with built-in security features such as hardware-based keys, secure boot, updated firmware, regular patching, and network slicing isolation.

Solving threat landscapes in the 6G-IoT ecosystem using AI integration and advanced principles ensures compliance and user privacy.

4.6 Economic, Market, and Ecosystem Insights

Economic, market, and ecosystem insights convergence has impacted the broader environment of business, technology, and society. When 6G and IoT technologies come together, value, trends, and collaborative dynamics emerge.

4.6.1 Market Forecasts and Investment Trends

Market insights denote identifying industry demand, growth trends, and customer needs based on 6G-IoT technology. The global market for emerging technologies and business models is continuously evolving by influencing technological advancements, consumer behavior, and global events. One of the most focused trends is sustainability and digitalization across multiple sectors. Green investment plays a pivotal role in promoting sustainable industrial development. Green investment aids in the growth of environmentally friendly industries. Their study emphasizes that green investment supports eco-friendly industrial development and helps accelerate the upgrading and transformation of industrial structures. This underscores the significance of green investment in driving sustainable economic growth [45].

Technological progress is another critical factor influencing market dynamics. Technological development promotes long-term industrial development, indicating that advancements in technology are essential for the evolution of industrial structures. The short- and long-run estimates attached to Research and Development (R&D) expenditures are positively significant, confirming that increasing R&D expenditures helps improve the industrial structure.

Market forecasts indicate significant growth in sectors like AI, renewable energy, biotechnology, and digital transformation. Investment trends reveal a great focus on technology-driven industries, particularly in fintech, health tech, and sustainable solutions. Research shows that the adoption of AI is set to be the most crucial factor in shaping market dynamics in the upcoming years. The growth is driven by the increasing need for automation, data-driven decision-making, and cost optimization across industries, and investments in green technologies are projected to reach 10 trillion by 2030.

4.6.2 Ecosystem Collaboration and Policy Frameworks

Ecosystem collaboration and policy frameworks play an important role in shaping innovation and sustainability across sectors. These frameworks enable inclusive engagement between governments, private sectors, academia, and local communities to cocreate solutions for complex challenges. The four steps of the collaboration process include planning, meeting, matching, and implementing; however, the process is not necessarily one.

Public policy is also crucial in facilitating this transformation. A framework for analyzing the role of policy in transforming innovation systems into innovation ecosystems is presented in [46]. This framework highlights policy guides transitions toward more integrated and dynamic systems. Moreover, successful ecosystem collaboration requires not only strategic coordination but also transparent communication and trust among stakeholders. As ecosystems evolve, adaptive policies that support open innovation and community engagement become increasingly essential for sustainable progress.

4.6.3 Monetization Strategies and Business Models

In the evolving landscape of digital transformation, monetization strategies and business models have also become pivotal in driving innovation and sustainability. Organizations are gradually moving toward platform-based modes to utilize network effects and generate value. The platform business model is characterized by delivering benefits through facilitating exchanges between two or more interdependent parties, usually consumers and producers [47].

Also, the adoption of digital technologies has catalyzed the emergence of innovative revenue streams. Digital transformation not only facilitates the creation of novel business models but also improves efficiency, ultimately driving increased revenues and building stronger customer relationships. By using data analytics, AI, and automation, organizations can tailor their offerings to meet changing consumer demands, thereby maximizing their market potential and establishing deeper connections with their user network.

However, these advancements also have challenges. The quick technological change requires businesses to constantly adjust their monetization strategies to remain competitive [48].

4.7 Societal Impacts and Ethical Considerations

The adoption of modern technologies, particularly AI, into various fields has far-reaching effects on society, affecting areas such as employment, privacy, and governance. AI technologies have the potential to significantly impact employment, privacy, and decision-making processes.

Ethical considerations are crucial in ensuring that technological advancements benefit society as a whole. Ethical AI research engages with broader legal, social, and economic concerns, requiring interdisciplinary governance approaches that manage process with ethical responsibility. Moreover, there is a growing emphasis on inclusivity and justice. The debate over algorithmic fairness, bias, and accountability has moved beyond technical concerns to engage with broader sociopolitical issues, particularly regarding AI's role in reinforcing systemic inequalities [49].

4.8 Conclusion

It can be said that 6G is not just an evolution of 5G but a game changer in the world of technology that combines communication with sensing, intelligence, and seamless global connectivity. It is the right time to regulate the THz band to support multi-Gbps and Tbps communication links. Incorporating 6G with the IoT provides a revolutionary step toward the development of ecosystems that are intelligent, fully autonomous, and interconnected. This collaboration is expected to revolutionize a number of industries, including Industry 4.0, smart cities, autonomous transportation, and healthcare. But in order to realize this vision, key operational and technical issues must be resolved, especially those related to real-time data processing, network scalability, energy consumption, and strong security. Through advanced and sophisticated topologies, decentralized edge computing, and AI-enabled administration, 6G provides the fundamental capabilities needed to support the upcoming generation of IoT applications. Key enabling technologies, sustainability, standardization, ethical AI governance, and resilient design approaches can flourish effectively in 6G and IoT convergence.

References

1 Akyildiz, I.F., Jornet, J.M., and Han, C. (2014). Terahertz band: next frontier for wireless communications. *Physical Communication* 12: 16–32.

2 Saad, W., Bennis, M., and Chen, M. (2019). A vision of 6G wireless systems: applications, trends, technologies, and open research problems. *IEEE Network* 34 (3): 134–142.

3 Pan, C., Ren, H., Wang, K. et al. (2021). Reconfigurable intelligent surfaces for 6G systems: principles, applications, and research directions. *IEEE Communications Magazine* 59 (6): 14–20.

4 Basar, E., Alexandropoulos, G.C., Liu, Y. et al. (2024). Reconfigurable intelligent surfaces for 6G: emerging hardware architectures, applications, and open challenges. *IEEE Vehicular Technology Magazine*.

5 Basharat, S., Hassan, S.A., Pervaiz, H. et al. (2021). Reconfigurable intelligent surfaces: potentials, applications, and challenges for 6G wireless networks. *IEEE Wireless Communications* 28 (6): 184–191.

6 Huo, Y., Lin, X., Di, B. et al. (2023). Technology trends for massive MIMO towards 6G. *Sensors* 23 (13): 6062.

7 Dala Pegorara Souto, V., Dester, P.S., Soares Pereira Facina, M. et al. (2023). Emerging MIMO technologies for 6G networks. *Sensors* 23 (4): 1921.

8 Wang, Z., Zhang, J., Du, H. et al. (2024). A tutorial on extremely large-scale MIMO for 6G: fundamentals, signal processing, and applications. *IEEE Communications Surveys & Tutorials* 26 (3): 1560–1605.

9 Xu, J., Kishk, M.A., and Alouin, M.S. (2023). Space-air-ground-sea integrated networks: modeling and coverage analysis. *IEEE Transactions on Wireless Communications* 22 (9): 6298–6313.

10 Nguyen, D.C., Ding, M., Pathirana, P.N. et al. (2021). 6G internet of things: a comprehensive survey. *IEEE Internet of Things Journal* 9 (1): 359–383.

11 Sethi, P. and Sarangi, S.R. (2017). Internet of things: architectures, protocols, and applications. *Journal of Electrical and Computer Engineering* 2017: 9324035.
12 Hosseinzadeh, M., Hemmati, A., and Rahmani, A.M. (2022). 6G-enabled internet of things: vision, techniques, and open issues. *Computer Modeling in Engineering and Sciences* 133 (3): 509.
13 Zhou, Z., Chen, X., Li, E. et al. (2019). Edge intelligence: paving the last mile of artificial intelligence with edge computing. *Proceedings of the IEEE* 107 (8): 1738–1762.
14 Park, J., Samarakoon, S., Bennis, M., and Debbah, M. (2019). Wireless network intelligence at the edge. *Proceedings of the IEEE* 107 (11): 2204–2239.
15 A. Dorri, S. S. Kanhere, and R. Jurdak. (2016). "Blockchain in Internet of Things: Challenges and Solutions," *arXiv preprint arXiv:1608.05187*. Available: https://arxiv.org/abs/1608.05187.
16 Cu, V.L., Burie, J.C., Ogier, J.M., and Liu, C.L. (2019). Hiding security feature into text content for securing documents using generated font. *2019 International conference on document analysis and recognition (ICDAR)*, pp. 1214–1219. IEEE.
17 Mosbah, A.B., Hammami, S.E., Moungla, H. et al. (2019). Enhancing device-to-device direct discovery based on predicted user density patterns. *Computer Networks* 151: 245–259.
18 Perera, C., Zaslavsky, A., Christen, P., and Georgakopoulos, D. (2013). Context aware computing for the internet of things: a survey. *IEEE Communications Surveys & Tutorials* 16 (1): 414–454.
19 Alwakeel, A.M. and Alnaim, A.K. (2024). Network slicing in 6G: a strategic framework for IoT in smart cities. *Sensors* 24 (13): 4254.
20 Zhou, F. and Cheriet, M. (2021). Intelligent network slicing management and control for 6G mobile networks. In: *6G Wireless Communications and Mobile Networking* (ed. X. Xie, B. Rong, and M. Kadoch), 189–207. Bentham Science Publishers.
21 Mei, J., Wang, X., and Zheng, K. (2020). An intelligent self-sustained RAN slicing framework for diverse service provisioning in 5G-beyond and 6G networks. *Intelligent and Converged Networks* 1 (3): 281–294.
22 Ammar, S., Lau, C.P., and Shihada, B. (2024). An in-depth survey on virtualization technologies in 6G integrated terrestrial and non-terrestrial networks. *IEEE Open Journal of the Communications Society* 5: 3690–3734.
23 Wang, J., Liu, J., Li, J., and Kato, N. (2023). Artificial intelligence-assisted network slicing: network assurance and service provisioning in 6G. *IEEE Vehicular Technology Magazine* 18 (1): 49–58.
24 Dangi, R., Choudhary, G., Dragoni, N. et al. (2023). 6G mobile networks: key technologies, directions, and advances. *Telecom* 4 (4): 836–876. https://doi.org/10.3390/telecom4040037.

25 Rieke, N., Hancox, J., Li, W. et al. (2020). The future of digital health with federated learning. *npj Digital Medicine* 3 (1): 119.

26 Dipierro, S. and Valdinoci, E. (2021). Description of an ecological niche for a mixed local/nonlocal dispersal: an evolution equation and a new Neumann condition arising from the superposition of Brownian and Lévy processes. *Physica A: Statistical Mechanics and its Applications* 575: 126052.

27 Li, T., Sahu, A.K., Talwalkar, A., and Smith, V. (2020). Federated learning: challenges, methods, and future directions. *IEEE Signal Processing Magazine* 37 (3): 50–60.

28 Yang, Q., Liu, Y., Chen, T., and Tong, Y. (2019). Federated machine learning: concept and applications. *ACM Transactions on Intelligent Systems and Technology (TIST)* 10 (2): 1–19.

29 Zhang, Z., Xiao, Y., Ma, Z. et al. (2019). 6G wireless networks: vision, requirements, architecture, and key technologies. *IEEE Vehicular Technology Magazine* 14 (3): 28–41.

30 Giordani, M., Polese, M., Mezzavilla, M. et al. (2020). Toward 6G networks: use cases and technologies. *IEEE Communications Magazine* 58 (3): 55–61.

31 Kumar, A., Masud, M., Alsharif, M.H. et al. (2025). Integrating 6G technology in smart hospitals: challenges and opportunities for enhanced healthcare services. *Frontiers in Medicine* 12: 1534551.

32 Strinati, E.C., Barbarossa, S., Gonzalez-Jimenez, J.L. et al. (2019). 6G: the next frontier: from holographic messaging to artificial intelligence using subterahertz and visible light communication. *IEEE Vehicular Technology Magazine* 14 (3): 42–50.

33 Li, L. (2023). A survey on intelligence-endogenous network: architecture and technologies for future 6G. *Intelligent and Converged Networks* 5 (1): 53–67.

34 Letaief, K.B., Shi, Y., Lu, J., and Lu, J. (2021). Edge artificial intelligence for 6G: vision, enabling technologies, and applications. *IEEE Journal on Selected Areas in Communications* 40 (1): 5–36.

35 Ahad, A., Jiangbina, Z., Tahir, M. et al. (2024). 6G and intelligent healthcare: taxonomy, technologies, open issues and future research directions. *Internet of Things* 25: 101068.

36 Lang, P., Tian, D., Han, X. et al. (2024). Towards 6G vehicular networks: vision, technologies, and open challenges. *Computer Networks* 257: 110916.

37 Ghildiyal, Y., Singh, R., Alkhayyat, A. et al. (2023). An imperative role of 6G communication with perspective of industry 4.0: challenges and research directions. *Sustainable Energy Technologies and Assessments* 56: 103047.

38 Kim, N., Kim, G., Shim, S. et al. (2024). Key technologies for 6G-enabled smart sustainable city. *Electronics* 13 (2): 268.

39 Kumar, R., Gupta, S.K., Wang, H.C. et al. (2023). From efficiency to sustainability: exploring the potential of 6G for a greener future. *Sustainability* 15 (23): 16387.

40 Mohanta, K. and Al-Rubaye, S. (2024). Towards 6G satellite–terrestrial networks: analysis of air mobility operations. *Electronics* 13 (14): 2855.

41 Mozaffari, M., Lin, X., and Hayes, S. (2022). Toward 6G with connected sky: UAVs and beyond. *IEEE Communications Magazine* 59 (12): 74–80.

42 Wang, M., Zhu, T., Zhang, T. et al. (2020). Security and privacy in 6G networks: new areas and new challenges. *Digital Communications and Networks* 6 (3): 281–291.

43 Jahid, A., Alsharif, M.H., and Hall, T.J. (2023). The convergence of blockchain, IoT and 6G: potential, opportunities, challenges and research roadmap. *Journal of Network and Computer Applications* 217: 103677.

44 Nguyen, D.C., Pathirana, P.N., Ding, M., and Seneviratne, A. (2020). Blockchain for 5G and beyond networks: a state of the art survey. *Journal of Network and Computer Applications* 166: 102693.

45 Chen, M., Chen, R., Zheng, S., and Li, B. (2023). Green investment, technological progress, and green industrial development: implications for sustainable development. *Sustainability* 15 (4): 3808.

46 Zheng, X. and Cai, Y. (2022). Transforming innovation systems into innovation ecosystems: the role of public policy. *Sustainability* 14 (12): 7520.

47 Deng, T., Qiao, L., Yao, X. et al. (2022). A profit framework model for digital platforms based on value sharing and resource complementarity. *Sustainability* 14 (19): 11954.

48 Judijanto, L. (2025). The impact of digital transformation on business models: a bibliometric study. *The Eastasouth Management and Business* 3 (2): 255–267.

49 Runcan, R., Hațegan, V., Toderici, O. et al. (2025). Ethical AI in social sciences research: are we gatekeepers or revolutionaries? *Societies* 15 (3): 62.

5

Societal Implications of Synergizing 6G and IoT

V. Vidhya

Department of Artificial Intelligence & Data Science, Easwari Engineering College, Ramapuram, Chennai, India

5.1 Introduction

The impending fusion of sixth-generation (6G) wireless technology and the Internet of Things (IoT) promises to fundamentally alter societal structures across numerous domains. Advancing beyond 5G's capabilities, 6G is envisioned to deliver exceptionally high-speed data transfer, reaction times measured in sub-milliseconds, and deeply woven Artificial Intelligence (AI). These elements are expected to create networks that can self-adjust and respond intelligently to their surroundings [1]. Such capabilities will underpin the connection of a massive array of devices and enable immediate data analysis, unlocking innovative applications from sophisticated transport networks to fully immersive digital worlds.

In contrast to earlier network upgrades, 6G is set to do more than just boost speed and data handling capacity; it aims to redefine human interaction with digital platforms. By embedding intelligence at the network's edge and combining varied communication methods—including satellites, ground-based stations, and aerial platforms—6G will establish a pervasive, interconnected framework [2]. This framework will be the backbone for a broad IoT deployment affecting smart residences, urban centers, industrial operations, and medical care. From a societal perspective, the integration of 6G and IoT offers pathways to a better quality of life via enhanced service provision, ecological sustainability, and economic advancement. However, this progress also brings forth significant considerations regarding privacy, data security, ethical application, and fair access. Effectively navigating these issues will necessitate cooperative efforts between technology developers, regulatory bodies, and the wider community to encourage responsible technological progress [3].

Synergies of 6G and IoT, First Edition. Edited by Himanshi Babbar.

Published 2026 by John Wiley & Sons, Inc.

This chapter delves into the wide-ranging societal effects of this technological convergence. It examines how IoT, empowered by 6G, is likely to shape social inclusivity, healthcare services, educational methods, city life, employment trends, environmental protection efforts, cultural shifts, and governmental systems, providing a comprehensive outlook on the approaching digital age.

5.2 Bridging the Digital Divide

The disparity in digital access continues to be a major hurdle in achieving worldwide inclusive growth. Even with broad mobile phone use, many global populations, especially in rural, economically disadvantaged, and underserved areas, do not have consistent, fast internet. This absence of connectivity amplifies existing inequalities in schooling, medical services, job opportunities, and community involvement [4]. 6G network designs aim to tackle this with new architectures that combine terrestrial stations, low earth orbit (LEO) satellites, high-altitude platform stations (HAPS), and unmanned aerial vehicles (UAVs) into a multilayered, robust communication system. These hybrid networks could bring broadband to previously isolated regions by surmounting the geographical and financial obstacles that hinder conventional infrastructure [5].

IoT is instrumental in this by facilitating decentralized, local communication systems, such as mesh networks, where devices can exchange data directly or pass it along over short ranges without relying entirely on a central backbone. This is particularly crucial for remote or disaster-affected zones where centralized connectivity is often weak or absent.

Illustrative projects show the power of integrating 6G and IoT to narrow the digital access gap. For example, initiatives in Sub-Saharan Africa are using LEO satellite systems and IoT sensor arrays to provide remote medical consultations, agricultural oversight, and learning materials, thereby enabling communities to engage in the digital economy. In rugged, mountainous terrains, UAVs fitted with IoT sensors have facilitated environmental surveillance and emergency communication, proving the viability of scalable remote connectivity solutions [6]. The advantages of bridging the digital divide with 6G and IoT go beyond mere access; they nurture local innovation by supplying the necessary infrastructure for digital business ventures and skill enhancement. Nonetheless, to fully achieve this potential, obstacles like the cost of devices, the need for digital skills, and the creation of culturally appropriate content must be tackled through comprehensive policies and collaboration among diverse stakeholders [7] (Table 5.1).

Table 5.1 Key Societal Implications of 6G and IoT Integration

Area	Opportunities	Challenges	Examples/ Applications
Healthcare	Real-time remote monitoring, personalized treatment	Data privacy, access inequality	Telemedicine, AI-assisted diagnostics
Education	Immersive learning, remote access, and personalized education	Digital divide, content quality control	VR classrooms, AI tutors
Smart cities	Efficient resource management, improved public safety	Surveillance concerns, infrastructure costs	Smart traffic, energy grids, and waste management
Workforce & employment	Automation, new job creation, and remote work	Job displacement, skills gap, and worker privacy	Smart factories, virtual collaboration
Environmental sustainability	Real-time monitoring, resource optimization, and emission reduction	E-waste, unequal access to green tech	Smart grids, precision agriculture
Ethics & privacy	Data protection technologies, ethical AI integration	Data misuse, surveillance, bias	Privacy-by-design IoT, blockchain for data transparency

5.3 Transforming Healthcare Delivery

The synergy between 6G networks and IoT is poised to transform healthcare by enabling real time, widespread, and individualized medical support. Healthcare infrastructures worldwide face mounting pressures from aging demographics, the increase in chronic illnesses, and the need for more economical care. The 6G-IoT framework can meet these demands by supporting ongoing health surveillance, remote diagnostic capabilities, and AI-enhanced treatment strategies [8]. The extremely low delay and high dependability of 6G will facilitate critical functions like remote surgical procedures and telemedicine, where immediate communication is vital. IoT devices, including wearable monitors, implantable gadgets, and intelligent hospital tools, will produce large quantities of live health data. This information, securely and rapidly transmitted via 6G, can be processed by AI systems at the network edge or in the cloud to identify irregularities, forecast disease development, and suggest treatments.

Table 5.2 Impact of 6G-IoT on Healthcare Services

Healthcare Domain	6G-IoT Benefits	Potential Challenges	Examples
Remote patient monitoring	Continuous real-time data, rapid emergency response	Data security, connectivity in rural areas	Wearable biosensors, emergency alerts
Personalized medicine	AI-driven treatment plans, genomics integration	Data privacy, algorithm bias	AI diagnostic tools, gene therapy
Surgery and rehabilitation	Remote robotic surgery, VR rehabilitation	High infrastructure cost, latency issues	Telesurgery, VR rehab programs

For instance, individuals in isolated or less-served regions could receive expert medical advice through high-quality video connections powered by 6G. Continuous monitoring of vital signs allows for the early identification of health issues, potentially lowering hospital readmissions and improving the management of long-term conditions. The capacity to gather and interpret ongoing health information facilitates personalized medical approaches suited to an individual's genetic makeup and lifestyle [9]. Furthermore, smart ambulances equipped with IoT sensors and 5G/6G links can send patient information to hospitals en route, helping medical staff prepare for emergencies. In public health crises like pandemics, 6G-supported IoT devices can aid in contact tracing, real-time disease outbreak surveillance, and the effective allocation of medical supplies. Nevertheless, this healthcare evolution brings up serious questions about data privacy, security, and ethical conduct [10]. The sensitive nature of health information necessitates strong encryption, secure identity verification, and clear governance to safeguard patient confidentiality and prevent unauthorized use. Guaranteeing fair access to these sophisticated healthcare options is also essential to prevent widening current health outcome gaps (Table 5.2).

5.4 Revolutionizing Education and Learning

The combination of 6G and IoT technologies signals a new phase in education, characterized by highly engaging, customized, and widely available learning opportunities. Conventional educational frameworks, often limited by geographic constraints, scarce resources, and rigid timetables, can be revolutionized by the constant connectivity and intelligent settings offered by 6G-enabled IoT [11]. The substantial bandwidth and minimal delay of 6G will underpin sophisticated augmented reality (AR), virtual reality (VR), and mixed reality (MR) uses in education, fostering immersive learning spaces where students can engage with 3D representations, virtual labs, and instructors remotely in real time. IoT sensors integrated into classrooms and learning tools will track student involvement, well-being, and environmental factors to dynamically adjust teaching approaches.

Customized learning programs, driven by AI analysis of data from IoT devices, will cater to individual student capabilities, weaknesses, and preferred learning styles. This method enhances educational results by adapting content delivery and speed, promoting better comprehension and knowledge retention [12]. Learners in remote or disadvantaged areas will gain equal access to superior educational materials, thus narrowing the educational access gap.

Moreover, IoT devices can support collaborative learning by linking students globally for joint projects, virtual excursions, and cultural interactions. Smart campus initiatives will optimize energy consumption, safety measures, and resource allocation, leading to healthier and more sustainable learning environments. However, the widespread gathering of student data brings forth concerns regarding privacy, data protection, and ethical considerations. Educational bodies and policymakers must establish unambiguous regulations to safeguard student rights while using data to enhance learning experiences [13].

5.5 Smart Cities and Urban Development

The fusion of 6G networks and IoT technologies is set to reshape urban existence by facilitating the creation of smart cities that are productive, sustainable, and focused on their inhabitants. Swift global urbanization has created issues like traffic jams, pollution, dwindling resources, and insufficient infrastructure. 6G-powered IoT systems present viable answers to these issues through continuous data gathering, smart decision-making processes, and automated management systems.

In smart cities, a broad array of IoT sensors will monitor traffic conditions, air purity, energy usage, waste disposal, and public security. The ultralow latency and high data capacity of 6G networks allow for immediate communication between these sensors and centralized or edge-based AI systems, which then analyze the data to refine city operations [14]. For example, adaptive traffic light systems can lessen congestion and pollution by dynamically altering signal timings based on current traffic. Energy management also sees substantial improvements. Smart grids, utilizing 6G-IoT, can more effectively balance energy supply and demand, incorporate renewable energy, and quickly identify system faults, thereby improving reliability and lessening environmental impact. Similarly, IoT-driven waste collection systems can optimize collection routes and timing, cutting operational expenses and environmental harm.

Public safety and emergency services are improved by interconnected surveillance systems, intelligent street lighting, and sensor networks that quickly identify accidents, crimes, or environmental dangers. The high reliability of the network guarantees prompt notifications and coordinated actions among emergency responders and city authorities [15]. While smart city innovations offer a better quality of life, they also introduce concerns about data privacy, surveillance, and fair access. Governance structures need to strike a balance between fostering innovation and protecting citizens' rights to privacy and involvement in decision-making.

5.6 Workforce Evolution and Employment

The synergy of 6G and IoT technologies is expected to deeply impact the workforce and job market. While automation, AI, and interconnected devices boost productivity and open up new job types, they also disrupt existing roles and necessitate new competencies. Grasping these changes is essential for managing the societal effects of technological shifts [16]. The highly dependable, low-latency connectivity of 6G will speed up industrial automation via IoT-equipped smart factories, where robots, sensors, and AI systems work together seamlessly. This Industry 4.0 progression improves accuracy, efficiency, and adaptability in manufacturing, logistics, and supply chains. As routine physical and mental tasks become automated, the demand in the workforce will move toward roles requiring complex problem-solving, creativity, and technology oversight (Table 5.3).

Table 5.3 Workforce Transformation Due to 6G and IoT

Aspect	Positive Impacts	Negative Impacts	Mitigation Strategies
Job creation	New roles in AI, IoT management, and data analysis	Decline in routine manual jobs	Reskilling, vocational training
Work environment	Flexible remote work, enhanced collaboration	Increased surveillance, work-life boundary issues	Transparent policies, worker consent
Skills demand	Higher demand for digital literacy and technical skills	Skills gap, access inequality	Inclusive education, affordable training
Economic impact	Productivity gains, innovation boost	Economic inequality, regional disparities	Government support, inclusive policies

Remote and hybrid work arrangements will be increasingly supported by immersive 6G-powered virtual collaboration platforms and IoT devices that monitor worker health, ergonomics, and output. These technologies offer the prospect of improved work-life integration, less commuting, and the inclusion of geographically scattered talent. However, this shift could also increase job displacement risks for workers with fewer skills, making comprehensive retraining and skill-upgrading programs necessary [17]. Policymakers and organizations need to invest in education and continuous learning to provide workers with digital literacy, AI proficiency, and adaptability. Social support systems and inclusive labor practices will also be key in lessening inequalities.

Furthermore, ethical questions arise regarding workplace monitoring enabled by ubiquitous IoT sensors, particularly concerning privacy and worker independence. Transparent governance and worker involvement in decisions about technology adoption are crucial for establishing fair and human-focused work settings.

5.7 Environmental Sustainability and Climate Action

The integration of 6G and IoT technologies offers considerable potential for advancing environmental sustainability and tackling climate change. Urban growth, industrial activity, and rising energy needs persistently deplete natural resources and worsen environmental decline. The 6G-IoT framework provides robust tools for real-time monitoring, data-informed decision-making, and automated controls to encourage efficient resource utilization and diminish ecological footprints [16] (Table 5.4).

Table 5.4 Environmental Benefits and Concerns of 6G-IoT

Focus Area	Environmental Benefits	Concerns	Examples/ Technologies
Energy management	Efficient grid operation, renewable integration	Energy consumption of IoT devices	Smart grids, demand–response systems
Pollution monitoring	Real-time air and water quality sensing	Data accuracy, deployment costs	IoT sensors for pollutants, smart meters
Agriculture	Precision irrigation, reduced chemical use	Electronic waste, sensor disposal	IoT soil sensors, automated irrigation
Waste management	Optimized collection, increased recycling rates	Implementation costs, data privacy	Smart bins, route optimization apps

The massive connectivity and ultralow latency of 6G will facilitate extensive sensor networks for ongoing environmental surveillance, capable of tracking air and water quality, soil conditions, wildlife habitats, and carbon emissions with exceptional detail. IoT devices embedded in ecosystems and urban settings can gather and send data to AI-driven analytical platforms to spot anomalies, foresee environmental hazards, and refine mitigation efforts [18]. Smart grids, bolstered by 6G-enabled IoT, help integrate renewable energy sources, implement demand–response initiatives, and manage energy storage, thus lessening dependence on fossil fuels and improving grid stability. Smart agriculture utilizes IoT sensors to optimize watering, fertilization, and pest management, boosting yields while cutting down on resource waste and environmental damage.

Additionally, intelligent transportation systems lower emissions by optimizing travel routes and facilitating electric vehicle use through real-time management of charging stations. Waste management improves with IoT-equipped smart bins and recycling systems that make collection more efficient and increase recycling rates [13].

Despite these advancements, challenges persist in ensuring fair access to green technologies and managing the environmental impact of manufacturing and disposing of IoT devices. Principles of sustainable design and a circular economy must inform the lifecycle management of 6G-IoT hardware.

5.8 Smart Cities and Infrastructure: The 6G-IoT Revolution

5.8.1 Introduction

The smart city concept is quickly moving from a futuristic idea to a global imperative. As urban populations expand and cities become more intricate, there's a pressing need for intelligent infrastructure that boosts efficiency, sustainability, and overall quality of life. The synergy between 6G and IoT technologies is central to this change, heralding ultra-connected urban settings where everything, from street lighting to waste management, functions in a coordinated, adaptive, and intelligent way.

5.8.2 A New Urban Command System

At the core of this transformation is the development of a "digital nervous system" for the city. Similar to how the human nervous system gathers, processes, and acts on sensory input, a smart city powered by 6G and IoT continuously observes its surroundings and adjusts accordingly. IoT sensors integrated into roads, buildings, public services, and even personal wearable devices transmit a constant stream of data to cloud or edge computing platforms. Thanks to 6G's minimal latency and high bandwidth, these platforms can process data instantly, allowing for dynamic adjustments. For instance, traffic congestion can be swiftly mitigated by redirecting vehicles through adaptive traffic signals or by suggesting alternative routes via connected cars and mobile applications.

5.8.3 Automated Infrastructure and Proactive Maintenance

Traditional infrastructure management often depends on reactive fixes—addressing issues after they arise. In contrast, smart infrastructure is proactive. Bridges can be fitted with stress and vibration sensors that warn authorities about potential structural issues before they become hazardous. Water systems can detect leaks using pressure and flow sensors, thereby reducing waste and repair expenses. Electrical grids can manage loads more effectively, averting power outages during high-demand periods. This move toward predictive maintenance enhances public safety, cuts operational costs, and prolongs the lifespan of urban infrastructure. The crucial element here is not merely connectivity but the real-time analysis and coordination made possible by 6G-enabled networks.

5.8.4 Intelligent Urban Mobility and Traffic Systems

Transportation networks are among the most immediate and noticeable beneficiaries of smart infrastructure. IoT-enabled traffic lights can modify their timing based on current traffic flow, decreasing waiting times and fuel use. Public transport, taxis, and autonomous vehicles can communicate with the infrastructure to optimize routes and prevent collisions. Furthermore, 6G integration will enable capabilities beyond 5G's reach, such as real-time high-definition mapping for autonomous vehicles, simultaneous device-to-device communication in congested traffic, and swarm intelligence for drone-based delivery. The outcome is a safer, quicker, and more environmentally friendly movement throughout cities.

5.8.5 Waste Management and Environmental Oversight

Efficient waste collection is another vital component of smart urban systems. Sensor-equipped bins that indicate when they are full allow cities to optimize garbage collection routes, cutting down on unnecessary trips and fuel consumption. Air and water quality sensors distributed across the city enable detailed environmental monitoring, allowing for rapid responses to pollution events, chemical spills, or unusual weather phenomena. Beyond immediate monitoring, the historical data from these sensors informs predictive models, helping city planners create more sustainable urban environments. For example, identifying areas with persistent poor air quality could lead to the establishment of green zones, traffic limitations, or urban afforestation projects.

5.8.6 Energy Management and Smart Grids

Urban energy use significantly contributes to environmental issues and operational expenses. 6G-IoT infrastructure supports the creation of intelligent energy grids that can manage demand in real time. Smart meters in homes and businesses offer insights into usage habits and assist consumers in optimizing their energy consumption. Additionally, these grids more effectively integrate renewable energy sources, balancing inputs from solar, wind, and conventional sources with city-wide demand. Surplus energy generated during low-demand times can be rerouted or stored for later use, enhancing the resilience and efficiency of the power supply.

5.8.7 Safety, Surveillance, and Emergency Response

In smart cities, public safety becomes more predictive and responsive. Surveillance cameras integrated with AI algorithms can identify unusual behaviors or activities, automatically alerting law enforcement. Smart street lighting can adjust brightness according to pedestrian presence, improving safety while saving energy. Emergency services are significantly enhanced with smart systems that direct ambulances via the quickest routes or deploy drones to deliver critical supplies during natural disasters. These functions demand ultrareliable low-latency communication—precisely what 6G networks are engineered to deliver (Table 5.5).

Table 5.5 Key Benefits of 6G-IoT in Smart Cities

Domain	Impact of 6G-IoT Integration
Traffic & mobility	Adaptive traffic lights, autonomous navigation
Public safety	Predictive surveillance, emergency response optimization
Energy management	Smart grids, real-time consumption tracking
Waste & environment	Sensor-based waste collection, pollution monitoring
Infrastructure maintenance	Predictive maintenance, reduced downtime
Governance	Real-time citizen feedback, transparent service delivery

5.8.8 Citizen Interaction and Digital Governance

A smart city is more than just a network of machines; it is a place where people live, work, and connect. 6G and IoT technologies open new avenues for citizen engagement and participatory governance. Mobile applications can empower residents to report issues like potholes, broken streetlights, or vandalism in real time, strengthening feedback mechanisms and shortening response times. Moreover, with integrated data platforms, city administrations can provide tailored information to citizens—from weather alerts to transport updates and public health notices—thereby improving transparency and fostering trust in governmental institutions.

5.8.9 Privacy and Ethical Dilemmas

Despite the immense potential of smart cities, they also present considerable ethical and privacy challenges. The extensive collection of personal and behavioral data raises questions about consent, surveillance, and potential misuse. Who controls this data? How is it stored, shared, and secured? If not handled properly, these issues could undermine public confidence and slow adoption. It is crucial for city planners, technologists, and policymakers to collaborate on implementing ethical guidelines, strong cybersecurity measures, and privacy-by-design principles that protect individual rights while delivering collective advantages.

5.8.10 Economic Growth and Urban Innovation

The deployment of 6G and IoT infrastructure spurs local economic activity through job creation, support for startups, and attraction of international investment. Urban innovation centers and smart technology labs become hubs for research and entrepreneurship. Startups can utilize open city data to develop innovative services, such as mobility-as-a-service platforms, urban agriculture applications, or citizen safety networks. As cities become more efficient and desirable places to live, they attract talent and investment, fostering a positive cycle of economic development powered by intelligent infrastructure.

5.8.11 Implementation Hurdles

While the benefits are numerous, implementing smart city projects faces several obstacles. High initial investment, a shortage of technical expertise, fragmented regulatory environments, and existing legacy infrastructure are common difficulties. Furthermore, ensuring interoperability between devices and systems from different manufacturers remains a significant concern. To address these issues, cities require thorough strategic planning, public-private collaborations, and adherence to global standards for smart urban systems. Pilot programs, stakeholder consultations, and phased deployments are effective methods for building scalable and sustainable smart cities [19].

5.9 Agricultural Innovation and Rural Empowerment Through 6G-IoT Integration

5.9.1 Introduction

Agriculture, historically a traditional field, is undergoing a swift transformation driven by digital connectivity advancements. The convergence of 6G and IoT technologies is set to revolutionize farming methods, bolster food security, and elevate rural economies. By enabling smart farming practices, real-time crop surveillance, and autonomous agricultural machinery, 6G-IoT integration allows for data-informed decision-making even in the most isolated rural areas.

5.9.2 Precision Agriculture and Smart Farming

One of the most notable changes facilitated by 6G and IoT is precision agriculture. Farmers can deploy soil sensors, climate monitoring devices, and automated irrigation systems that gather real-time data on soil moisture, pH levels, nutrient content, and temperature. This data, transmitted almost instantaneously over 6G networks, permits highly localized field management, leading to optimized resource use and increased yields. Farming machinery, such as tractors and harvesters, is evolving into autonomous units capable of operating with minimal human oversight. Connected via 6G, these machines can coordinate with drones for aerial monitoring and optimize routes for planting, watering, or harvesting based on live analytical data.

5.9.3 Climate Adaptability and Crop Prediction

Unpredictable weather patterns resulting from climate change pose a significant risk to agriculture. With 6G's capacity to process vast amounts of data in real time, weather forecasts can be tailored to specific fields, assisting farmers in planning their activities more effectively. IoT devices distributed across farmland can feed data into climate models that predict rainfall, pest infestations, or temperature fluctuations, enabling farmers to take proactive measures. Crop forecasting tools, utilizing historical data and real-time satellite imagery, can predict yield outcomes with considerable accuracy. This assists farmers in making knowledgeable decisions regarding planting schedules, crop rotation, and marketing strategies.

5.9.4 Livestock Management

Beyond crop cultivation, 6G and IoT are also transforming livestock farming. Wearable sensors for animals can track health indicators like body temperature, movement patterns, and feeding habits. In cases of illness or injury, early alerts are sent to veterinarians or animal caretakers. These smart systems also facilitate automated feeding and milking processes, enhancing productivity while maintaining animal welfare [20]. Geo-fencing, made possible by satellite and IoT integration, enables farmers to monitor and manage animal movement over extensive areas, especially in open grazing systems. This technology improves security and lessens the risk of animal loss or disease transmission.

5.9.5 Water and Resource Management

The efficient use of water and energy is fundamental to sustainable agriculture. IoT-based smart irrigation systems analyze soil moisture levels and adjust water application accordingly. 6G connectivity ensures this data is processed and acted upon in real time, minimizing waste. Solar-powered sensors and pump controls further decrease reliance on nonrenewable energy sources. Additionally, farmers gain access to dashboards and mobile applications that display resource consumption, making conservation efforts more concrete and manageable (Table 5.6).

5.9.6 Market Access and Supply Chain Clarity

A persistent challenge in agriculture is the limited market access for rural farmers. With 6G networks extending to even remote areas, farmers can directly engage with digital marketplaces, compare pricing, and conduct secure transactions. Blockchain-based platforms, integrated via IoT, create transparent supply chains by tracking agricultural products from the farm to the consumer. Real-time tracking of perishable items improves logistics, cuts down on spoilage, and ensures

Table 5.6 Impact of 6G-IoT on Agriculture and Rural Development

Agricultural Function	6G-IoT Contribution
Crop management	Real-time soil monitoring, automated irrigation
Climate adaptation	Localized weather alerts, yield forecasting
Livestock monitoring	Health tracking, geo-fencing, and smart feeding
Resource efficiency	Precision water use, solar-powered systems
Market access	Digital sales platforms, transparent logistics
Rural upliftment	Education access, community apps, and fintech integration

timely delivery, benefiting both producers and consumers. Farmers can also access financial services, crop insurance, and training materials through 6G-connected mobile applications.

5.9.7 Empowerment of Rural Communities

Digital inclusion through 6G offers transformative advantages to rural populations. Educational materials, farming advice, and government support programs become accessible via smart devices. Community-based networks foster knowledge exchange, cooperative farming efforts, and collective problem-solving. Rural youth, exposed to smart technologies, are empowered to innovate, thereby bridging the divide between rural and urban areas.

5.10 Environmental Sustainability and Ecosystem Monitoring

5.10.1 Introduction

Preserving the environment is a pressing global concern. The synergy of 6G and IoT offers potent tools for real-time environmental surveillance, predictive ecosystem analysis, and the encouragement of sustainable practices across various sectors. By facilitating highly distributed sensor networks and edge-based data processing, this integration supports both regulatory enforcement and community-led conservation initiatives.

5.10.2 Real-Time Environmental Monitoring

With 6G's high capacity and low latency, extensive sensor networks can be deployed across urban and natural landscapes. These sensors measure air quality, water purity, soil conditions, radiation levels, and other ecological indicators in real time. Such data empowers municipalities, researchers, and citizens to respond to pollution events and monitor adherence to environmental regulations [20]. Smart grids, for instance, can use sensors to detect emissions from industrial areas. Automated alerts can be dispatched to environmental agencies, while public dashboards enhance transparency.

5.10.3 Forest and Wildlife Conservation

Illegal logging and poaching pose serious threats to biodiversity. 6G-connected drones and ground sensors can continuously monitor protected areas. Image recognition and motion detection technologies identify unauthorized activities or track animal movement patterns. These systems function with minimal human involvement, reducing the expense and expanding the scope of conservation efforts. Wildlife equipped with IoT tracking collars can be observed for migration routes, health conditions, and breeding behaviors. These insights inform breeding programs, habitat preservation strategies, and anti-poaching operations.

5.10.4 Smart Waste Management Systems

Urban waste presents a significant environmental challenge. Smart bins, connected via IoT and 6G, signal collection requirements based on fill levels and material types. Waste collection routes are dynamically optimized, thereby reducing fuel consumption and emissions. Furthermore, these systems help track recycling compliance and identify areas prone to illegal dumping. Data from these systems can also be used to encourage citizen participation by offering incentives for proper waste separation and recycling.

5.10.5 Renewable Energy and Grid Optimization

6G and IoT support the integration of renewable energy sources into national power grids. Sensors monitor outputs from solar and wind farms, while predictive analytics balance energy demand and supply [21]. In smart homes, appliances can react to real-time energy pricing, encouraging efficient use during off-peak hours. Microgrids in rural areas, coordinated by 6G-enabled control centers, ensure a continuous power supply even in remote regions. These systems lessen reliance on fossil fuels and enhance energy resilience.

5.10.6 Disaster Prevention and Ecosystem Restoration

Natural disasters such as floods, wildfires, and landslides can be predicted, and their impacts lessened, with the aid of IoT sensors and AI analytics. Real-time data from rain gauges, seismometers, and satellite imagery enable early warning systems to operate effectively. Emergency services can be deployed before disasters occur, saving lives and reducing damage. In degraded environments, remote sensors assess soil health and vegetation growth, guiding reforestation or wetland restoration efforts. Over time, data-driven restoration improves ecosystem services like carbon sequestration, water purification, and biodiversity (Table 5.7).

5.11 Ethical and Privacy Considerations

As 6G and IoT technologies become more deeply embedded in daily life, ethical and privacy considerations emerge as paramount societal issues. The widespread collection, transmission, and analysis of personal data introduce risks concerning consent, individual autonomy, surveillance, and data misuse. Tackling these

Table 5.7 Environmental Benefits of 6G-IoT Synergy

Environmental Domain	Impact of 6G-IoT Integration
Air and water monitoring	Real-time pollution detection, compliance enforcement
Wildlife protection	GPS tracking, drone surveillance, and poaching alerts
Waste management	Smart bins, route optimization, and recycling incentives
Renewable energy	Grid optimization, adaptive energy pricing
Disaster response	Early warning systems, satellite-driven coordination
Ecosystem restoration	Data-guided reforestation, habitat assessment

challenges demands a multifaceted approach that includes technology design, policy development, and public involvement. 6G-enabled IoT devices will produce enormous datasets containing sensitive personal, behavioral, and environmental details. Without strong privacy protections, this data could be exploited for unauthorized profiling, discrimination, or surveillance by corporations, governments, or malicious entities. Moreover, the vast scale and diverse nature of IoT ecosystems make it difficult to establish clear ownership, control, and accountability [22].

Ethical design principles, such as privacy-by-design and data minimization, must be incorporated into hardware, software, and network protocols from the outset. Advanced cryptographic methods, including homomorphic encryption and secure multiparty computation, can safeguard data confidentiality during transmission and processing. Decentralized systems using blockchain can improve transparency and trust by enabling verifiable audit trails. Legal and regulatory frameworks must evolve alongside technological advancements [23]. International collaboration is vital to align data protection laws, manage cross-border data transfers, and set ethical standards that uphold human rights and dignity. Public awareness initiatives and inclusive discussions are essential for promoting informed consent and empowering individuals to manage their digital presence.

Finally, AI-driven decision-making, increasingly integrated with 6G-IoT systems, must be transparent, understandable, and free from biases to ensure fairness and accountability. Ethical review boards and impact assessments can help guide responsible innovation and lessen unforeseen negative outcomes.

5.12 Conclusion and Future Directions

The synergy between 6G and IoT technologies is set to fundamentally transform society across various aspects—communication, healthcare, education, urban living, employment, the environment, and ethics. As discussed in previous sections,

6G's ultrahigh speed, massive connectivity, and extremely low latency, combined with the intelligent data collection and automation capabilities of IoT, will create unparalleled opportunities for innovation and improvements in quality of life.

However, this technological revolution also presents considerable challenges, including digital disparities, privacy vulnerabilities, workforce shifts, and environmental concerns. Addressing these requires concerted efforts from governments, industry, academia, and civil society to formulate inclusive policies, ethical guidelines, and resilient infrastructures. Investments in education, digital literacy, and equitable access must go hand-in-hand with technological deployment to ensure that no community is excluded.

Looking ahead, research and development will concentrate on further enhancing AI integration within 6G-IoT ecosystems, enabling autonomous decision-making at scale across different sectors. The advent of quantum communication, advanced materials for IoT devices, and decentralized network architectures may reshape network paradigms in the upcoming decades. Furthermore, fostering global cooperation on regulatory standards, data governance, and sustainability will be of utmost importance. Ultimately, the successful realization of 6G and IoT synergies offers a transformative vision for a connected, intelligent, and sustainable society that empowers individuals and communities while protecting fundamental rights and the planet.

References

1 Uusitalo, M.A. (2020). 6G networks: visions, requirements, and enabling technologies. *IEEE Communications Magazine* 58 (12): 90–96.

2 Mumtaz, S. (2021). 6G wireless systems: vision, requirements, challenges, insights, and opportunities. *Proceedings of the IEEE* 109 (7): 1166–1199.

3 Hossain, R. (2022). Societal challenges and opportunities with 6G technologies. *IEEE Consumer Electronics Magazine* 11 (6): 6–15.

4 International Telecommunication Union (ITU) (2021). *The State of Broadband 2021: Tackling Digital Inequality*. ITU Report.

5 Jing, Y. (2020). Integrating satellites into 6G networks for universal connectivity. *IEEE Network* 34 (3): 190–197.

6 Nkengafac, T. (2020). IoT-enabled mesh networks for rural connectivity: case studies. *IEEE Access* 8: 199345–199360.

7 Singh, R. (2020). Satellite-IoT integration for remote healthcare delivery. *IEEE Internet of Things Journal* 7 (9): 8800–8809.

8 Baccelli, F., Hahm, O., Günes, M., and Wählisch, M. (2021). Networking and computing challenges in 6G: a research agenda. *IEEE Communications Magazine* 59 (3): 14–20.

9 Zhang, L., Chen, M., and Zhang, H. (2022). Blockchain-based privacy preservation for IoT data sharing in 6G networks. *IEEE Internet of Things Journal* 9 (4): 2328–2341.

10 Chen, S., Hu, J., Shi, Y., and Zhao, L. (2020). 6G wireless networks: vision, requirements, architecture, and key technologies. *IEEE Vehicular Technology Magazine* 15 (4): 28–41.

11 Chowdhury, M.Z., Shahjalal, M., Hossain, M.A., and Rahman, M.J. (2020). 6G wireless communication systems: applications, requirements, technologies, challenges, and research directions. *IEEE Open Journal of the Communications Society* 1: 957–975.

12 Boulogeorgos, A., Faruk, N.R.H., Yang, H.C., and Karagiannidis, G.K. (2020). 6G wireless networks: applications, trends, technologies, and open research problems. *Electronics* 9 (10): 1634.

13 Mao, Y. (2021). A survey on societal impacts of 6G and IoT. *IEEE Access* 9: 47872–47892.

14 Yang, K., Wang, Y., and Chen, J. (2021). Privacy-preserving machine learning in 6G and IoT: opportunities and challenges. *IEEE Network* 35 (4): 106–113.

15 Yu, F.R. and Tang, H. (2020). Challenges and research opportunities for 6G networks. *IEEE Network* 34 (6): 212–217.

16 Yang, P. (2021). A survey of 6G wireless communications: emerging technologies. *IEEE Communications Surveys & Tutorials* 23 (2): 1–29.

17 Lin, S., Ma, M., and Wang, X. (2021). Sustainable development in 6G: enabling technologies and future challenges. *IEEE Wireless Communications* 28 (6): 36–42.

18 Huang, J., Qian, Y., and Ma, Y. (2021). AI-empowered 6G and IoT networks: from vision to reality. *IEEE Communications Magazine* 59 (6): 52–58.

19 Shi, Y., Zhao, J., and Alwakeel, M.M. (2022). 6G-enabled intelligent IoT: opportunities and challenges. *IEEE Internet of Things Journal* 9 (10): 7611–7625.

20 Sharma, S.K. and Wang, X. (2021). Towards massive machine type communications in 6G: a survey. *IEEE Communications Surveys & Tutorials* 23 (1): 114–140.

21 Poor, H.V. (2021). Privacy and security challenges in 6G and beyond networks. *Proceedings of the IEEE* 109 (2): 230–247.

22 Liu, X., Sun, Y., and Zhang, L. (2023). Security and privacy in 6G-enabled IoT networks: challenges and solutions. *IEEE Internet of Things Journal* 10 (4): 2555–2568.

23 Lee, J. (2021). Ethical frameworks for AI and IoT systems in future networks. *IEEE Internet of Things Journal* 8 (12): 9876–9885.

6

Enabling Technologies in 6G for VIoT and Autonomous Driving

Ishan Kumar

Department of Computer Science and Engineering, Chitkara University, Rajpura, Punjab, India

6.1 Introduction

The global automotive industry is undergoing a radical transformation with the convergence of connectivity, automation, and artificial intelligence (AI). At the heart of this transformation lies the concept of the Vehicular Internet of Things (VIoT), which integrates vehicles, infrastructure, and cloud services into a unified, intelligent transportation ecosystem. However, as vehicular systems grow more autonomous and data-driven, they encounter technological limitations related to communication latency, reliability, localization accuracy, and computational responsiveness—especially in complex, high-mobility environments.

Fifth-generation (5G) networks, while a significant step forward, still fall short in supporting ultralow latency, real-time decision-making, and decentralized intelligence required by autonomous vehicles. This gap is increasingly critical as applications such as emergency maneuvering, cooperative perception, and vehicle platooning become mainstream. The need for a more robust, intelligent, and scalable communication infrastructure has led to the emergence of sixth-generation (6G) wireless networks, envisioned to address these exact demands.

This chapter delves into how 6G, with its native integration of artificial intelligence (AI), ultrareliable low-latency communication (URLLC), sub-centimeter localization, and hierarchical computing models, can unlock the full potential of autonomous driving and VIoT systems. In particular, the focus is on enabling technologies such as edge computing, fog-based orchestration, cooperative localization, and sensor fusion, which collectively redefine the vehicular experience.

Synergies of 6G and IoT, First Edition. Edited by Himanshi Babbar.

Published 2026 by John Wiley & Sons, Inc.

By offering a comprehensive examination of recent innovations, architectural frameworks, and real-world use cases—including patent-based implementations—this chapter provides readers with a forward-looking perspective on the evolution of vehicular communication. It not only identifies existing gaps in current technologies but also proposes a cohesive roadmap for integrating 6G into the vehicular domain. The following sections build upon this foundation to explore the capabilities, challenges, and future directions of 6G-enabled VIoT and autonomous mobility systems.

6.1.1 Background and Problem Statement

Vehicular communication has evolved significantly, yet limitations in connectivity, latency, and computational delay persist, impeding safe and efficient autonomous driving. Current 4G and 5G networks cannot deliver deterministic communication guarantees required by mission-critical automotive applications. Key problems include unstable network connectivity in high-speed environments, limited localization precision in complex terrains, and inefficient centralized data processing models. For instance, systems like the wireless communication and collision avoidance method described in ref. [1] demonstrate how real-time radar sensor data fusion and V2V communication can prevent collisions. However, such systems demand ultralow-latency and highly reliable communication links that current technologies struggle to consistently provide, especially in dense urban or high-mobility scenarios.

6.1.2 Research Gap

While 5G introduced enhanced Mobile Broadband (eMBB) and URLLC, it falls short in scenarios demanding real-time distributed intelligence, sub-centimeter localization, and ultrahigh reliability. There is a clear need for a paradigm shift toward a network architecture capable of ubiquitous edge intelligence (Edge AI) and multilayer computing support. Existing studies rarely focus on the integration of edge-fog-cloud infrastructure with 6G in vehicular contexts.

6.2 Literature Review

Recent work by Saad et al. [2] outlines the fundamental vision of 6G, highlighting its potential for sub-ms latency and AI-native operations. Khan et al. [3] and Liu et al. [4] discussed cooperative perception and high-mobility challenges. However, they stop short of detailing the architectural shifts involving edge and fog computing. Zhou et al. [5] and Mao et al. [6] emphasize the promise of edge AI but do not elaborate on vehicular-specific adaptations. The collision avoidance and wireless vehicular communication system [1] offers a concrete example of integrating radar sensor data with V2V communication for autonomous safety, which aligns with the emerging 6G-enabled edge-fog computing paradigms but remains underrepresented in the literature. Therefore, a comprehensive model leveraging edge-fog-cloud collaboration for VIoT in a 6G landscape remains underexplored.

6.2.1 Novelty of the Research

This chapter uniquely synthesizes the interplay between 6G features and vehicular technologies, with a special focus on how distributed AI, via edge and fog layers, transforms decision-making in real-time. It highlights novel sensor fusion mechanisms, cooperative localization strategies, and hierarchical computation architectures that optimize the behavior of autonomous vehicles.

6.3 Methodology

This chapter employs a descriptive-analytical approach, integrating findings from contemporary literature, simulation data, and architectural modeling to explore the potential of 6G networks in vehicular ecosystems. Comparative tables and use-case diagrams are utilized to evaluate 6G capabilities against 5G benchmarks. Emphasis is placed on edge computing scenarios and fog-based orchestration layers designed for vehicular environments.

6.3.1 Problems Faced by Current Vehicular Systems

1) **Inconsistent Network Connectivity:** High-speed vehicular mobility often leads to network disconnections and frequent handovers, affecting continuous communication.
2) **Latency-Sensitive Applications:** Applications such as emergency braking and intersection navigation require ultralow latency.
3) **High Bandwidth Demand:** Emerging services like cooperative perception and HD map sharing demand more bandwidth than current systems can provide.
4) **Localization Issues:** In tunnels or dense urban environments, GPS signals become unreliable, risking the precision needed for autonomous maneuvers [7].

These limitations highlight the need for more advanced network capabilities, which 6G aims to deliver.

6.4 Characteristics of 6G Networks for Vehicular Communication

6G networks are built to overcome the connectivity, latency, and positioning limitations that hamper current vehicular systems. They are inherently **AI-native**, support **URLLC**, and enable **sub-centimeter-level localization**—all crucial for autonomous driving and VIoT use cases.

AI-native 6G networks

The AI-native nature of 6G signifies that artificial intelligence is not an add-on but an integral component of the network architecture. AI is embedded across all layers—radio access, core network, edge, and application layers—to enable self-learning, self-optimization, and self-healing capabilities. For vehicular systems, this allows real-time adaptation to high mobility, dynamic traffic density, and varying channel conditions. AI-native operation supports predictive handovers, intelligent resource allocation, and autonomous network management, which are essential for scalable, reliable, and context-aware VIoT and autonomous driving environments.

Ultra-Reliable Low-Latency Communication (URLLC)

URLLC is critical for mission- and safety-critical vehicular applications that require deterministic communication guarantees. In autonomous driving, decisions such as emergency braking, collision avoidance, cooperative maneuvering, and platooning must be executed within microseconds with near-zero packet loss. 6G-enabled URLLC significantly reduces end-to-end latency (sub-millisecond or even sub-0.1 ms) while achieving extremely high reliability ($\geq$99.999%). This ensures that control messages and sensor data are delivered instantaneously and consistently, preventing catastrophic failures in high-speed and dense traffic scenarios.

Sub-centimeter-level localization

Sub-centimeter-level localization represents a major advancement over traditional GPS-based positioning, which is often unreliable in tunnels, urban canyons, or dense environments. By leveraging terahertz/mmWave signals, joint communication and sensing (JCAS), cooperative localization, and AI-assisted sensor fusion, 6G enables ultra-precise positioning and orientation awareness. This level of accuracy is crucial for lane-level navigation, autonomous overtaking, intersection management, and pedestrian safety. For VIoT systems, precise localization also enhances cooperative perception and coordination among vehicles and infrastructure, leading to safer, more efficient, and fully autonomous mobility.

6.4.1 Ultrareliable Low-Latency Communication (URLLC)

URLLC is a foundational pillar of 6G, providing latency as low as **0.1 ms** and **reliability above 99.999%**. This is critical for real-time autonomous driving applications, such as:

- Lane-keeping and emergency braking.
- Cooperative collision avoidance.
- Traffic light communication and control.
- Remote operation of vehicles in hazardous conditions.

5G's latency (~1 ms) and reliability (99.99%) are not sufficient for mission-critical vehicular operations where **instant decision-making** is mandatory.

Key Enablers of URLLC in 6G:

- **Massive MIMO and beamforming** for efficient spectrum use [2].
- **AI-driven spectrum sharing** to avoid congestion [8].
- **Intelligent reflecting surfaces (IRS)** for maintaining LoS in urban environments [9].
- **Multi-connectivity and predictive handovers** to reduce disruptions [3] (Figure 6.1 and Table 6.1).

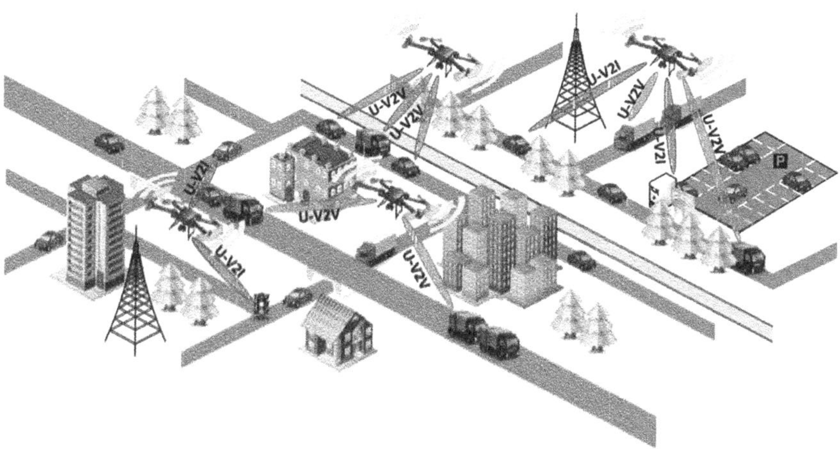

Figure 6.1 UAV-assisted edge computing framework in 6G-enabled networks.
Source: IEEE / UAV-assisted network architecture for 6G vehicular edge computing / https://www.researchgate.net/publication/352801803_UAV-Assisted_Vehicular_Edge_Computing_for _the_6G_Internet_of_Vehicles_Architecture_Intelligence_and_Challenges Accessed on July 07, 2025.

Table 6.1 Comparison of 5G and 6G URLLC Metrics Use Cases Enabled by URLLC

Feature	5G	6G
Latency	~1 ms	~0.1 ms
Reliability	99.99%	99.999%
Mobility support	<500 km/h	Up to 1000 km/h
Use-case example	V2X Messaging	Remote driving

- **Platooning**: Vehicles nearby exchange data in real-time to maintain tight formations [3].
- **Remote Driving**: Operators can control vehicles in hazardous zones without delay [4].
- **Cooperative Perception**: Fusion of sensors from multiple vehicles enhances environment perception [9].

Challenges:

- **Line-of-Sight Obstruction**: Can be mitigated by using IRS [8].
- **Interference in Dense Areas**: AI-powered spectrum allocation dynamically reduces interference [10].

6.4.2 High-Precision Positioning and Localization

High-precision positioning with **centimeter-level accuracy** is a cornerstone of 6G-enabled VIoT systems. This capability is indispensable for advanced vehicular functions such as **lane-level navigation**, **pedestrian avoidance**, **autonomous overtaking**, and **intersection management**. Existing Global Navigation Satellite System (GNSS) technologies like GPS are insufficient in complex environments due to **multipath interference**, **signal blockages** in tunnels or urban canyons, and **low update rates** [7, 11].

6G enhances localization by combining multiple advanced technologies, including:

- **Millimeter-wave (mmWave)** and **Terahertz (THz)** frequency signals, which provide high-resolution spatial awareness due to their short wavelengths [3, 8].
- **Inertial Navigation Systems (INS)**, **LiDAR**, and **Visual Odometry**, which capture precise motion and environment data [11, 12].

- **Ultra-Wideband (UWB)** and **Cooperative V2X Communication** for redundancy and augmentation in sensor-compromised scenarios [4, 13].

One of the most effective techniques in this domain is **Visual-Inertial SLAM (Simultaneous Localization and Mapping)**, which integrates camera data with inertial measurement unit (IMU) measurements. When deployed on edge devices, it enables vehicles to navigate complex, GPS-denied areas in real time [12]. Similarly, **Joint Communication and Sensing (JCAS)** allows the same 6G infrastructure to be used for both data transmission and environmental perception, significantly reducing latency and hardware redundancy [10].

6G also empowers **cooperative localization**, where vehicles dynamically share location data with infrastructure (e.g., Road Side Units [RSUs] and fog nodes) to refine their own position estimates. This **data fusion** reduces error margins in crowded or signal-challenged environments and improves overall fleet safety [4, 6, 14].

Patent-backed research [1] outlines a dynamic localization and communication system that uses radar-based environment scanning integrated with V2V messaging. It illustrates how real-time **environmental feedback** can adjust both vehicle positioning and communication routing, leading to robust and adaptable positioning accuracy.

These innovations are closely tied to 6G's goals of enabling **context-aware, resilient vehicular systems** [8, 15].

Additionally, **AI-powered edge and fog computing** units enhance localization by processing vast amounts of real-time sensor data with minimal latency [16–18]. Techniques such as **predictive localization using federated learning (FL)**, **sensor reliability scoring**, and **adaptive path reconstruction** are being deployed to strengthen precision even during **temporary signal loss or occlusion** [5, 14, 19] (Table 6.2).

- **JCAS**: Combines network signals with radar-like sensing [20].
- **Cooperative Localization**: Uses multi-vehicle data for enhanced positioning.
- **Visual-Inertial SLAM**: Fuses camera and IMU data for real-time mapping [12].

Table 6.2 5G vs. 6G Localization Comparison of Emerging Methods

Metric	5G	6G
Accuracy	~1 m	1–10 cm
Update rate	10 Hz	>100 Hz
Environmental resilience	Moderate	High (incl. Tunnels, indoor)

6.5 Edge Computing and Fog Computing for Vehicular Applications

6.5.1 Edge Intelligence and Distributed Data Processing

Edge computing in 6G architecture plays a pivotal role in enabling real-time responsiveness by bringing computational resources closer to the data source, whether it's embedded within a vehicle, roadside infrastructure, or traffic control units. This distributed architecture is indispensable for **latency-sensitive vehicular applications** such as **object detection**, **lane departure alerts**, **adaptive cruise control**, and **collision avoidance** [6, 16].

Unlike traditional cloud models that suffer from core network congestion and centralized delays, **edge computing minimizes round-trip latency** by processing data locally.

This is crucial in high-mobility environments where even milliseconds of delay could jeopardize safety-critical decisions [2, 15].

Building upon edge computing, **Edge AI** integrates **machine learning** and **AI inference** capabilities directly into edge nodes. With the help of lightweight models, these nodes can perform the following functions:

- Real-time object classification,
- Hazard recognition,
- Anomaly detection (e.g., sudden driver behavior shifts),
- Local route optimization [14, 17, 18] (Figure 6.2).

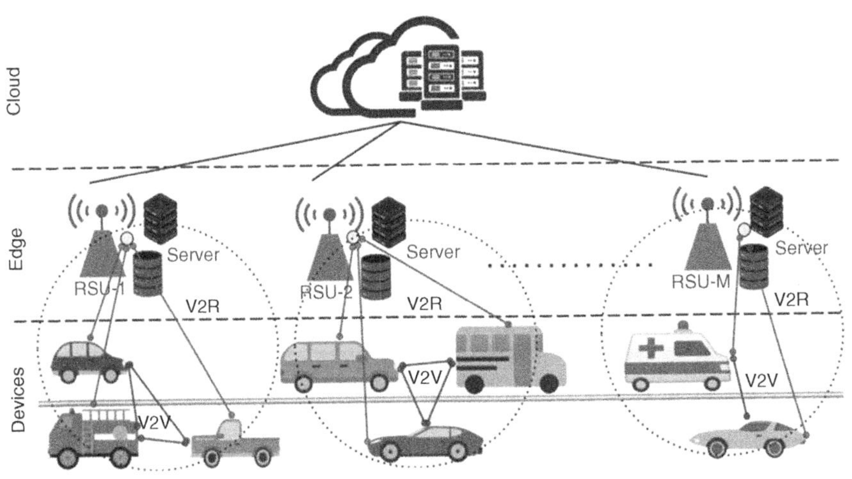

Figure 6.2 High-level layered architecture of a 6G intelligent network system. *Source:* Adapted from High-level overview of the 6G architecture.

This enables **context-aware, autonomous decision-making** directly at the vehicle or roadside level, bypassing cloud reliance. For instance, when a pedestrian is detected stepping onto a road, an edge node can trigger an emergency braking system faster than centralized cloud logic would allow [5, 13].

The use of **FL** and **distributed neural architectures** allows multiple edge nodes to collaboratively learn and adapt while preserving data privacy. These models are especially beneficial for environments with variable vehicle densities, changing road layouts, or unpredictable weather conditions [21].

Further enhancing this vision is the **patent-based hierarchical model** detailed in ref. [1], which outlines a **smart edge AI deployment framework** for urban transportation grids. In this model:

- **Tier-1 edge nodes** are deployed within vehicles.
- **Tier-2 nodes** are integrated into traffic infrastructure like RSUs.
- **Tier-3 fog orchestrators** supervise regional data traffic and inference tasks.

The system dynamically reallocates computation based on **vehicular density**, **mobility patterns**, and **contextual urgency**, ensuring both **resource efficiency** and **service continuity**, especially in congestion-prone areas or during peak traffic hours.

Key Benefits of Edge Intelligence in VIoT:

- **Ultralow Latency:** Enables real-time decision-making for safety-critical tasks (e.g., emergency braking, obstacle avoidance) [15, 16].
- **Decentralized Processing:** Reduces load on central/cloud systems and ensures local autonomy [6, 22].
- **Data Privacy and Security:** Raw sensor data remains local; only model updates are shared (via FL) [17, 21].
- **Scalability:** Edge nodes can be scaled up across infrastructure with modular AI services [14, 23].
- **Personalized Analytics:** Enables individual vehicle behavior analysis, such as **driver monitoring**, **drowsiness detection**, or **aggressive driving patterns** [13, 18] (Figure 6.3 and Table 6.3).

6.5.2 Fog-Based Decision-Making for Autonomous Vehicles

Fog computing serves as an essential **intermediary layer** between edge nodes and centralized cloud systems, enabling **hierarchical processing**, **context-aware inference**, and **collaborative decision-making** across distributed vehicular networks. In the context of autonomous vehicles, **fog nodes**—often deployed in **RSUs**, traffic lights, or regional gateways—aggregate data from multiple vehicles, analyze traffic dynamics, and disseminate time-critical commands [6, 15].

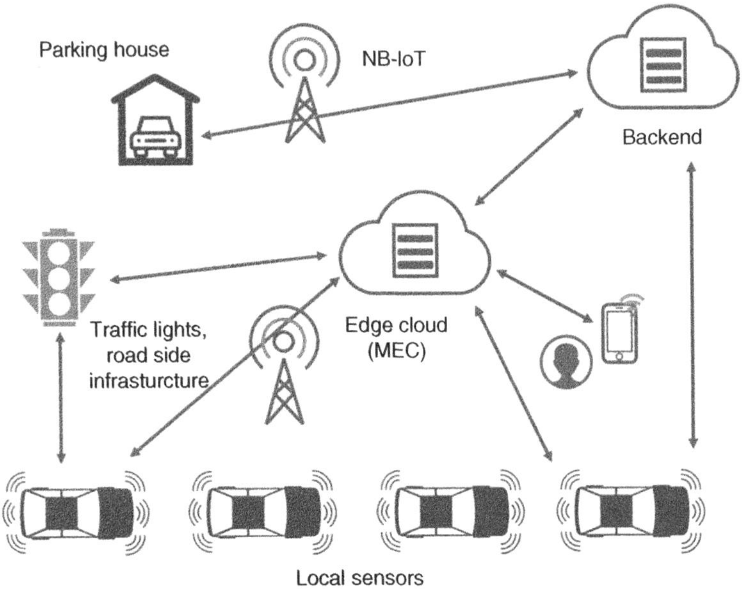

Figure 6.3 Edge computing architecture integrated with autonomous vehicle systems. *Source:* Architecture of autonomous vehicles with edge computing / IEEE / CC BY 4.0.

Table 6.3 Edge vs. Cloud Capabilities in Vehicular Systems Applications

Attribute	Edge Computing	Cloud Computing
Latency	Milliseconds	Seconds
Location	Onboard/near-source	Remote datacenter
Decision time	Real-time	Delayed
Use cases	Emergency braking and lane change	Traffic analysis and training models

This **multilayered architecture** enhances vehicular autonomy by reducing reliance on long-latency cloud infrastructure. For example, during peak hours or urban congestion, fog nodes can manage localized traffic through:

- Dynamic **signal adjustment**.
- **Route optimization**.

- Real-time **incident response coordination.**
- **Platooning synchronization**, where a series of vehicles move cohesively with minimal delay [4, 23].

Key Functionalities Enabled by Fog Nodes:

- **Distributed Consensus:** Vehicles communicate with fog controllers to agree on shared routes or actions.
- **Cross-Vehicle Learning:** Sensor data from one vehicle (e.g., object detection or braking event) can inform others in the vicinity [17].
- **Traffic Flow Prediction:** Using regional AI models, fog nodes forecast traffic congestion patterns and advise rerouting [8, 19].

Fog computing also plays a pivotal role in maintaining **data integrity, security, and availability**. By integrating **blockchain** and **zero-trust communication protocols**, fog-based systems can **authenticate data packets**, **prevent tampering**, and **log communication events** in real time [21]. This is particularly important for safety-critical systems such as **emergency vehicle prioritization**, **hazard alert broadcasting**, or **vehicle recall warnings**.

The **hierarchical orchestration system** proposed in the Chitkara University patent [1] demonstrates a practical implementation of fog intelligence. The model includes the following:

- **Vehicle-level agents** (edge).
- **RSU-based fog controllers**.
- **Cloud-based long-term model training**.

It dynamically assigns resources based on environmental triggers such as **vehicle density**, **weather conditions**, or **event urgency**, ensuring **service continuity and low-latency communication** even during partial infrastructure degradation or disasters.

Advantages of Fog-Based Decision-Making in 6G VIoT:

- **Low Latency and High Reliability:** Local processing prevents delays due to backhaul congestion [2, 16].
- **Scalability:** Can be deployed across regions without overloading cloud networks [6, 15].
- **Resilience to Infrastructure Failures:** Redundant fog nodes can maintain operations even during outages.
- **Enhanced Security:** Integration with blockchain and localized authentication improves trustworthiness [21].

Use Cases in Smart Transportation:

- **Cooperative Adaptive Cruise Control (CACC):** Fog nodes coordinate speed and spacing across a platoon [4, 17].
- **Emergency Routing:** It prioritizes ambulance/fire truck movements by rerouting surrounding vehicles [13].
- **Hazard Alert Distribution:** Broadcast hazard events (e.g., black ice and accidents) in real-time to vehicles within range [1, 9].
- **Dynamic Traffic Signal Control:** Adjust signal timing based on traffic flow predictions [8, 15].

With the introduction of **6G's AI-native and URLLC features**, fog-based vehicular ecosystems become not only **reactive**, but also **predictive and autonomous**. Fog nodes can host **machine learning models** trained in the cloud and execute them locally for faster response, adapting to both real-time events and forecasted conditions [5, 10, 11].

These developments position fog computing as a **foundational layer** in building **resilient, scalable, and intelligent 6G-powered smart transportation systems** (Figure 6.4).

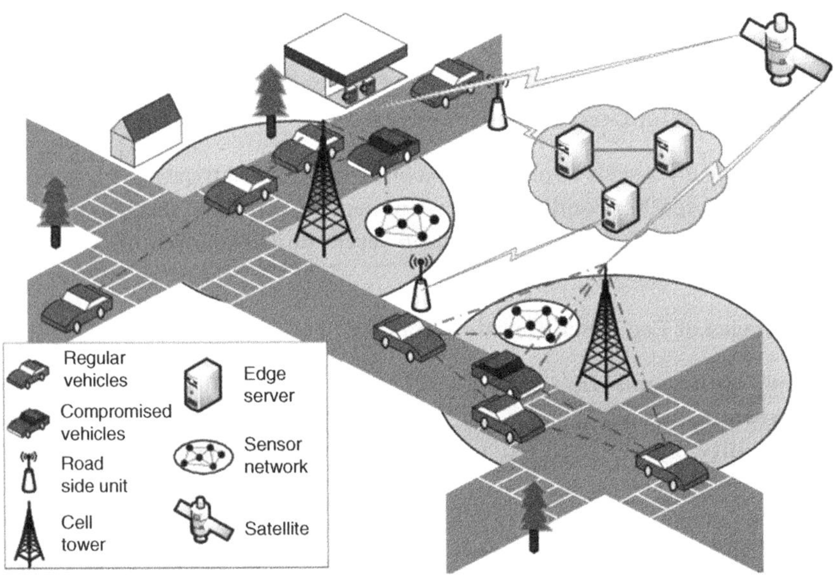

Figure 6.4 TRM architecture for trust-based vehicular communication in 6G. *Source:* An example of TRM architecture for 6G-supported VANET / IEEE / CC BY 4.0.

Three-Tier Fog Architecture:

1) **Vehicle Edge Layer**: Initial sensing and local computation.
2) **Fog Node Layer**: Regional RSUs aggregate and analyze data.
3) **Cloud Layer**: Performs long-term storage and model training.

Use Cases:

- Cooperative Adaptive Cruise Control (CACC).
- Dynamic Traffic Signal Adjustment.
- Emergency Response Coordination.

Advantages:

- Lower latency than cloud-only solutions.
- Greater scalability and resilience.
- Better support for heterogeneous device collaboration [6].

6.6 Conclusion

6G stands to redefine vehicular communication and autonomous driving through its unique set of characteristics—URLLC, high-precision localization, and native AI integration. The chapter underscores the inadequacy of current systems in supporting time-sensitive, safety-critical vehicular functions and introduces distributed computing paradigms to resolve these issues. By unifying edge and fog computing within the 6G framework, this work lays the foundation for robust and scalable VIoT systems that ensure safety, efficiency, and real-time responsiveness.

References

1 Chitkara University (2024). Wireless vehicular communication and collision avoidance system. Indian Patent Application No. 202411021682.

2 Saad, W., Bennis, M., and Chen, M. (2019). A vision of 6G wireless systems: applications, trends, technologies, and open research problems. *IEEE Network* 34 (3): 134–142.

3 Khan, L.U., Yaqoob, I., Imran, M., and Guizani, M. (2020). 6G wireless systems: a vision, architectural elements, and future directions. *IEEE Access* 8: 147029–147044.

4 Liu, Y., Wang, L., Chen, M., and Zhang, Y. (2021). Cooperative perception for connected autonomous vehicles: architectures, challenges, and solutions. *IEEE Network* 35 (2): 144–151.
5 Zhou, Z., Chen, X., Zhang, E. et al. (2019). Edge intelligence: paving the last mile of artificial intelligence with edge computing. *Proceedings of the IEEE* 107 (8): 1738–1762.
6 Mao, Y., Zhang, J., and Letaief, K.B. (2017). Mobile edge computing: survey and research outlook. *IEEE Communications Surveys & Tutorials* 19 (4): 2322–2358.
7 Zafari, F., Papapanagiotou, I., and Christidis, K. (2019). A survey of indoor localization systems and technologies. *IEEE Communications Surveys & Tutorials* 21 (3): 2568–2599.
8 Zhang, H., Liu, N., Chu, X. et al. (2020). Machine learning enabled dynamic spectrum sharing for 6G. *IEEE Wireless Communications* 27 (5): 24–31.
9 Wu, Q. and Zhang, R. (2019). Towards smart and reconfigurable environment: intelligent reflecting surface aided wireless network. *IEEE Communications Magazine* 58 (1): 106–112.
10 Jiang, W., Han, B., and Schotten, H.D. (2021). Joint communication and sensing in 6G: a review. *IEEE Access* 9: 137204–137222.
11 Li, X., Liu, C., and Cui, H. (2022). High-precision positioning based on UWB and IMU fusion for vehicle localization. *IEEE Sensors Journal* 22 (3): 2341–2351.
12 Li, S. and Wang, H. (2022). Visual-inertial SLAM technologies for autonomous driving. *Sensors* 22 (5): 1769.
13 Park, J., Kim, J., and Lee, H. (2020). Reliable and low-latency communication for VIoT: challenges and solutions. *IEEE Internet of Things Journal* 7 (12): 11793–11803.
14 Zhang, K., Mao, Y., Leng, S. et al. (2021). Edge intelligence in 6G: vision, enabling technologies, and applications. *IEEE Wireless Communications* 28 (3): 29–35.
15 Chiang, M. and Zhang, T. (2016). Fog and IoT: an overview of research opportunities. *IEEE Internet of Things Journal* 3 (6): 854–864.
16 Chen, M., Yang, J., Hu, L. et al. (2021). Edge AI: on-demand accelerating deep neural network inference via edge computing. *IEEE Transactions on Wireless Communications* 20 (1): 177–191.
17 Tang, J., Ren, J., Zhang, D. et al. (2020). Road-aware edge AI: toward energy-efficient edge AI for autonomous driving. *IEEE Internet of Things Journal* 7 (5): 4021–4032.
18 Liu, Y., Zhang, Y., and Zhang, H. (2022). Toward edge computing for autonomous driving: architectures and challenges. *IEEE Internet of Things Magazine* 5 (1): 34–39.
19 Sun, Y., Peng, M., Zhou, Y. et al. (2020). Application of machine learning in wireless networks: key techniques and open issues. *IEEE Communications Surveys & Tutorials* 21 (4): 3072–3108.

20 Taleb, T., Samdanis, K., Mada, B. et al. (2020). On multi-access edge computing: a survey of the emerging 5G network edge cloud architecture and orchestration. *IEEE Communications Surveys & Tutorials* 19 (3): 1657–1681.

21 Chen, M., Zhao, Y., and Zhang, Y. (2021). AI-native 6G systems: vision, enabling technologies, and applications. *IEEE Network* 35 (4): 24–31.

22 Shi, W., Cao, J., Zhang, Q. et al. (2016). Edge computing: vision and challenges. *IEEE Internet of Things Journal* 3 (5): 637–646.

23 Dastjerdi, A.V. and Buyya, R. (2016). Fog computing: helping the internet of things realize its potential. *Computer* 49 (8): 112–116.

7

Trends and Forecasts in 6G and IoT

Betshrine R. Jibinsingh[1] *and Leema Nelson*[2]

[1] *Ramanujan Computing Centre, College of Engineering Guindy, Anna University, Chennai, India*
[2] *Centre for Cybersecurity Research and Education, Chitkara University, Rajpura, Punjab, India*

7.1 Introduction: The Dawn of Pervasive Interconnectivity

The unceasing momentum of technological evolution is steering humanity toward a future where the distinctions between the digital and physical dimensions are increasingly blurred, leading to their inextricable amalgamation. Spearheading this profound transformation are the nascent sixth-generation (6G) wireless networks and the continuously broadening Internet of Things (IoT) [1, 2]. Even as fifth-generation (5G) networks are still navigating their deployment and commercialization stages, intensive research and strategic envisioning for 6G are vigorously underway, promising a suite of capabilities that will substantially overshadow those of current-generation networks. More than a mere incremental enhancement, 6G is conceptualized as a fundamental paradigm recalibration, moving toward an "Intelligent Network of Everything." This future network is anticipated to be characterized by extreme performance metrics, ubiquitous embedded intelligence, and truly pervasive connectivity [3, 4]. This chapter endeavors to forecast the developmental trajectory of these technologies. It will delve into the emerging patterns, project future application landscapes, and underscore the cornerstone innovations that define the 6G and IoT ecosystem. Our examination will encompass the foundational technological pillars, potential transformative use cases, synergistic interactions across diverse technological disciplines, and the wider strategic frameworks essential for actualizing a future that is not only intelligent but also sustainable and inclusive for all. The abbreviations used in this chapter are in Table 7.1.

Synergies of 6G and IoT, First Edition. Edited by Himanshi Babbar.

Published 2026 by John Wiley & Sons, Inc.

Table 7.1 Abbreviations and Their Full Forms

Abbreviation	Full Form
3GPP	3rd-Generation Partnership Project
5G	Fifth Generation
6G	Sixth Generation
AI	Artificial Intelligence
AR	Augmented Reality
BANs	Body Area Networks
BCI	Brain–Computer Interfaces
CAV	Connected and Autonomous Vehicles
DLT	Distributed Ledger Technologies
DSS	Dynamic Spectrum Sharing
ETSI	European Telecommunications Standards Institute
FL	Federated Learning
HAPS	High-Altitude Platform Stations
IEEE	Institute of Electrical and Electronics Engineers
IoBNT	Internet of Bio-Nano-Things
IoT	Internet of Things
ITU	International Telecommunication Union
LEO	Low Earth Orbit
MEC	Mobile Edge Computing
ML	Machine Learning
Mmwave	Millimeter-Wave
MR	Mixed Reality
NTN	Non-Terrestrial Networks
OWC	Optical Wireless Communication
PPP	Public–Private Partnerships
QKD	Quantum Key Distribution
QRC	Quantum-Resistant Cryptography
SDO	Standardization Development Organizations
STEM	Science, Technology, Engineering, and Mathematics
Tbps	Terabits per Second
THz	Terahertz
UAV	Unmanned Aerial Vehicle
URLLC	Ultrareliable Low-Latency Communication

Table 7.1 (Continued)

Abbreviation	Full Form
V2X	Vehicle-To-Everything
VLC	Visible Light Communication
VR	Virtual Reality
XAI	Explainable AI
XR	Extended Reality
ZSM	Zero Touch Network and Service Management

7.1.1 Charting the Course for Tomorrow: Anticipating Key Developments

The voyage toward mature 6G networks and a fully realized IoT ecosystem is distinguished by an accelerated pace of technological breakthroughs and dynamically shifting societal expectations and demands. A thorough comprehension of nascent trends, coupled with the ability to make well-grounded forecasts, is indispensable for adeptly navigating this intricate and rapidly evolving technological terrain. This section is dedicated to exploring the primary drivers and projected evolutions that are poised to mold the near-to-mid-term outlook for 6G and IoT. By doing so, it aims to lay the groundwork for understanding the transformative applications and innovative use cases that are expected to surface in this new era.

7.1.1.1 Identifying Emergent Patterns and Future Projections in 6G and IoT

The transition from 5G to 6G, alongside the progressive maturation of the IoT, is propelled by a confluence of significant technological undercurrents and evolving societal paradigms. Current forecasts paint a picture of a future characterized by an unparalleled density of connectivity, a colossal explosion in data volumes, and the seamless integration of sophisticated intelligence into the very fabric of networked systems. Figure 7.1 illustrates the diverse trends, technologies, applications, and considerations pertinent to 6G and IoT landscapes.

The Advent of the Terahertz Spectrum and Ultrahigh-Capacity Networks

A predominant and defining trend for the 6G era involves the ambitious exploration and strategic utilization of significantly higher frequency bands, with a particular emphasis on the Terahertz (THz) spectrum, typically spanning from 0.1 to 10 THz [1, 2]. While 5G technology marked an advancement by extending into millimeter-wave (mmWave) bands (reaching up to 300 GHz), 6G sets its sights

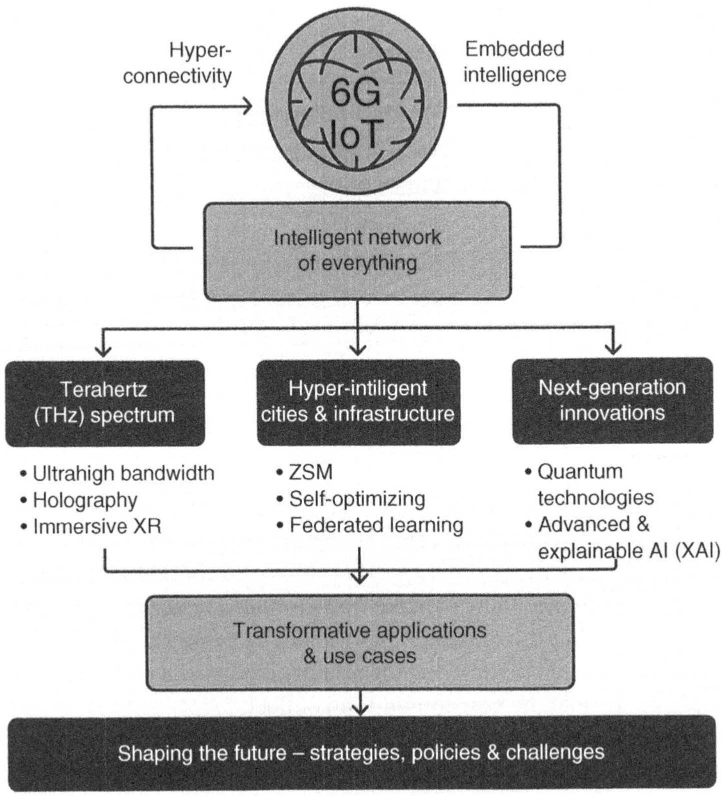

Figure 7.1 Various trends, technologies, applications, and considerations pertinent to 6G and IoT landscapes.

on unlocking the enormous untapped bandwidth potential residing within THz frequencies.

The objective is to achieve unprecedented multi-terabits per second (Tbps) data rates [1, 5]. Such extraordinary throughput is deemed essential for adequately supporting extremely bandwidth-intensive applications, including, but not limited to, true holographic communication and immersive virtual or augmented reality experiences with extreme visual resolution. Projections indicate that THz communication, despite encountering substantial challenges such as severe signal propagation loss and significant atmospheric absorption [1, 5], will likely form a foundational element of 6G infrastructure, at least for short-range communication links demanding exceptionally high capacity. Current research endeavors are concentrated on pioneering advanced materials, such as graphene-based components for transceivers, developing highly efficient transceiver architectures, and devising sophisticated beamforming techniques to effectively counteract these inherent physical limitations [1, 5]. Furthermore, the integration of THz systems with optical wireless communication (OWC) technologies, notably visible light communication (VLC) [1], is emerging as a significant trend aimed at delivering ultrahigh-capacity indoor networking solutions.

The Rise of AI-Native and Cognitive Network Architectures

Artificial intelligence (AI) and machine learning (ML) are projected to be intrinsically woven into the foundational architecture of 6G networks, signifying a departure from the supplementary, add-on approach commonly observed in 5G systems [6, 7]. The vision for 6G networks is that they will be "AI-native," meaning AI/ML algorithms will comprehensively govern network operations, spanning from physical layer optimizations like channel estimation and beam management to network-level functions such as dynamic resource allocation, intelligent traffic steering, proactive security measures, and automated service orchestration [6, 8]. This overarching trend also encompasses the anticipated ascendance of "Zero Touch Network and Service Management" (ZSM) [1], a paradigm designed to enable networks that are self-configuring, self-healing, self-optimizing, and capable of self-evolution without extensive human intervention. Cognitive architectures will further empower these networks to intelligently sense, learn from, and adapt to their operational environment and specific user requirements in real time. This adaptability is expected to culminate in unparalleled levels of operational efficiency and deeply personalized user experiences. Federated learning (FL) is identified as a crucial enabling trend in this domain, facilitating distributed AI model training across multiple nodes while meticulously preserving the privacy of the underlying data [7, 9].

The Expansion of Edge Computing and Distributed Intelligence Frameworks

The prevailing trend toward the decentralization of computational intelligence is anticipated to gain further momentum with the advent of 6G and the continued expansion of IoT ecosystems. Edge computing, which encompasses established paradigms like mobile edge computing (MEC) and fog computing, will assume a role of paramount importance. It will be critical for processing the enormous volumes of data generated by a multitude of IoT devices, much closer to their point of origin or data source [3, 9]. This proximity offers several key advantages, including a significant reduction in communication latency, conservation of valuable backhaul bandwidth, and an enhancement of data privacy and security. 6G networks are envisioned to foster a "computing continuum," a seamless spectrum of computational resources stretching from the individual device level, through the network edge, and up to centralized cloud infrastructures. Forecasts strongly suggest a future populated by AI-empowered edge nodes. These nodes will be equipped with the capability for sophisticated real-time analytics, autonomous decision-making, and dynamic service provisioning tailored to immediate needs [7, 9]. Concurrently, distributed ledger technologies (DLT), such as blockchain, are increasingly trending as solutions for secure and decentralized data management and transaction processing at the network edge. These technologies are expected to underpin a variety of applications, including the coordinated operation of unmanned aerial vehicle (UAV) swarms and the establishment of secure marketplaces for IoT-generated data [10, 11].

Analyzing Growth Trajectories and Global Projections for Connectivity
Global forecasts consistently point toward an exponential surge in the number of connected IoT devices, with estimates suggesting that this figure could reach hundreds of billions, or potentially even trillions, within the 6G era [1, 3]. Correspondingly, data traffic is predicted to escalate dramatically, potentially reaching zettabytes per month on a global scale [1, 12]. This unprecedented growth necessitates a massive upscaling of network capacity, a significant increase in connection density (with targets aiming for greater than 10^7 devices per square kilometer) [1, 5], and a substantial enhancement in processing power across the network infrastructure. Prominent standardization bodies, including the International Telecommunication Union (ITU) with its IMT-2030 vision and the 3rd-Generation Partnership Project (3GPP) through its Release 20 and subsequent specifications, are actively defining the developmental timeline. Initial commercial deployments of 6G technology are widely anticipated to commence around the year 2030 [1, 13]. Major global research initiatives, such as the 6G Flagship program in Finland, the Hexa-X project in Europe, and various national programs in countries like the United States, South Korea, Japan, and China, are at the forefront of driving innovation in this domain. These initiatives are backed by forecasts of significant public and private sector investment [1, 13]. Furthermore, the deployment of low Earth orbit (LEO) satellite mega-constellations, exemplified by projects like Starlink, OneWeb, and Kuiper, represents a key evolving trend. These constellations are poised to provide ubiquitous global coverage and are expected to integrate seamlessly with terrestrial 6G networks, thereby extending connectivity to remote and underserved regions [14, 15].

7.1.1.2 Envisioning Future Applications and Transformative Use Cases

The sophisticated and advanced capabilities inherent in 6G technology, combined with a fully matured IoT ecosystem, are set to unlock a wide array of novel applications while simultaneously transforming existing ones. This technological synergy is poised to profoundly impact diverse sectors across society and industry, ushering in new paradigms of interaction, operation, and service delivery.

The Dawn of Fully Immersive Realities and Holographic Communications
6G networks are expected to serve as the critical enabler for truly immersive extended reality (XR) experiences, an umbrella term that encompasses virtual reality (VR), augmented reality (AR), and mixed reality (MR) [16, 17]. This includes the realization of holographic telepresence, a technology where high-fidelity three-dimensional (3D) representations of individuals can interact dynamically in real time. Such applications will demand extraordinary data rates, potentially in the Tbps range, coupled with ultralow, sub-millisecond latencies to ensure seamless and natural interaction [1, 5]. The Metaverse, conceived as a persistent, shared, and interactive collection of 3D virtual worlds, is projected to evolve into a mainstream platform. It will facilitate a wide range of activities, including social interaction, collaborative work, entertainment experiences, and commercial transactions, all of which will be heavily reliant on the robust capabilities offered by 6G [1, 17]. Furthermore, haptic communication, often referred to as the Tactile Internet, will allow users to experience the sensation of touch and receive physical feedback remotely. This capability is anticipated to revolutionize numerous applications, such as remote surgical procedures, advanced collaborative robotics, and immersive skill training programs [1, 3].

Revolutionizing Smart Infrastructure and Urban Living Environments The synergy between 6G and IoT is poised to propel the evolution of smart cities toward the creation of hyper-intelligent urban environments, characterized by unprecedented levels of automation, efficiency, and responsiveness [16, 18]. This transformation will encompass AI-driven traffic management systems capable of optimizing flow and reducing congestion [18, 19]; highly optimized energy distribution through advanced smart grids (often termed Smart Grids 2.0) that seamlessly integrate massive renewable energy sources and accommodate energy prosumers [11]; the deployment of fully autonomous public transportation systems; and the delivery of highly personalized city services tailored to individual citizen needs. The concept of Digital Twins—virtual replicas of cities and their critical infrastructure—will enable real-time monitoring, complex simulations for planning and optimization, and predictive maintenance scheduling to prevent failures [1, 2]. Ubiquitous sensor networks, deployed throughout the urban landscape, will provide granular data streams crucial for continuous environmental monitoring, enhancing public safety and security, and enabling the efficient utilization of resources. The convergence of communication, sensing, and localization capabilities (often referred to as 3CLS—Communication, Computation,

Control, and Localization/Sensing) will be a fundamental technological pillar underpinning these advancements [1, 2].

Enabling Autonomous Mobility and Intelligent Transportation Systems Connected and autonomous vehicles (CAVs) are anticipated to achieve widespread adoption, their operation critically supported by the ultrareliable low-latency communication (URLLC) capabilities and high-precision positioning services intrinsic to 6G networks [1, 16]. This category includes not only fully autonomous passenger cars and commercial trucks but also sophisticated drones or UAVs [1, 2] and autonomous maritime vessels. Coordinated swarms of collaborative robots and UAVs will be deployed to perform complex and multifaceted tasks in sectors such as logistics and supply chain management, precision agriculture, advanced surveillance operations, and rapid disaster response [2]. 6G will be instrumental in supporting comprehensive Vehicle-to-Everything (V2X) communication. This will enable vehicles to interact seamlessly and intelligently with each other, with roadside infrastructure, with pedestrians, and with the broader network, thereby significantly enhancing road safety and overall traffic efficiency [1]. Beyond terrestrial applications, high-speed rail networks and even futuristic concepts like space tourism and deep-sea exploration are forecasted as potential use cases for 6G, demanding extreme coverage extension and robust support for high-mobility scenarios [1, 5].

Pioneering Advanced Healthcare and Remote Patient Monitoring The advent of 6G technology is set to revolutionize the healthcare sector through a myriad of innovative applications. These include remote surgical procedures (telesurgery), which necessitate haptic feedback for tactile sensation and extreme precision for delicate operations [20, 21]. Personalized medicine will become more attainable, driven by continuous health monitoring facilitated by wearable sensors and implantable Internet of Bio-Nano-Things (IoBNT) [1, 22], coupled with AI-powered diagnostic tools that can analyze complex medical data with remarkable speed and accuracy. Intelligent healthcare systems, built upon this technological foundation, will enable proactive health management strategies and facilitate the early detection of diseases, leading to improved patient outcomes. Personalized body area networks (BANs) will continuously collect vast quantities of physiological data from individuals. This data can then be processed at the network edge to provide real-time health insights and alerts [1, 22]. The profound integration of AI within 6G networks will be crucial for analyzing intricate medical datasets, supporting sophisticated remote patient care models, and addressing the healthcare needs of aging populations and individuals residing in underserved or remote geographical areas [20, 21, 23].

7.2 Peering into the Future: Innovations on the Technological Horizon

As the development of 6G and the IoT ecosystem continues its relentless march forward, a fresh wave of technological innovations is poised to fundamentally redefine the existing technological landscape. These forthcoming advancements, which span from novel fundamental communication paradigms to highly sophisticated AI integrations and intricate cross-domain synergies, will be instrumental in unlocking and realizing the full, transformative potential of a hyper-connected and pervasively intelligent future. This section aims to explore these next-generation technologies and the convergent pathways that are expected to shape the innovations emerging on the horizon, offering a glimpse into the capabilities that will define tomorrow's digital world.

7.2.1 Emerging Next-Generation Technologies: Quantum Communication and Advanced AI Integration

The ambitious pursuit of unparalleled performance metrics, ironclad security, and pervasive intelligence within the framework of 6G and IoT is acting as a powerful catalyst, driving intensive research into revolutionary technologies that aim to extend capabilities beyond the limitations of classical paradigms. This includes explorations into the quantum realm for communication and the deeper, more intrinsic fusion of AI into network operations.

7.2.1.1 The Advent of Quantum-Secure Networks and Novel Encryption Methodologies

A highly significant innovation anticipated on the horizon is the seamless integration of quantum communication principles into network architectures, with the primary objective of ensuring unconditional security for data transmission [1, 24]. Quantum key distribution (QKD) is rapidly emerging as a leading technology designed to provide robust protection against the looming threat posed by quantum computers, which are theoretically capable of breaking current industry-standard cryptographic algorithms [1, 2]. It is envisioned that 6G networks may natively incorporate QKD mechanisms for the secure transmission of highly sensitive data, a feature of particular importance for critical national infrastructure, governmental communications, and enterprise data requiring the highest levels of confidentiality. Concurrently, research is actively exploring the development of quantum-resistant cryptography (QRC), also known as post-quantum cryptography, as an alternative or complementary approach.

Beyond security, there is also significant investigation into the potential for a future "Quantum Internet." Such a network could offer entirely novel capabilities that extend far beyond secure communication, potentially including distributed quantum computing resources and significantly enhanced quantum sensing applications [1, 2]. However, substantial challenges persist concerning the practical aspects of deployment, including limitations in range, the cost of implementation, and the complexities of integrating these quantum technologies with existing classical network infrastructures.

7.2.1.2 Embedding AI at the Network Core and Extending It to the Extreme Edge

The integration of AI into network systems is set to become significantly more profound and intrinsic, with AI models evolving from merely optimizing network functions to actively participating in the design and dynamic adaptation of various aspects of the 6G network itself [6, 7]. This paradigm shift includes the development of AI-native air interfaces, where AI/ML algorithms are expected to replace many traditional signal processing blocks, leading to more adaptive and efficient wireless communication [1, 6]. Furthermore, AI will drive the sophisticated orchestration of end-to-end network slices, tailoring network resources dynamically to meet specific application requirements. In this context, Explainable AI (XAI) will become crucially important for ensuring trust, transparency, and accountability in these increasingly complex AI-driven systems [1, 7]. This is especially pertinent for critical IoT applications deployed in sensitive domains such as healthcare and autonomous systems, where understanding the decision-making process of AI is paramount. Innovations in advanced ML techniques, including FL for privacy-preserving distributed model training [25, 26], transfer learning for leveraging knowledge across different tasks and domains [6, 7], and reinforcement learning for autonomous decision-making and control [6, 7], will enable more efficient and privacy-conscious distributed intelligence. This intelligence will span the entire network, from the centralized core to the extreme edge, potentially even enabling on-device AI capabilities within individual IoT sensors [7, 9].

7.2.1.3 Exploring Neuromorphic and Bio-inspired Computing Paradigms

To effectively manage the immense computational demands associated with pervasive AI in 6G and IoT environments while simultaneously addressing critical concerns regarding energy efficiency, significant innovations in computing hardware architectures are anticipated. Neuromorphic computing, a field that draws inspiration from and aims to mimic the highly efficient parallel architecture of the human brain, along with other bio-inspired computing paradigms, holds the promise of delivering orders-of-magnitude improvements in power efficiency specifically for AI-related workloads [1, 2]. These advanced computing technologies could potentially be integrated into edge devices and network infrastructure components. Such integration would enable more sophisticated on-device intelligence, thereby reducing the reliance on energy-intensive, centralized cloud processing. This shift is particularly relevant and beneficial for energy-constrained IoT devices, which often operate under strict power budgets, allowing them to perform complex tasks locally with minimal energy consumption.

7.2.1.4 Pioneering Brain–Computer Interfaces and the Cognitive Internet of Things

A truly futuristic and transformative innovation involves the prospect of directly interfacing human cognition with the digital world through advanced brain–computer interfaces (BCIs) [1]. While this field is currently in its nascent stages of research and development, the ultralow latency and exceptionally high bandwidth characteristics promised by 6G technology could potentially provide the necessary support for rudimentary BCI applications. Such capabilities could usher in entirely new forms of human–machine interaction and foster the development of a "cognitive IoT," where interconnected devices are capable of responding directly to human thought, intent, or even nuanced cognitive states. The development and deployment of BCI technologies naturally raise profound ethical, societal, and privacy-related questions that will require careful consideration and robust governance frameworks. Nevertheless, BCIs represent a long-term horizon for innovation where 6G could facilitate a significantly deeper and more intuitive integration of human intelligence with the surrounding digital environment, thereby extending the conceptual boundaries of the Internet of Senses and human–computer symbiosis.

7.2.2 Harnessing Cross-Domain Synergies and Convergence with Allied Technologies

The overall impact and transformative potential of 6G and IoT will be substantially amplified through strategic synergies and deep convergence with other rapidly advancing technological domains. This interdisciplinary fusion is expected to give rise to a new generation of novel, converged solutions that leverage the strengths of multiple technologies to address complex challenges and unlock unprecedented capabilities, thereby accelerating innovation across various sectors.

7.2.2.1 Leveraging Blockchain for Enhanced IoT Security, Transparency, and Decentralization

The convergence of DLT, most notably blockchain, with the advanced capabilities of 6G and the expansive reach of IoT is poised to create new and robust paradigms for secure, transparent, and decentralized data management, reliable identity verification processes, and the automated execution of contractual agreements through smart contracts [10, 11]. This amalgamation is particularly relevant and beneficial for massive-scale IoT deployments that inherently require trustless interactions among a diverse array of stakeholders. It can also ensure secure data provenance and traceability for complex supply chains, and facilitate decentralized energy trading mechanisms within next-generation smart grids [10, 11]. However, significant challenges remain, particularly concerning the scalability and energy consumption associated with many current blockchain technologies. These limitations are actively being addressed by ongoing research and development into newer, more efficient, and lightweight DLT designs tailored for resource-constrained environments [27, 28].

7.2.2.2 Fostering Synergies with Renewable Energy Systems and Intelligent Smart Grids

6G and IoT technologies are set to play a pivotal and transformative role in accelerating the global transition toward more sustainable energy systems [11, 27]. The deep convergence of these communication and sensing technologies with advanced smart grid functionalities will enable highly efficient management of distributed renewable energy sources, such as solar and wind power. It will also facilitate the implementation of real-time demand-response mechanisms, allowing for dynamic adjustments in energy consumption, and support the sophisticated optimization of energy storage solutions. AI-powered IoT platforms, operating within the framework of 6G networks, will enable predictive maintenance schedules for critical energy infrastructure, thereby minimizing downtime and improving reliability. Furthermore, these platforms will foster peer-to-peer energy trading ecosystems, promoting more resilient, democratized, and decentralized energy landscapes. Concurrently, energy harvesting technologies are expected to converge more closely with the design of low-power IoT devices, with the ultimate aim of achieving self-sustaining sensor networks that require minimal or no external power sources [27].

7.2.2.3 Integrating IoT with Advanced Satellite Communication Networks

The burgeoning synergy between terrestrial 6G networks and non-terrestrial networks (NTNs), particularly those involving constellations of LEO satellites, is on track to ensure truly global and ubiquitous IoT coverage [14, 15]. This convergence is of vital importance for a wide range of applications situated in remote geographical areas, maritime environments, and aeronautical domains. By bridging connectivity gaps, it will enable seamless and uninterrupted communication for IoT devices, irrespective of their physical location on the planet. Future innovations in this sphere will concentrate on developing sophisticated inter-working protocols for smooth handover between terrestrial and satellite segments, implementing dynamic resource management strategies across these integrated networks, and designing ultralow-power IoT chipsets capable of direct-to-satellite communication. These advancements build upon initiatives already seen in some 5G NTN development efforts, aiming for even greater integration and efficiency in the 6G era.

7.2.2.4 Interfacing with Biotechnology: The Rise of the Bio-Internet of Things

A cutting-edge frontier of innovation lies in the progressive convergence of IoT with the rapidly advancing field of biotechnology, a synergy that is giving rise to the concept of the IoBNT [1, 22]. This emerging domain involves the development and deployment of microscopic devices and sophisticated sensors that are designed to operate safely and effectively within biological systems, including the human body. Such technologies hold immense promise for applications like advanced, continuous health monitoring; highly targeted and precise drug delivery mechanisms; and gaining an unprecedented understanding of intricate biological processes at a molecular or cellular scale. The unique capabilities of 6G networks, particularly in handling a massive density of connected devices and enabling precise, reliable communication with nano-scale entities (potentially utilizing THz frequencies or even molecular communication techniques), could pave the way for these transformative biomedical applications. However, the realization of IoBNT's potential necessitates significant further research and careful navigation of paramount ethical considerations and regulatory frameworks.

7.3 Actively Shaping the Future: Strategies and Policies for a Connected World

The realization of the ambitious and transformative vision underpinning 6G and ubiquitously deployed IoTs demands more than just groundbreaking technological breakthroughs. It necessitates the implementation of concerted, well-defined strategies aimed at fostering sustainable and inclusive innovation. These must be complemented by supportive and adaptive policy frameworks, alongside robust collaborative initiatives that span academia, industry, and government. This section outlines key strategic approaches and policy considerations designed to ensure that the future shaped by the confluence of 6G and IoT is not only technologically advanced but also broadly beneficial to society as a whole, addressing potential challenges proactively.

7.3.1 Crafting Strategies for Sustainable Development and Inclusive Innovation

As 6G and IoT technologies continue their rapid proliferation, their design, development, and deployment processes must be meticulously guided by foundational principles of sustainability and inclusivity. This proactive approach is essential to maximize the societal benefits of these powerful technologies while simultaneously mitigating potential negative impacts on the environment, social equity, and individual rights.

7.3.1.1 Championing Green Networking Principles and Low-Power IoT Solutions

A primary and critical strategic imperative is the prioritization of energy efficiency across the entire lifecycle and ecosystem of 6G and IoT technologies [1, 27]. This commitment involves the comprehensive adoption of "green networking" principles, which include designing inherently energy-efficient hardware components such as chipsets, antennas, and base stations [1, 27]. It also entails developing power-saving communication protocols and leveraging AI-driven resource management systems that can dynamically adjust energy consumption based on real-time demand and network conditions [6, 28]. Furthermore, promoting the widespread use of renewable energy sources to power network infrastructure is a key component of this strategy. Specifically, for the IoT domain, the development and deployment of ultralow-power communication technologies and innovative energy harvesting techniques are crucial. These advancements aim to enable long-lasting, battery-less, or entirely self-sustaining IoT devices [27], thereby minimizing their environmental footprint and reducing long-term operational costs. The ambitious concept of "Zero-Energy IoT," where devices harvest all required energy from their ambient environment, represents a significant and active research direction in this pursuit [27].

7.3.1.2 Systematically Bridging the Global Digital Divide for Equitable Access

While offering immense potential, 6G and IoT technologies also carry the risk of either exacerbating or, if managed thoughtfully, alleviating the existing global digital divide. A consciously crafted strategy for inclusive innovation must therefore place a strong emphasis on extending affordable, reliable, and high-quality connectivity to currently underserved and remote regions across the globe [1, 14]. This multifaceted approach involves strategically leveraging NTNs, such as LEO satellite constellations and high-altitude platform stations (HAPS) [14, 15], to reach areas where terrestrial infrastructure is unfeasible or economically prohibitive. It also requires the development of cost-effective terrestrial infrastructure solutions tailored for rural or low-density population areas, and the codesign of IoT applications specifically addressing the unique needs of these communities, for instance, in sectors like agriculture, healthcare access, and education. Complementary to these technological solutions, policies aimed at promoting digital literacy, affordability, and equitable access to services are essential to ensure that the benefits of 6G and IoT are widely shared.

7.3.1.3 Establishing Robust Ethical Frameworks and Promoting Responsible AI

The pervasive and deep integration of AI within 6G networks and across the IoT landscape necessitates the urgent development and implementation of robust ethical frameworks. These frameworks must comprehensively govern critical aspects such as data privacy, information security, algorithmic bias, transparency, and accountability [1, 9]. Strategies centered on "Responsible AI" are paramount and should include the active development of XAI techniques. XAI aims to render AI-driven decisions transparent, interpretable, and understandable to humans, which is crucial for building trust and enabling oversight [1, 7]. The implementation of advanced privacy-enhancing technologies (PETs), such as FL for decentralized model training [7, 9] and homomorphic encryption for computation on encrypted data, will be vital for protecting user information. Furthermore, establishing clear, enforceable guidelines for data governance and AI ethics is essential. A critical focus must be placed on ensuring fairness and actively preventing non-discrimination in AI algorithms to prevent the inadvertent amplification or perpetuation of existing societal biases through these powerful new technologies [1, 7].

7.3.1.4 Fostering Inclusivity in Design Processes and Deployment Strategies

Innovation processes themselves must be inherently inclusive, actively involving a diverse range of stakeholders, including end users, various community representatives, industry players, academic researchers, and governmental bodies throughout the entire lifecycle of 6G and IoT solutions, from initial design to final deployment [1, 2]. This co-creation approach helps to ensure that the developed technologies are genuinely aligned with real-world user needs, societal values, and ethical considerations. Accessibility must be treated as a core design principle from the outset, ensuring that 6G services and IoT devices are designed to be usable by people with diverse abilities, backgrounds, and technical literacies. The promotion and adoption of open standards and interoperable platforms can further foster a more inclusive technological ecosystem. This approach helps to prevent vendor lock-in, encourages broader participation in innovation, and stimulates a more competitive and vibrant marketplace.

7.3.2 Formulating Policy Recommendations and Fostering Collaborative Initiatives

The creation of supportive and adaptive policy environments, coupled with the establishment of robust and far-reaching collaborative initiatives, is indispensable for successfully navigating the complex transition into the 6G and IoT era. Such frameworks are essential to harness the full potential of these technologies while mitigating risks and ensuring equitable benefit distribution on a global scale.

7.3.2.1 Harmonizing Regulatory Approaches and Advancing Spectrum Governance

Effective and forward-looking spectrum governance stands as a paramount requirement for the successful rollout and operation of 6G networks [1, 13]. This encompasses the timely and efficient allocation of new spectrum bands, particularly in the challenging but capacious THz range. Crucially, it also involves harmonizing spectrum policies at an international level to facilitate seamless global roaming for users and devices, and to achieve necessary economies of scale in equipment manufacturing and service deployment. The implementation of flexible spectrum management techniques, such as dynamic spectrum sharing (DSS) and AI-driven spectrum sensing capabilities [1, 6], will necessitate correspondingly adaptive and agile regulatory frameworks that can accommodate these new paradigms. Policy-making must also proactively address complex issues related to data sovereignty, the legalities of cross-border data flows, and the assignment of liability within intricate IoT ecosystems that often involve multiple actors and jurisdictions. Coordinated standardization efforts, spearheaded by organizations like the ITU, 3GPP, and European Telecommunications Standards Institute (ETSI), remain critical for ensuring global interoperability and a cohesive technological landscape [1, 13].

7.3.2.2 Strengthening Public–Private Partnerships in Research and Development

The sheer scale of investment required and the inherent complexity of the research and development efforts for 6G and advanced IoT solutions necessitate the formation of strong, synergistic public–private partnerships (PPPs) [1, 13]. Governments can play a crucial role in fostering innovation by strategically funding precompetitive research initiatives, establishing shared testbeds and living labs where new technologies can be trialed in real-world or near-real-world conditions, and creating incentives that encourage robust collaboration between industry leaders, academic institutions, and dedicated research organizations. International collaborative efforts, exemplified by flagship programs such as the European Union's HexaX initiative [1, 13] and various national 6G programs worldwide, are vital for pooling diverse resources, sharing critical knowledge and expertise, and collectively developing globally aligned solutions that benefit from a wider range of perspectives and contributions.

7.3.2.3 Promoting Global Standards and Interoperability Frameworks

The development and widespread adoption of open and globally recognized standards are absolutely crucial for ensuring the seamless interoperability of 6G networks and diverse IoT devices originating from different vendors and deployed across various geographical regions [1, 13]. Standardization development organizations (SDOs) such as the ITU-R (with its IMT-2030 focus for 6G), 3GPP, the ETSI, and the Institute of Electrical and Electronics Engineers (IEEE) play a pivotal and indispensable role in this process. Policies should actively encourage broad participation in these SDOs from all stakeholders. Furthermore, promoting the adoption of open interfaces and modular architectures is key to fostering a competitive, dynamic, and innovative ecosystem, thereby avoiding the pitfalls of market fragmentation and proprietary vendor lock-ins that can stifle progress and limit consumer choice.

7.3.2.4 Investing in Education, Skills Development, and Workforce Readiness

The profound transition toward a 6G-enabled world and an AI-driven IoT will inevitably demand a workforce equipped with a new and evolving set of advanced skills. Expertise will be particularly sought after in areas such as AI/ML, data science and analytics, cybersecurity, emerging quantum technologies, and advanced network engineering and operations [1, 2]. Policy initiatives must therefore focus on comprehensively reforming educational curricula at all levels to align with these future needs. This includes promoting STEM (Science, Technology, Engineering, and Mathematics) education from an early age and supporting targeted upskilling and reskilling programs designed to prepare the current and future workforce for the highly skilled jobs of tomorrow. Fostering interdisciplinary research and education programs is also of significant importance, as addressing the complex socio-technical challenges and opportunities of the 6G era will require collaborative approaches that bridge traditional disciplinary boundaries.

7.4 Conclusion: Embracing a Hyper-connected, Intelligent Future

The ongoing journey toward the sixth generation of wireless technology (6G) and the full realization of a pervasive IoT represents far more than a simple technological progression; it signifies a fundamental reshaping of how humanity connects, computes, interacts with information, and engages with the surrounding world. This chapter has endeavored to chart the anticipated trends and provide informed forecasts, highlighting the immense and multifaceted potential for innovation across a broad spectrum of applications. These range from hyper-personalized immersive experiences that blur the lines between physical and digital reality to globally interconnected intelligent infrastructure that optimizes resources and enhances quality of life. Key technological enablers, including the strategic and efficient utilization of the THz spectrum, the deep and intrinsic integration of AI-native architectures into network design and operation, and the widespread proliferation of distributed intelligence facilitated by edge computing, are poised to overcome current technological limitations and unlock capabilities of an unprecedented scale and scope.

Innovations looming on the horizon, such as the development of quantum-secure communications to safeguard data integrity and the deployment of advanced AI paradigms for enhanced automation and insight, coupled with powerful cross-domain synergies across various technological fields, promise to further accelerate this profound transformation. However, the successful and equitable realization of this envisioned future is contingent upon our collective ability to proactively and effectively address significant challenges. These challenges relate primarily to ensuring environmental sustainability, promoting digital inclusivity and bridging divides, guaranteeing robust security, and protecting user privacy in an increasingly data-rich environment. Consequently, strategic initiatives that focus on developing green networking solutions, actively working to bridge the global digital divide, establishing comprehensive and robust ethical frameworks for the responsible deployment of AI, and promoting collaborative, open standards are of paramount importance.

Supportive policy recommendations and sustained global cooperation in areas such as research and development, spectrum governance, and human capital development through education and skills training will be critical navigational tools in managing the complexities inherent in this new era. Ultimately, the future shaped by the convergence of 6G and IoT holds the compelling promise of a more intelligent, efficient, sustainable, and interconnected world. Yet, its successful and equitable actualization will depend critically on our collective foresight, meticulous strategic planning, unwavering commitment to responsible innovation, and the shared will to harness these transformative technologies for the betterment of all humankind.

References

1 De Alwis, C., Kalla, A., Pham, Q.V. et al. (2021). Survey on 6G frontiers: trends, applications, requirements, technologies and future research. *IEEE Open Journal of the Communications Society* 2: 836–886.

2 Quy, V.K., Chehri, A., Quy, N.M. et al. (2023). Innovative trends in the 6G era: a comprehensive survey of architecture, applications, technologies, and challenges. *IEEE Access* 11: 39824–39844.

3 Nguyen, D.C., Ding, M., Pathirana, P.N. et al. (2022). 6G Internet of things: a comprehensive survey. *IEEE Internet of Things Journal* 9 (1): 359–383.

4 Hosseinzadeh, M., Hemmati, A., and Rahmani, A.M. (2022). 6G-Enabled internet of things: vision, techniques, and open issues. *CMES Computer Modeling in Engineering and Sciences* 133 (3): 509–556.

5 Tariq, F., Khandaker, M.R., Wong, K.K. et al. (2020). A speculative study on 6G. *IEEE Wireless Communications* 27 (4): 118–125.

6 Noman, H.M.F., Hanafi, E., Noordin, K.A. et al. (2023). Machine learning empowered emerging wireless networks in 6G: recent advancements, challenges and future trends. *IEEE Access* 11: 83017–83051.

7 Mahmood, M.R., Matin, M.A., Sarigiannidis, P., and Goudos, S.K. (2022). A comprehensive review on artificial intelligence/machine learning algorithms for empowering the future IoT toward 6G era. *IEEE Access* 10: 87535–87562.

8 Abd Elaziz, M., Al-qaness, M.A., Dahou, A. et al. (2024). Evolution toward intelligent communications: impact of deep learning applications on the future of 6G technology. *Wiley Interdisciplinary Reviews: Data Mining and Knowledge Discovery* 14 (1): e1521.

9 Ferrag, M.A., Friha, O., Kantarci, B. et al. (2023). Edge learning for 6G-enabled internet of things: a comprehensive survey of vulnerabilities, datasets, and defenses. *IEEE Communications Surveys & Tutorials* 25 (4): 2654–2713.

10 Liang, W., Zhang, L., and Kadoch, M. (2022). 6G IoT tracking-and ma- chine learning-enhanced blockchained supply chain management. *Electronics* 12 (1): 40.

11 Mithul Raaj, A.T., Balaji, B., Sai Arun Pravin, R.R. et al. (2024). Intelligent energy management across smart grids deploying 6G IoT, AI, and blockchain in sustainable smart cities. *IoT* 5 (3): 560–591.

12 Shayea, I., Alhammadi, A., El-Saleh, A.A. et al. (2022). Time series forecasting model of future spectrum demands for mobile broadband networks in Malaysia, Turkey, and Oman. *Alexandria Engineering Journal* 61 (10): 8051–8067.

13 Khan, B., Mihovska, A., Prasad, R., and Velez, F.J. (2021). Trends in standardization towards 6G. *Journal of ICT Standardization* 9 (3): 327–348.

14 Vaezi, M., Azari, A., Khosravirad, S.R. et al. (2022). Cellular, wide-area, and non-terrestrial IoT: a survey on 5G advances and the road toward 6G. *IEEE Communications Surveys & Tutorials* 24 (2): 1117–1174.

15 Shayea, I., El-Saleh, A.A., Ergen, M. et al. (2024). Integration of 5G, 6G and IoT with Low Earth Orbit (LEO) networks: opportunity, challenges and future trends. *Results in Engineering* 23: 102409.

16 Das, M., Nag, A., Hassan, M.M. et al. (2024). Synergy of 6G technology and IoT networks for transformative applications. *International Journal of Communication Systems* 37 (14): e5869.

17 Zawish, M., Dharejo, F.A., Khowaja, S.A. et al. (2024). AI and 6G into the metaverse: fundamentals, challenges and future research trends. *IEEE Open Journal of the Communications Society* 5: 730–778.

18 Chauhan, D., Sharma, S., and Singh, V. (2023). Roadmap from 4G to 6G in smart cities impact and challenges. *International Journal of Wireless and Microwave Technologies (IJWMT)* 13 (6): 39–50.

19 Wang, F., Zhu, M., Wang, M. et al. (2021). 6G-Enabled short-term forecasting for large-scale traffic flow in massive IoT based on time-aware locality-sensitive hashing. *IEEE Internet of Things Journal* 8 (7): 5321–5331.

20 Kumar, A., Masud, M., Alsharif, M.H. et al. (2025). Integrating 6G technology in smart hospitals: challenges and opportunities for enhanced healthcare services. *Frontiers in Medicine* 12: 1534551.

21 Mohod, S.B., Ingole, K.R., Raghuwanshi, P. et al. (2025). Next-gen medical intelligence: Fuzzy logic-driven expert systems for clinical decision-making. *International Journal of Environmental Sciences* 11 (21s): 2025.

22 Dao, N.N. (2023). Internet of wearable things: advancements and benefits from 6G technologies. *Future Generation Computer Systems* 138: 172–184.

23 Anto Bennet, M., Prince Samuel, S., Mustare, N.B. et al. (2025). Efficient cervical cancer diagnosis using supervised machine learning: a performance perspective. In: *Proceedings of the IEEE International Conference on Compute, Control, Network & Photonics (ICCCNP), Bangalore, India*, 1–6. https://doi.org/10.1109/ICCCNP63914.2025.11233345.

24 Rachel, B., Kavali, S., Raghuwanshi, P. et al. (2025). Revolutionizing healthcare informatics with fuzzy logic: Smarter data, smarter decisions. *International Journal of Environmental Sciences* 11 (21s): 2025.

25 Jibinsingh, B.R. and Nelson, L. (2025). FL-WOSP: Federated learning with walrus optimization for sepsis prediction using MIMIC-III physiological and clinical data. *Pattern Recognition* 173: 112813. https://doi.org/10.1016/j.patcog.2025.112813.

26 Kumar, K.S., Nelson, L., and Jibinsingh, B.R. (2025). Systematic review of privacy-preserving federated learning in decentralized healthcare systems. *Franklin Open* 13: 100440. https://doi.org/10.1016/j.fraope.2025.100440.

27 Khan, T., Veedu, S.N.K., Rácz, A. et al. (2024). Toward 6G zero-energy internet of things: standards, trends, and recent results. *IEEE Communications Magazine* 62: 82–88.

28 Sefati, S.S., Haq, A.U., Craciunescu, R. et al. (2024). A comprehensive survey on resource management in 6G network based on internet of things. *IEEE Access* 12: 113741–113784.

8

Advancing Connectivity: The Role of Satellite IoT in 6G Networks

Garima Chopra

Department of Electronics and Communication, Chitkara University Institute of Engineering and Technology, Chitkara University, Rajpura, Punjab, India

8.1 Introduction

The Internet of Things (IoT) has impacted the lives of many, ranging from industries to everyday use. Many new technologies and use cases are integrating with time to support the network with considerably better communication devices and extended lifetime. Lately, IoT has positioned itself as one of the most crucial technologies for future applications. A massive amount of data is being collected by various sensors, impacting valuable decisions, aiding in object tracking, proposing more innovative healthcare solutions, and reducing the operational cost of industries and homes through automation. The key objectives of data collected through devices in the IoT ecosystem are to ensure reliability and optimize processes with cooperation and federation as key features. The introduction of emerging technologies like machine learning and federated learning will contribute to an added advantage to the IoT infrastructure.

Recent analyses predict that connected IoT devices will surpass a total of 83 billion by 2024, which was 35 billion in 2020. The industrial IoT sector (IIoT), covering manufacturing, retail, and agriculture, is expected to represent more than 70% of all IoT connections by 2024, with a projected 180% increase in IIoT units over a duration of four years [1]. To extend IoT benefits to regions where traditional coverage is impractical as a result of technical or economic barriers, telecom providers and businesses are inclining toward integrated global solutions. IoT's role becomes critical in remote areas lacking cellular connectivity, serving industries like transportation (maritime, aviation, and rail), logistics, energy production (solar, oil, and gas), offshore monitoring, smart utilities, agriculture,

Synergies of 6G and IoT, First Edition. Edited by Himanshi Babbar.

Published 2026 by John Wiley & Sons, Inc.

environmental tracking, and mining. Satellite systems are increasingly vital, filling gaps where terrestrial networks are unviable, such as isolated land regions (e.g., forests) or oceanic zones, by integrating with ground-based IoT infrastructure. These technologies enable connectivity in locations previously excluded from the digital ecosystem, ensuring operational continuity for sectors reliant on real-time data in challenging environments. In massive machine-type communication (mMTC), the fifth-generation (5G) wireless communications and their enabled application scenarios are interlinked with a vast number of devices with the involvement of base stations (BSs). The BSs in mMTC establish a connection and manage it with the devices. Due to deployment constraints and commercial difficulties, it sometimes becomes impractical to deploy access points (APs) to ensure connections for the remote areas such as oceans, forests, hilly areas, and deserts. These limitations provide us with the coverage loss in case of terrestrial IoT.

Satellite IoT involves leveraging orbiting satellites to link IoT devices in areas where ground-based networks (like cellular or Wi-Fi) are absent, unstable, or too costly to deploy. These systems rely on satellites positioned at varying altitudes, that is, low-Earth (closer to the planet), medium-Earth, or geostationary (fixed over one location), to guarantee connectivity across the globe. This approach addresses gaps in hard-to-reach zones such as open oceans, rugged terrains, polar regions, or dense forests. By merging with existing terrestrial IoT frameworks, satellite-enabled solutions support real-time functions like monitoring cargo ships at sea, tracking wildlife in isolated ecosystems, managing irrigation in rural farms, or overseeing equipment in remote oil fields and solar farms. Benefits include near-universal reach, adaptability for industries with dispersed operations (e.g., shipping, mining, and emergency services), and resilience in disaster-prone areas. The potential drawbacks involve delays in data transmission (especially with geostationary satellites), limited battery life for field devices, and higher expenses for satellite-ready hardware. The emerging trends like compact nanosatellites, energy-efficient protocols, and hybrid networks (satellite and 5G/NB-IoT) are making satellite IoT essential for global, uninterrupted connectivity. This enables seamless data flow even in remote and inaccessible regions.

Satellite companies are actively pursuing opportunities to expand or solidify their role in the IoT landscape. Meanwhile, to reduce the gap between ground infrastructure and satellites, the telecom industry, equipment vendors, and agile startups are innovating and establishing adaptive terrestrial wireless technologies such as long-range networks to serve as bridges between ground infrastructure and satellites. Success centers around stakeholders like industry leaders, regulatory groups, and space organizations harmonizing standards and aligning priorities to develop practical, complementary systems. Efforts focus on building upon current IoT frameworks to ensure compatibility with existing terrestrial systems, a critical step for rapid adoption within evolving communication ecosystems. The landscape is dynamic, blending proven solutions with emerging innovations, where the pace of change demands flexible, unified approaches to meet both current and future connectivity demands.

There are two distinct categories for existing long-range wireless IoT technologies, such as mobile cellular networks and low-power wide-area network (LPWAN) innovations [2–4]. Mobile networks operated by national mobile network operators (MNOs) utilize radio access technologies (RATs) such as 2G (GPRS), 3G (UMTS), and 4G (LTE). Although initially developed for human-focused services such as voice and data transmission, recent advancements in 3GPP standards have increasingly supported IoT connectivity [5]. In particular, mMTC targets high-density IoT scenarios by accommodating relaxed latency and reliability requirements. Nevertheless, research indicates that existing 4G infrastructure has scalability limitations—for instance, a typical BS can efficiently serve only a few thousand devices while maintaining access latency under 100 ms and ensuring high success rates for data transmission [6]. Pushing latency down to ultralow levels (around 10 ms) would require significantly reducing the number of devices per BS, which is not practical. To overcome these limitations, 5G standardization efforts within 3GPP have emphasized improvements in mMTC support. The 5G New Radio (NR) air interface introduces ultrareliable low-latency communication (URLLC), tailored for time-critical IoT applications that involve moderate traffic loads per BS [7]. This progression signals a broader industry movement toward hybrid frameworks that integrate legacy networks with the evolving demands of IoT (Figure 8.1).

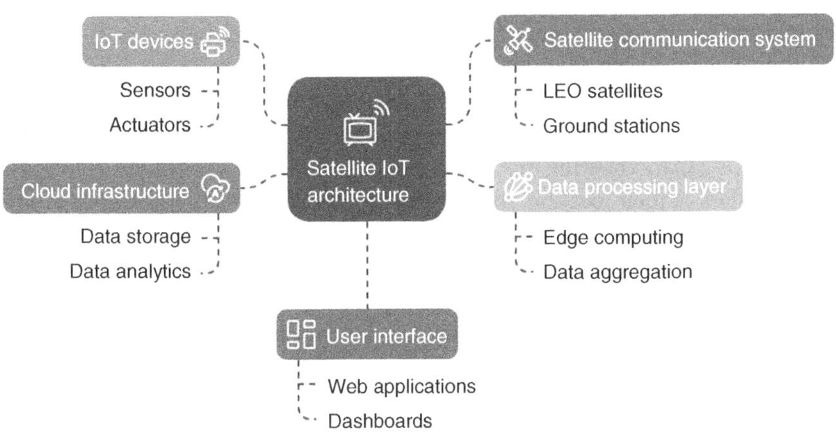

Figure 8.1 Architecture of satellite IoT.

8.1.1 Applications and Use Cases

IoT has rapidly emerged as a key technology in many different fields, from cutting-edge methods to automated control, real-time data collection, and monitoring. Wearable technology and remote monitoring technologies enable continuous patient tracking in the healthcare industry, supporting early diagnosis and more realistic treatment strategies. Whereas in the agriculture sector, IoT-based sensor systems collect the data on important environmental parameters such as soil moisture, temperature, and crop conditions, supporting precision farming techniques that enhance productivity and resource efficiency. With the initiative of smart cities, urban development has also seen significant benefits by employing IoT devices to manage traffic, track pollution levels, and regulate energy practice, all helping to create safer and more sustainable living environments. In IIoT links, the machinery and embedded sensors allow predictive maintenance, rationalize production workflows, and decrease operational interruptions, adding an advantage to the setup. Moreover, Satellite IoT (Sat-IoT) has gained traction as a resource for delivering connectivity to remote regions. By leveraging low earth orbit satellite (LEO) satellites and advanced communication networks, this technology spreads the influence of IoT to areas missing traditional infrastructure, benefiting sectors like maritime logistics and emergency response. Overall, these developments highlight the significant role of IoT in increasing efficiency, innovations, and informed decision-making for various applications.

8.1.2 Motivation

Integrating satellites into IoT ecosystems is largely driven by the demand for reliable, global connectivity, particularly in areas that are underserved or inaccessible by conventional terrestrial networks. While urban and suburban regions are typically well-covered by ground-based infrastructure, remote areas such as rural landscapes, oceans, deserts, and mountainous zones often lack sufficient connectivity due to infrastructural and economic limitations. Satellite communication bridges this gap by ensuring that IoT devices can maintain continuous connectivity regardless of their location, thus supporting consistent data transmission on a global scale.

Recent technological progress has also contributed to the viability of satellite IoT. Innovations such as LEO satellite constellations and compact, cost-efficient platforms like CubeSats have significantly lowered barriers to adoption. These advanced systems offer quicker deployment timelines, reduced communication latency, and enhanced performance, making them well-suited for a wide range of IoT use cases, including environmental sensing, asset tracking, and emergency response. Moreover, their design aligns well with the low-power operational requirements of typical IoT devices, further supporting scalable and energy-efficient implementations.

In conclusion, the motivation behind using satellites for IoT stems from their ability to provide reliable, scalable, and cost-effective connectivity solutions. Satellite IoT addresses the limitations of terrestrial networks and supports the growing number of connected devices to enable a more resilient global IoT infrastructure.

8.1.3 Organization

The chapter is organized as follows. Section 8.2 describes the hybrid network architecture. Section 8.3 discusses the solutions for remote and rural areas, keeping in view the strengths and future prospects of 6G networks, along with possible use cases for different scenarios. Section 8.4 illustrates the future research directions for Sat-IoT, followed by a conclusion in Section 8.5.

8.2 Hybrid Network Architectures for Seamless Connectivity

Terrestrial long-range IoT has expanded application opportunities by surpassing the constraints of short-range sensor networks. This progress is particularly significant for sectors like transportation, oil and gas, agriculture, and mining, where devices are often spread across vast, remote areas [8]. These industries benefit from IoT's ability to automate remote maintenance and control, enhancing efficiency and system resilience even when traditional network infrastructure is impractical or too costly to deploy [9]. Therefore, there is a need for an alternative to overcome the practical impossibilities of the terrestrial infrastructure. Satellite networks provide an alternative to sparse terrestrial infrastructures by offering backhaul capabilities for sensor gateways, enabling connectivity in remote areas. This approach, often termed the Internet of Remote Things or Internet of Everything Everywhere, leverages satellite solutions for cost-effective, global IoT deployments [10, 11]. This section outlines the network architecture of the Sat-IoT system. It consists of different parts, that is, equipment on the terrestrial area, deployment on the sea area, and space area. The terrestrial area consists of sensors, ground BSs, IoT service control planes, etc. The sea area, meanwhile, includes cargo ships, etc., and all are communicating with the space network and establishing communication with the terrestrial network as well. On the other hand, the space region is bifurcated into different zones depending upon the height of the hovering device, such as satellites and drones, to maintain direct as well as indirect communication (as shown in Figure 8.2).

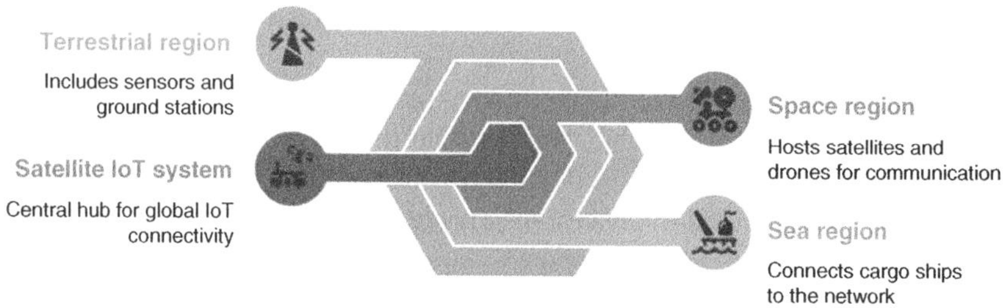

Figure 8.2 Satellite IoT ecosystem.

8.2.1 Architecture of Sat-IoT

The architecture of a Sat-IoT system is composed of several crucial elements working together to ensure communication over vast and often remote areas. At the ground level, machine-type devices (MTDs) are scattered across isolated locations like ships in the ocean or sparsely populated rural areas. These devices generate data that needs to be transmitted to centralized systems for processing and decision-making. The specialized satellites forming a network or constellation orbit above the MTDs. These satellites are specifically designed to support IoT operations by covering wide geographic areas, where terrestrial infrastructure fails to provide the services. They act as relay points, picking up data from MTDs and forwarding it toward ground infrastructure/stations or back to the MTDs in the case of downlink communication.

In some scenarios, especially when there's a dense grouping of IoT devices (like on a cargo ship), an intermediate gateway may be deployed. This gateway collects data from nearby devices and transmits it to the satellite, streamlining communication. It also handles messages coming from satellites and directs them to the appropriate MTDs. This setup helps manage large numbers of devices more efficiently. The storage information transmitted through satellites must reach the intended site of the central system used for analytics and real-time response purposes. This is achieved through ground stations distributed across different regions. These ground stations communicate with the satellite constellation to receive and send information, acting as the final terrestrial interface between the satellite layer and the cloud-based IoT services.

Satellite-IoT systems are typically structured around two primary communication paradigms. The first, referred to as the *indirect-to-satellite* model, involves MTDs transmitting their data to a local gateway, which then relays the information to a satellite. This method is advantageous in regions where infrastructure supports gateway deployment and allows for the use of standard IoT communication protocols. The second model, known as *direct-to-satellite*, enables MTDs to communicate with satellites without intermediary gateways. This configuration is particularly suited for isolated or mobile environments where deploying fixed gateways is not feasible, making it ideal for real-time global applications such as environmental surveillance or mobile asset tracking.

Using inter-satellite links (ISLs), the IoT network with satellites is able to exchange the desired information with each other, including facilities like onboard data routing and reducing reliance on ground-based infrastructure. Additionally, the buffered data is stored temporarily by satellites and forwarded once the ground stations or the intended destinations are made available, which guarantees flexibility for data delivery.

Another important development in Sat-IoT architecture is the incorporation of edge computing. The local data analysis can be performed by placing processing capabilities close to the data source, either at the gateway level or onboard satellite systems. This approach minimizes the volume of data transmitted over satellite links, reduces end-to-end (E2E) latency, and supports latency-sensitive services like emergency notifications or live system monitoring.

8.2.2 Interoperability and Standard Protocols

8.2.2.1 Importance of Interoperability in Satellite IoT

Interoperability is a fundamental issue for operational Sat-IoT systems. In a Sat-IoT context, devices differ widely in their hardware, software, and communication protocols. As these systems grow, facilitating smooth interaction among diverse devices and platforms becomes increasingly important. This enables data from different vendors to be integrated through the devices, creating an interconnected ecosystem. Moreover, Sat-IoT systems often need to interface with terrestrial networks like LPWAN, LTE, or 5G, needing border-crossing protocols for domain data interaction. Especially for remote monitoring, maritime surveillance, and disaster response activities, the Sat-IoT greatly restricts the consistency as a foundation, which limits its utilities.

8.2.2.2 Problems for Achieving Interoperability

Sat-IoT interoperability presents profound system design issues. The main reason is the lack of uniformity across the network parts. Satellite links, for instance, stand apart from terrestrial networks in terms of latency, bandwidth, and reliability. All these factors demand different or new designs at the very least. Also, the IoT ecosystem today is a puzzle dominated by multiple standards and proprietary technologies, which drives up cost and market risks (Table 8.1).

Table 8.1 Standards for NTN & IoT Infrastructure

Standard	Organization	Description
ETSI TS 103 267	ETSI	Protocols for satellite-based M2M communication systems.
3GPP release 17/18	3GPP	Support for non-terrestrial networks (NTN) and 5G IoT integration.
DVB-S2X	DVB Project	Enhanced spectral efficiency for satellite IoT communication.
CCSDS standards	CCSDS	Data communication and telemetry standards for space systems.
IEEE 802.15.4	IEEE	Standard for LR-WPANs, adapted for satellite sensor networks.
ITU-R recommendations	ITU	Frequency regulation and interference management for satellite IoT.

8.2.2.3 Standard Protocol in Sat-IoT

Several communication protocols have been adapted or developed specifically to support Sat-IoT systems. On the physical and data link layers, protocols like DVB-S2/S2X are commonly used for efficient forward-link transmission in satellite communications. For space communications, the Consultative Committee for Space Data Systems (CCSDS) offers standardized protocols that are now being tailored for IoT applications. At higher layers, lightweight application protocols such as MQTT-SN and CoAP have gained popularity due to their minimal overhead and suitability for constrained environments. Additionally, protocols such as LoRaWAN have proven effective in LEO satellite deployments for low-data-rate IoT services. With the emergence of 5G, the 3GPP has started to incorporate support for non-terrestrial networks (NTNs) by extending existing protocols like NB-IoT to work over satellite links, thereby enabling broader coverage and seamless integration.

8.2.2.4 Standardization Bodies and Initiatives

Numerous global entities are working toward creating and developing procedures for Sat-IoT. Notable contributions have been made by the 3rd-Generation Partnership Project (3GPP), which is actively supporting the integration of satellite communications within its NTNss framework, specifically in Release 17 and later versions. Also actively participating in the field is ETSI

(European Telecommunications Standards Institute), which focuses on the integration of protocols for user IoT systems and satellites. The Internet Engineering Task Force (IETF) has crafted numerous protocols that are lightweight and compatible with IP, including 6LoWPAN and CoAP, which can be utilized in Sat-IoT systems. Lastly, the International Telecommunication Union (ITU) also contributes tremendously to the universal spectrum and satellite communication practices as well as standard etiquette between differing regions and operators (Figure 8.3).

8.2.2.5 Research Areas

With the growing demand for IoT infrastructure in different terrains, it is expected to design a unified architecture that helps to establish a smooth transition between the terrestrial networks and satellite infrastructure. One potential avenue is the integration of artificial intelligence (AI) for real-time protocol modification and dynamic interoperability, permitting optimal routing and protocol selection based on prevailing network conditions. Another important research focus is the lightweight secure communication protocols, which embrace the limitations of satellite-linked cross communication, such as high latency and unpredictable connectivity. Research is also being done on robustness and reliability using Delay-Tolerant Networking (DTN) and edge computing. All these efforts will aid in the framework development of robust, scalable, and secure infrastructure integrating Sat-IoT, which can be utilized in remote monitoring and sensing, agriculture, environmental surveillance, and emergency management.

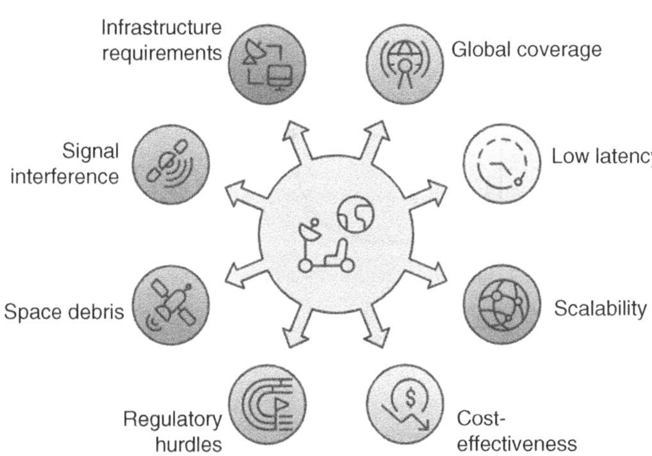

Figure 8.3 Pros and cons of LEO satellites for IoT applications.

8.2.2.6 Role of Edge Computing for Sat-IoT

In designing satellite-based IoT networks, edge computing operates as a key enabler. In highly remote areas with low bandwidth where user-plane data requires local processing, an indirect-to-satellite architecture must be employed. This configuration usually contains an edge computing server that is collocated with the local gateway. The edge server performs preprocessing and filtering of data so that it has less burden for the satellite communication link (refer to Figure 8.4).

Such an architecture is particularly beneficial for IoT islands, isolated networks with limited or intermittent connectivity, and where local decision-making must occur quickly, yet connectivity to cloud-based IoT services is still required. By shifting key network functions to the edge, the system can maintain real-time responsiveness and lower latency, which are critical for time-sensitive applications like remote monitoring, predictive maintenance, or disaster response.

Additionally, the core network design must evolve to support this decentralization by including functions like session management, security handling, and parts of the control plane closer to the edge. This architecture ensures robust E2E system performance by decentralizing network functions, which is vital for maintaining service continuity in satellite IoT deployments with latency and bandwidth constraints (Figure 8.4).

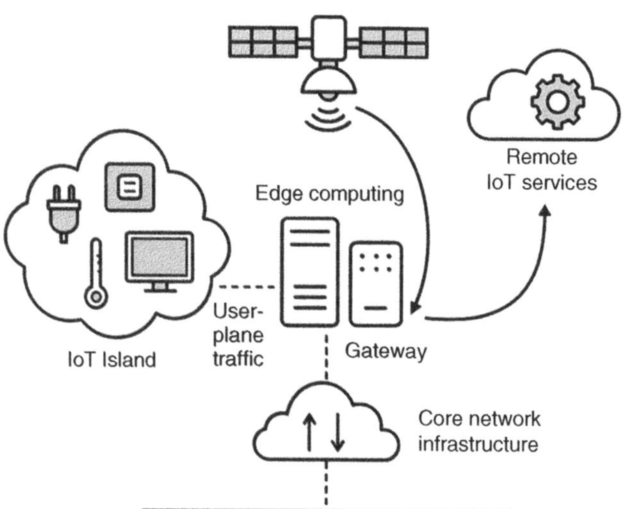

Figure 8.4 Architecture of indirect-to-satellite communication for Sat-IoT.

8.3 6G Connectivity Solutions for Remote and Rural Areas

Remote and rural regions often remain underserved by conventional communication infrastructure due to geographical, economic, and logistical constraints. The 6G promises to provide global connectivity and ultralow latency and present an opportunity to bridge the digital divide. This segment explores the enabling technologies, architectures, and strategies that can allow 6G to connect remote and rural societies.

8.3.1 Connectivity Limitations in Underserved Areas

Due to limitations in the terrestrial networks, there are several limitations in the rural environment during the deployment phase. Certain performance limitations in this regard are outlined as follows:

- **Infrastructure Sparsity**: The high operational expenditure (OPEX) and capital expenditure (CAPEX) impede the scalability of the backhaul links supported by optic fibers or deployment of terrestrial BSs, particularly for spacious and densely populated regions.
- **Topographical Challenges**: It becomes difficult to install the devices physically in the tough terrains such as dense mountains, vegetation, and larger water bodies.
- **Energy Availability**: There are certain challenges for off-grid areas that limit the performance, such as consistent power supplies and deploying networking equipment without additional need for infrastructure for storage or energy harvesting.
- **Backhaul Constraints**: The absence of reliable, high-capacity backhaul links limits the feasibility of high-throughput services, especially in latency-sensitive use cases such as telemedicine or disaster response.

8.3.2 Satellite-IoT Integration for Ubiquitous Coverage in 6G Networks

In satellite communication, LEP constellations appear as a convincing substitute for the connectivity gap in rural areas. These systems offer global footprint capabilities and reduced signal propagation delay compared to traditional geostationary (GEO) satellites. When integrated with IoT architectures, satellite-enabled 6G networks can provide continuous monitoring and intelligent control in remote areas across verticals such as:

- **Agricultural Informatics**: Precision farming through real-time sensing of soil moisture, pest activity, and climatic conditions.
- **Offshore and Industrial Monitoring**: Data acquisition from oil rigs, mining operations, and power distribution nodes.
- **Disaster Management and Emergency Response**: Coordinated sensor networks and communication links for rapid damage assessment and resource deployment.

These implementations are further enhanced through standardized NTN frameworks being specified by the 3GPP for 5G and extended into 6G [12].

8.3.3 MEC-Enabled Satellite Networks: Edge Intelligence at Scale

Mobile edge computing (MEC), when integrated with satellite infrastructure, enables real-time data processing and analytics at or near the point of data generation. Three principal deployment architectures facilitate satellite-MEC integration:

1) **Computing-in-Forward-Link (CIF)**: MEC servers deployed on aerial platforms (e.g., unmanned aerial vehicles (UAVs) and high altitude platforms (HAPs)) relay data to and from satellites, offering ultralow-latency computing in localized regions.
2) **Computing-on-Orbit (COO)**: Satellites equipped with on-board MEC modules support direct computation in space, enabling wide-area low-latency services with minimal reliance on ground infrastructure.
3) **Computing-After-Feeder-Link (CAF)**: Edge servers located at ground gateway stations handle computational tasks post satellite transmission, suitable for delay-tolerant, computation-intensive workloads.

Each structure presents unique trade-offs between latency, energy efficiency, and computational capacity, depending on application-specific quality of service (QoS) requirements [13].

8.3.4 Bridging the Digital Divide with Satellite Connectivity

One of the fundamental promises of 6G is to enable truly ubiquitous connectivity, particularly in regions where terrestrial infrastructure is impractical or economically unfeasible. Satellite networks, especially those based on LEO constellations, offer a scalable and cost-effective approach to bridging the digital divide by delivering broadband access to underserved and geographically isolated communities. Unlike geostationary satellites, LEO systems operate at altitudes between 500 and

2000 km, enabling lower latency, enhanced link reliability, and greater spectral efficiency, which are essential for interactive services and real-time data applications [12]. This capability is particularly vital in supporting digital services such as remote education, e-healthcare, and e-governance in rural regions where fiber deployment remains constrained. The 3GPP's NTN framework, extended from 5G into 6G, is being designed to harmonize satellite communication with terrestrial mobile networks, thereby enabling seamless interoperability across network domains [13]. Moreover, the convergence of satellite communications with MEC and AI technologies is expected to enhance local data processing capabilities and optimize content delivery at the network edge—a critical feature for latency-sensitive services in isolated areas [14]. As demonstrated in recent literature, such integrated architectures not only enhance network resilience and scalability but also play a crucial role in democratizing digital access globally [15].

8.3.5 Use Cases in Remote Monitoring, Agriculture, and Disaster Management

The integration of satellite communication technologies with IoT frameworks, especially under the 6G architecture, has unlocked new dimensions of intelligent connectivity across sectors that are inherently distributed, infrastructure-constrained, or prone to environmental unpredictability. This section elaborates on three critical application domains: remote environmental monitoring, smart agriculture, and disaster management, where satellite-enabled 6G networks play a transformative role in ensuring coverage, continuity, and computational intelligence (ref Table 8.2).

Table 8.2 Comparison of Key Performance Metrics for Representative 6G Use Cases in Remote and Rural Connectivity Scenarios [14, 16, 17]

Use Case	Data Rate (Mbps)	Latency (ms)	Reliability (%)	Coverage (km)	Mobility Support
Precision agriculture	10	50	99.9	5–10	Low
Telemedicine	100	<10	99.99	50–100	Medium
Remote education	50	100	99.5	50–100	Low
Disaster recovery	20	<100	99.9	100+	High
IoT sensor networks	1	1000	99.0	100+	Static
Satellite IoT	0.1	600	99.5	>10,000	High

8.3.5.1 Remote Environmental Monitoring

Remote sensing and environmental surveillance are among the most compelling use cases for satellite-IoT integration. The vast and inaccessible terrains, such as oceans, mountain ranges, and polar regions, require persistent observation for ecological research, climate modeling, and conservation efforts. In such scenarios, terrestrial network deployments are often either impractical or economically unjustifiable. Leveraging constellations of LEO satellites, 6G networks can interconnect thousands of ground-based and airborne sensor nodes across these remote areas, enabling applications such as real-time meteorological data collection, wildfire detection, glacial melt monitoring, and air pollution tracking [12].

Energy-efficient LPWAN technologies like NB-IoT and LoRaWAN, combined with indirect-to-satellite and direct-to-satellite connectivity modes, ensure that MTDs deployed in these regions can transmit data at infrequent intervals with minimal power consumption. Furthermore, these systems can be enhanced with onboard satellite edge processing capabilities enabled by MEC to filter, compress, or prioritize data before it is forwarded to centralized processing hubs, thus reducing communication overhead and response latency [13].

8.3.5.2 Smart and Precision Agriculture

Sat-IoT supported by 6G infrastructure has significantly benefitted the developing countries and rural landscape, especially for agricultural practices, which is the pillar for the development. The continuous feedback from in-field deployed sensors provides reliable support to smart farming by enabling inputs about fertilizers, pesticide levels, and moisture levels. Satellite networks encompass this intellect in areas where cellular connectivity is sparse or unpredictable. With support from MEC-enabled UAVs or HAPs, farmers can benefit from real-time data analytics on soil moisture, pH level, nutrient status, pest infestation, and microclimatic conditions [15].

With the aid of localized gateways and links through satellite communication links, the infrastructure is supported with E2E connectivity to the sensors deployed in fields, cloud-based agricultural analytics platforms, and the farmer. Moreover, the incorporation of AI with MEC servers organized in close proximity to agricultural fields enables predictive modeling for crop yields, irrigation scheduling, and anomaly detection, permitting hands-on and optimized farming practices. This architecture not only enhances productivity and sustainability but also increases food security by safeguarding the effective use of scarce resources like water and land.

8.3.5.3 Disaster Management and Emergency Response

The natural disaster situations, such as earthquakes, floods, wildfires, and tsunamis, frequently result in the failure of terrestrial communication infrastructures, generating blind spots in organization and rescue processes. However, Sat-IoT supported with 6G can provide enormous assistance during the tough situations where a timely response is required to save the lives of populations. They permit the deployment of autonomous drones, sensor-embedded robots, and alternative communication terminals to collate the data, relay situational updates, and direct response teams in real-time scenarios [18].

One of the prominent advantages of satellite integration is its ability to uphold connectivity in spite of local infrastructure, thus guaranteeing operational continuity. For example, LEO satellites enabled by MEC can process visual and thermal imagery captured by UAVs to measure damage zones, recognize survivors, and prioritize rescue missions. The usage of ISLs and onboard storage also permits "data ferrying" when ground communication is provisionally inaccessible, enhancing the strength of the network.

Besides, satellite multicast and broadcast capabilities can circulate early warnings, evacuation orders, and public safety directions to a larger population within seconds, possibly saving lives. In addition to real-time sensitivity, post-disaster recovery planning can also benefit from satellite-IoT systems through continuous monitoring of infrastructure integrity for a longer duration, resource distribution, and environmental recovery [13].

8.4 Future Research Directions

The conjunction of satellite communication, IoT, and 6G architecture opens several possibilities for future research expected to enhance the system intelligence, energy efficiency, and operational security. It is quite evident from the advantages of Sat-IoT systems that they have an extended role in global infrastructure. The three key research directions shaping their evolution are the following.

8.4.1 AI-Driven Satellite Routing and Network Orchestration

AI and ML are anticipated to play a pivotal role in the composition of satellite networks. It is anticipated that future research is in desperate need to explore AI-based solutions for dynamic beam steering, predictive satellite handovers, and adaptive routing protocols to optimize network throughput and latency under highly dynamic conditions [14]. In this regard, the federated learning-based approach can enable the distributed approach and provide intelligence to satellite nodes, thereby reducing the unnecessary data exchange, preserving the privacy of the information, and enhancing real-time decision-making [15].

8.4.2 Green and Energy-Efficient Sat-IoT Protocols

Sustainability is a growing concern for spaceborne IoT deployments, particularly as LEO constellations continue to expand. Novel approaches such as lightweight MAC protocols, DTN, and low-power edge computing are essential to reduce power consumption in both satellites and ground-based devices. Research is also needed to integrate solar energy harvesting, dynamic sleep-wake cycles, and carbon-aware satellite operation strategies into Sat-IoT frameworks to align with global sustainability objectives [19].

8.4.3 Security and Privacy for Non-Terrestrial IoT Networks

The open and heterogeneous nature of satellite-IoT architectures introduces serious vulnerabilities related to data integrity, authentication, and secure access control. Given the limitations in processing and energy resources, future work should focus on lightweight cryptographic schemes, intrusion detection systems, and blockchain-based trust mechanisms adapted for satellite and edge computing environments. These methods must account for high-latency, intermittent connectivity, and multi-operator interoperability constraints.

8.5 Conclusion

As the demand for ubiquitous connectivity and intelligent services accelerates, especially in rural and remote environments, the integration of satellite communication into 6G architectures becomes not just beneficial but essential. This chapter has highlighted how Sat-IoT, empowered by LEO constellations, MEC, and AI-driven analytics, extends the frontier of network coverage beyond the limitations of terrestrial systems. Through technical discussions on minimal satellite-MEC structures and standardized protocols, we examined how satellite-enabled 6G networks can provide reliable, low-latency, and scalable connectivity across various verticals from agriculture and environmental monitoring to emergency response. The challenges of deployment, including latency, energy constraints, and protocol interoperability, remain significant but are actively being addressed through hybrid architectures and global standardization initiatives. Moving forward, the continued evolution of Sat-IoT will rely on collaborative innovation across industry, academia, and regulatory bodies. With intelligent edge computing, standardized interoperability, and a robust hybrid network backbone, 6G-enabled satellite IoT is poised to transform connectivity for the next generation, making inclusive and resilient communication networks a global reality.

References

1 Rothmuller, M. and Barker, S. (2020). IoT–the internet of transformation 2020. Basingstoke, U.K., Juniper Res., White Paper.

2 *GPRS; Overall Description of the GPRS Radio Interface; Stage 2 (Release 15); Revision 15.3.0*, document TSG RAN WG3Meeting #43.064, 3GPP, Gothenburg, Sweden, Mar. 2020.

3 *UTRAN Overall Description (Release 16), Revision 16.0.0*, document TSG RAN WG3Meeting #25.401, 3GPP, Gothenburg, Sweden, Jul. 2020.

4 *E-UTRA and E-UTRAN; Overall Description; Stage 2; Revision 16.1.0*, document TSG RAN WG3Meeting #36.300, 3GPP, Gothenburg, Sweden, Apr. 2020.

5 Agiwal, M., Roy, A., and Saxena, N. (2016). Next generation 5G wireless networks: a comprehensive survey. *IEEE Communications Surveys and Tutorials* 18 (3): 1617–1655.

6 Castagno, P., Mancuso, V., Sereno, M., and Marsan, M.A. (2020). A simple model of MTC flows applied to smart factories. *IEEE Transactions on Mobile Computing* 20 (10): 2906–2923.

7 *NR; Overall Description; Stage-2; Revision 16.1.0*, document TSG RAN WG3Meeting #38.300, 3GPP, Gothenburg, Sweden, Apr. 2020.

8 Conti, Matteo. (2021). "Satellite Systems in the Era of 5G Internet of Things.".

9 Ayoub, W., Samhat, A.E., Nouvel, F. et al. (2019). Internet of mobile things: overview of LoRaWAN, DASH7, and NBIoT in LPWANs standards and supported mobility. *IEEE Communications Surveys & Tutorials* 21 (2): 1561–1581.

10 De Sanctis, M., Cianca, E., Araniti, G. et al. (2016). Satellite communications supporting internet of remote things. *IEEE Internet of Things Journal* 3 (1): 113–123.

11 Palattella, M. R. and Accettura, N. Enabling Internet of Everything Everywhere: LPWAN with Satellite Backhaul, *2018 Global Information Infrastructure and Networking Symposium (GIIS)*, Thessaloniki, Greece, 2018, pp. 1–5.

12 Centenaro, M., Costa, C.E., Granelli, F. et al. (2021). A survey on technologies, standards and open challenges in satellite IoT. *IEEE Communications Surveys and Tutorials* 23 (3): 1693–1720.

13 Lin, Y., Feng, W., Wang, Y. et al. (2024). Satellite-MEC integration for 6G internet of things: minimal structures, advances, and prospects. *IEEE Open Journal of the Communications Society* 5: 3886–3903.

14 Giordani, M., Polese, M., Mezzavilla, M. et al. (2020). Toward 6G networks: use cases and technologies. *IEEE Communications Magazine* 58 (3): 55–61.

15 Vaezi, M., Azari, A., Khosravirad, S.R. et al. (2022). Cellular, wide-area, and non-terrestrial IoT: a survey on 5G advances and the road toward 6G. *IEEE Communications Surveys and Tutorials* 24 (2): 1117–1174.

16 Zhou, J., Wang, R., Shihada, B., and Alouini, M.-S. (2024). End-to-end uplink performance analysis of satellite-based IoT networks: a stochastic geometry approach. *IEEE Open Journal of the Communications Society* 5: 4036–4045.

17 Haque, A.B., Zihad, M.O.M., and Hasan, M.R. (2023). 5G and internet of things—integration trends, opportunities, and future research avenues. In: *5G and Beyond*, 217. Springer.

18 Mozaffari, M., Saad, W., Bennis, M. et al. (2019). A tutorial on UAVs for wireless networks: applications, challenges, and open problems. *IEEE Communications Surveys and Tutorials* 21 (3): 2334–2360.

19 Rinaldi, F., Maattanen, H.L., Torsner, J. et al. (2020). Non-terrestrial networks in 5G & beyond: a survey. *IEEE Access* 8: 165178–165200.

9

Emerging IoT Scenarios Enhanced by Machine Learning in 6G

Renu Popli[1], *Deepanshi Batra*[2], *Rajeev Kumar*[3], *and Ashutosh Sharma*[4]

[1] *Department of Computer Science and Engineering, Institute of Engineering and Technology, Chitkara University, Rajpura, Punjab, India*
[2] *Department of Computer Science and Engineering, Ambala College of Engineering and Applied Research Devsthali, Kurukshetra University, Kurukshetra, Ambala, Haryana, India*
[3] *Chitkara University Research Innovation and Network, Institute of Engineering and Technology, Chitkara University, Rajpura, Punjab, India*
[4] *Department of Computer Science, Henan University of Science and Technology, Luoyang, Henan, China*

9.1 Introduction

The Internet of Things (IoT) has transformed from a niche innovation into an essential backbone for today's society, allowing trillions of devices and volumes of data to be shared and processed across the world. An increasing range and quantity of connected devices are driving greater pressure on the current wireless communication infrastructure. Fifth-generation (5G) networks have already made significant progress in solving some of these issues, providing faster data transfer, real-time communication, and larger bandwidths. The massive growth of IoT applications calls for an evolution to networks that offer significantly greater levels of connectivity, intelligence, and adaptability [1].

Developers are working on sixth-generation (6G) network technology that'll deliver remarkable speeds and latencies while priming the network for the intelligent embedding of machine learning (ML) components. As a result, IoT systems will be able to shift from collecting data only after an event occurs to making decisions and preempting events based on their surrounding environment. ML will allow IoT systems to continuously improve their own performance, act independently, and offer tailored solutions in near real-time [2, 3].

We examine the scenarios where integrating 6G, IoT, and ML technologies will achieve the greatest results. We analyze how 6G, IoT, and ML will revolutionize

Synergies of 6G and IoT, First Edition. Edited by Himanshi Babbar.

Published 2026 by John Wiley & Sons, Inc.

urban infrastructure, healthcare, industry, agriculture, and digital experiences, considering the technical and societal obstacles that must be resolved. We seek to guide the exploration of IoT applications and services in the 6G era by mapping out this transition.

9.2 Emerging IoT Scenarios Enhanced by Machine Learning in 6G

The integration of 6G networks and ML will dramatically transform IoT in multiple industries. As 5G networks are being rolled out around the world, efforts to develop the sixth generation of wireless technology are now well underway to overcome existing challenges and enable new levels of advancement [4]. The chapter analyzes how the convergence of ML and 6G networks will improve the performance and effectiveness of IoT applications, driving innovation and positive change in different sectors.

The transition to 6G goes beyond mere network enhancements. It represents a significant move toward new approaches to network designs and device interplay. Current 5G networks are enabling innovative IoT solutions, but there are obstacles to broadly introducing IoT applications in areas such as healthcare and manufacturing [5].

6G networks feature breakthrough improvements essential for the advancement of IoT in the future. The massive data rates, sub-microsecond latency, and incredible density of connections offered by 6G networks form the basis for completely revolutionary IoT environments. As a result, ML-based analytics will be able to handle and interpret huge amounts of input data from IoT devices in real-time, allowing for faster and more efficient decisions.

6G networks will leverage terahertz (THz) wave technology to transmit information more efficiently and at higher speeds, and novel security features such as blockchain and quantum encryption will ensure secure data transmission over the internet. This allows for innovative solutions to problems that were not feasible using previous-generation wireless systems [6].

9.2.1 IoT Evolution with 6G

6G development will trigger drastic changes in IoT technologies. IoT devices will be extremely responsive and will react to our thoughts and movements instead of displaying data. This will likely spawn the emerging fields known as the Internet of Senses and the Internet of Reality. Biological interfaces will replace traditional user interfaces in IoT applications.

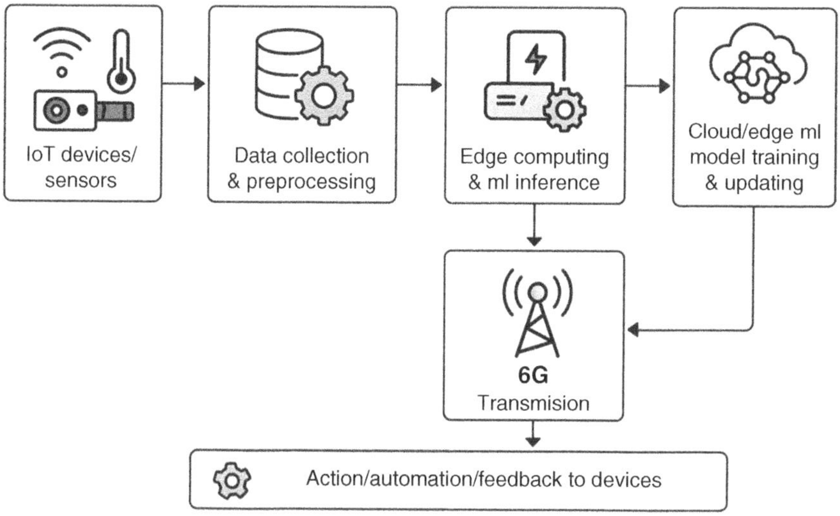

Figure 9.1 Workflow of ML-enhanced IoT data processing in 6G networks.

6G significantly addresses broader problems such as security, privacy, complexity, and sovereignty for IoT systems. Equipped with advanced features, IoT systems will operate independently, evolve as needed, and deliver individualized and tailored services that adjust to changing circumstances [7].

9.2.2 Machine Learning Framework for 6G-Enabled IoT

Machine intelligence will harness the powerful capabilities of 6G connectivity to unfold smart and fully autonomous IoT ecosystems [8]. Unlike earlier standards, 6G networks are being built with inbuilt capabilities for ML, as shown in Figure 9.1.

9.3 AI/ML Approaches for 6G Networks

Extensive research in artificial intelligence (AI)-based resource management for 6G has revealed a number of ML techniques that will power the development of future IoT platforms:

a) **Deep Learning (DL):** School teaching assists students in understanding concepts faster and helps educators assess their abilities better.
b) **Federated Learning (FL):** IoT devices are able to learn and train cooperatively without risking privacy by working with local data. Research demonstrates that FL enables IoT devices to collectively train a shared model

without exchanging both collective data and computational resources. This is especially important in healthcare, smart homes, and any field where privacy issues are paramount.

c) **Reinforcement Learning (RL):** Real-time network management tasks, including resource allocation, routing, and provisioning, can be automated by RL algorithms.
d) **Evolutionary Learning (EL):** EL provides solutions to challenging optimization tasks in 6G telecommunications, including antenna design, network optimization, and spectrum allocation.

Bringing together these techniques enables networks to dynamically respond to environmental fluctuations, forecast incoming events, and independently take actions aimed at achieving optimal efficiency and quality of service [8].

9.3.1 Edge Computing and Distributed Intelligence

The widespread adoption of edge computing in 6G networks makes possible the application of ML to IoT devices. Studies have shown that combining edge computing with 6G networks allows IoT applications to deliver exceptional performance across fields, including autonomous driving, smart cities, healthcare, and manufacturing. The distribution of intelligence across cloud, edge, and device layers is depicted in Figure 9.2.

Advancing IoT networks by deploying computational resources at the edge efficiently tackles key hurdles in these networks, as outlined below:

a) **Latency Reduction:** Processing data closer to its source enables nearly instantaneous responses in demanding scenarios involving autonomous systems, smart factories, and more.
b) **Bandwidth Optimization:** Despite having less data transferred across the network, edge computing allows devices to dedicate bandwidth to critical data exchange.
c) **Enhanced Privacy:** By performing sensitive data processing at the edge, risks of data theft and unauthorized access are greatly reduced during its journey and storage at centralized locations.
d) **Energy Efficiency:** Limiting the amount of data that is sent over networks helps IoT devices last longer on battery and become more practical for deployments in challenging environments.

6G distributed intelligence goes beyond the boundaries of standard edge computing [9]. According to current studies, FL effectively allows devices to jointly improve a collective model without transferring user data. As a result, both the security and performance of the system improve. As a result, the network becomes an adaptive knowledge-sharing ecosystem that preserves the confidentiality of data.

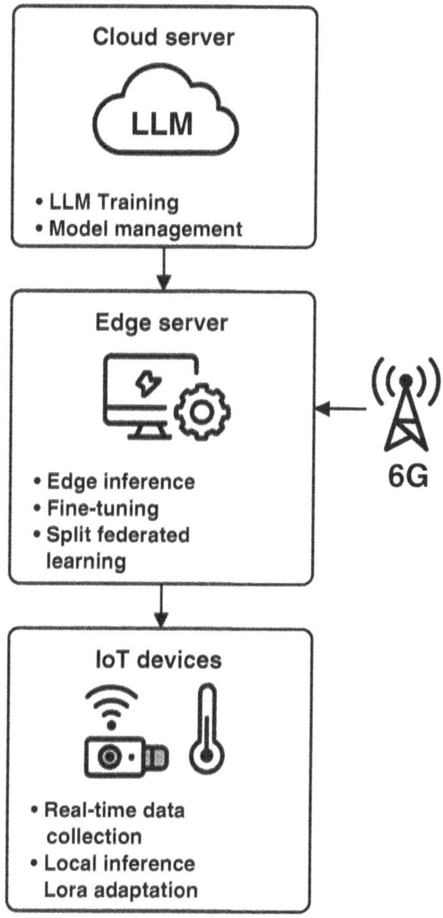

Figure 9.2 Distribution of intelligence across cloud, edge, and device layers.

9.3.2 LLM-Empowered IoT Architecture

LLMs (Large Language Models) are driving revolutionary changes in how IoT systems are designed and operated in the context of 6G networks. Research suggests that integrating LLMs into IoT architectures of 6G networks can enable greater levels of autonomy in IoT systems while supporting sophisticated applications.

LLMs significantly improve the ability of IoT systems to process and comprehend language, understand situational information, and make intelligent choices. It allows users to interact with IoT devices and services in a more natural and straightforward way using voice, text, or gestures [10].

There are several benefits to deploying LLMs in conjunction with 6G networks at the edge:

- Leveraging LLMs improves how machines interpret the data received from sensors.

- Human interactions with the system are more intuitive because of the use of conversational language interfaces.
- Making decisions with information from both the past and the present.
- Adjusting the system's operations without the need for external input in the event of unknown situations.

9.4 Smart Cities and Urban Infrastructure

Smart cities showcase a wide range of applications enabled by 6G IoT that fully leverage the power of ML. Bringing together a wide range of technologies and sensors in urban areas opens up the possibility for major enhancements in productivity, environmental friendliness, and well-being [11]. Table 9.1 represents key IoT scenarios in smart cities, which are enhanced by 6G and ML.

9.4.1 Intelligent Transportation Systems

Advancements in 6G, IoT, and ML are driving sweeping changes in the structure and management of transportation in smart cities. Widespread use of networks with extremely fast and responsive connections will allow tools for efficient and timely management of all aspects of transportation.

Advanced ML techniques analyze millions of real-time data points collected by distributed sensors, allowing for precise predictions of traffic flow and leading to more efficient traffic control. As an example, intelligent traffic lights adjust themselves according to traffic flow, lessening traffic jams and resulting in cleaner air

Table 9.1 Key IoT Scenarios in Smart Cities Enhanced by 6G and ML

Aspect	Description	Benefits
Intelligent transportation systems [12]	Real-time coordination between vehicles, infrastructure, and pedestrians using 6G and ML	Reduced congestion and emissions, and improved travel times
Environmental monitoring [13]	Continuous monitoring of air quality, noise, and water quality with predictive ML analytics	Proactive pollution management and environmental hazard prediction
Public safety and emergency response [14]	Detection of emergencies and threats using video, acoustic, and environmental sensors with ML	Optimized emergency response and reliable communication during disasters

and shorter commute times. This move requires traffic management to move from a reactive to a proactive approach.

Smart cities can expect significant improvements in traffic management, environmental oversight, and public safety, thanks to 6G networks. 6G's extraordinary richness of connections will enable extensive sensor networks to monitor urban traffic networks in great detail.

Vehicle-to-everything (V2X) communication opens up a range of capabilities for autonomous vehicles, such as working together to find routes, perceiving the environment cooperatively, and moving as a group. ML models process information collected from many vehicles and sensors to improve how traffic is distributed throughout a city.

9.4.2 Environmental Monitoring and Management

6G-driven IoT networks empowered with ML hold a key solution to implementing environmental sustainability in cities. Large numbers of sensors collecting information about air pollution, noise, water quality, and other aspects of the environment produce an ongoing data stream for assessment.

ML algorithms can detect the causes of pollution, forecast environmental risks, and direct focused response measures using past trends and current data. Using data collected from air quality sensors together with weather and traffic information, cities can identify areas prone to air pollution well ahead of time and take preemptive actions.

Linking these monitoring systems with other city assets enables the development of comprehensive urban management strategies. As an example, ML algorithms can help synchronize traffic, transit, and building functions to lessen air pollution in specific areas.

9.4.3 Public Safety and Emergency Response

Applications for public safety are among the most important examples of 6G-connected IoTs improved with AI. These innovative technologies working together provide a holistic view of risks to public safety.

AI is able to identify abnormal behavior or developments that could potentially lead to heavy casualties as a result of a disaster, crime, or public health hazard. Acoustic sensors can alert authorities to gunshots, smashed windows, or other signals of potential emergencies. Video surveillance picks up on unsafe or unusual activities such as injuries or fire.

6G network technologies ensure that first responders always remain connected during emergencies. ML algorithms assist with efficient response to emergencies by forecasting the probability of incidents worsening, determining the quickest paths for emergency response, and collaborating among involved organizations.

The uninterrupted dependability and broad connectivity of 6G networks allow these systems to remain operational even in the face of disaster or crisis. Edge computing allows safety applications to maintain operation independently of connections to central networks.

9.5 Healthcare and Telemedicine

Healthcare is among the fields that can be transformed the most using 6G-empowered IoTs in conjunction with ML. Integrating continuous monitoring, real-time analysis, and remote intervention will bring revolutionary changes to the way healthcare is provided, enabling predictive, tailored, and more widely distributed care. Table 9.2 shows application scenarios used in healthcare leveraging the benefits of IoT, ML, and 6G.

9.5.1 Remote Patient Monitoring

6G networks will enable remote monitoring systems to shift from concentrating on a patient's basic health metrics to more complete health assessments. 6G networks will make it possible for large numbers of medical IoT devices to be deployed and continually send data from patients.

Thanks to ML processing, the system can identify early signs of illness long before they show up as regular symptoms. Some algorithms, on the other hand, might spot signs in a person's changing heart rhythms, breathing, or activity that could predict an impending heart problem or recognize alterations in speech or cognition that point to neurological issues.

Table 9.2 Key Healthcare Applications Enhanced by 6G IoT and ML

Application	Description	Benefits
Remote patient monitoring [15]	Continuous health data collection and ML analysis for early detection	Improved patient care and real-time monitoring
Smart hospitals and telesurgery [16]	Coordinated hospital systems and ultralow latency remote surgery	Increased hospital efficiency and surgical precision
Personalized medicine and AI diagnostics [17]	Individualized treatment plans using ML on multimodal health data	More effective treatments and earlier diagnosis

Research has shown that 6G's ultralow latency, massive device connectivity, and high data throughput can significantly improve patient care, real-time monitoring, and the efficiency of hospital operations. These features allow healthcare providers to conduct high-resolution monitoring of a larger number of patients at a reduced cost.

Healthcare applications stand to gain the most from the privacy-safeguarding features of FL. Patients retain ownership of their data and simultaneously advance the accuracy of predictive models by conducting computations on their own devices.

9.5.2 Smart Hospitals and Telesurgery

6G and ML will transform medical institutions into examples of smooth operations, high productivity, and exact precision. Research shows that a 6G-powered smart hospital model allows technology systems to exchange information effortlessly, promoting intelligent decisions and improving resource utilization.

Integrating these technologies results in an integrated hospital infrastructure that functions seamlessly in all areas, such as patient care, diagnostics, record keeping, team communication, and facility operations. ML algorithms are able to improve how patients are routed, staff are assigned tasks, equipment is utilized, and supplies are ordered using past data analysis and the present situation [18].

Remote surgeries performed over long distances present the highest challenge for 6G networks, which must deliver consistent, near real-time responsiveness during the procedure. The incredibly low latency offered by 6G will make it possible for surgeons to receive real-time haptic feedback while performing operations from any location, equaling the accuracy achieved by on-site surgery.

ML technologies can improve these systems by offering decisions, correcting for network fluctuations, and automating elements of surgeries. Example applications include using algorithms to correct motions from the surgeon's hands and the network to achieve a greater level of accuracy than that achieved by a human alone.

9.5.3 Personalized Medicine and AI Diagnostics

The confluence of health monitoring data, genetic data, environmental influences, and behavior facilitates the evolution of precise medicine on an entirely new scale. ML algorithms can predict how different combinations of these factors contribute to specific health outcomes, allowing for personalized care plans.

For example, AI algorithms can personalize therapy choices by predicting which patients with a specific illness will experience the best outcomes from different treatments. AI systems leveraging a wide range of input data are better able to identify diseases earlier and more reliably than conventional diagnostic methods.

The reliable and secure transfer of rich medical information between devices is made possible by the high-capacity 6G networks. Some analysis is done directly at the edge of the network, while more sophisticated computations can take place in a cloud environment.

9.6 Industrial Automation and Smart Factories

6G, IoT, and ML are expected to transform the industrial sector by providing revolutionary new ways to automate, optimize, and agilely manage manufacturing and logistics activities.

9.6.1 Intelligent Manufacturing Systems

Future smart factories using 6G technology and ML will reach new heights of automation, changeability, and efficiency. A study shows that 6G will be the foundation of effective industrial automation, allowing for instantaneous and accurate transfer of information throughout the production process. 6G will support the fast and seamless transfer of massive amounts of data for automation and AI tasks throughout the entire production process.

The security and responsiveness of 6G allow for the precise synchronization and regulation of manufacturing tasks as they happen. These algorithms use information about product quality, energy usage, materials, and machines to make the best choices for each task [19].

One example is the use of ML that continuously adjusts settings based on production line data to produce top-quality goods with the least amount of resources. Similarly, algorithms can spot early signs of quality problems and allow for quick corrective measures.

The vast connectivity options offered by 6G allow for the widespread installation of sensors in factories, providing real-time insights into every part of the production process. This integration unites previously separate entities and maximizes overall factory efficiency.

9.6.2 Predictive Maintenance and Quality Control

Advanced methods of predictive maintenance are made possible by ML, which uses sensor data to identify signs of impending equipment malfunctions. 6G networks with their ultrafast speeds and large number of connected devices allow more machines to be monitored in much greater detail than before.

ML algorithms use patterns found in vibration, temperature, sound, and other metrics to predict the failure of equipment well before it happens. This allows maintenance to be scheduled when production is stopped, reducing the number of operational disruptions and preventing disruptions that affect several systems at once.

Quality control systems will also shift from random sampling to continuous monitoring. ML algorithms use data from sensors to identify tiny changes in quality markers promptly, allowing adjustments to operating values before flaws arise. It moves quality control efforts from a reactive approach to a proactive one.

Some experts suggest that the use cases for manufacturing industries would benefit greatly from the development of 6G. This type of monitoring allows for ongoing improvement of industrial processes using data collected in the present moment.

9.6.3 Digital Twins and Simulation

The development of advanced digital twins will be heavily influenced by the potential of 6G and ML technologies. Research teams believe that digital twin technology is crucial for unifying 6G and IoE (Internet of Everything) by offering a model representing physical objects that enables smooth collaboration between the real and virtual environments.

They allow operators to try out modifications on a virtual existing before making them in the real one, improving performance, forecasting results, and preventing difficulties. Predicting the behavior of physical systems under different conditions is enhanced by using ML to analyze historical data from the corresponding digital twin simulations [20].

A production line's digital twin can be used to understand how modifying parameters influences aspects such as product quality, energy use, and production capacity. This virtual exploration of various options allowed them to choose the best approach without any unnecessary risks.

6G's remarkable bandwidth and low latency support seamless, real-time synchronization between physical systems and their digital counterparts, no matter how many surveillance sensors gather vast data streams. Processing requests requiring immediate action takes place at the network's edge, while demanding simulations tap into cloud computing when necessary.

9.7 Agricultural Applications and Food Production

Tackling the needs of a larger population, dealing with climate change, and preserving resources may be achieved by leveraging 6G-powered IoT systems with ML to enable precision farming, automate farm work, and optimize resource use.

Precision agriculture uses real-time data intelligence to analyze and customize how resources are applied at specific locations within a field. Sensors can be placed densely in agricultural environments thanks to 6G's vast ability to connect, providing a detailed understanding of soil quality, plant well-being, environmental changes, and other important attributes.

ML models infer insights from the fusion of different data sources and provide tailored advice on seeding, irrigation, fertilization, and pest control. Sample applications include using algorithms to combine data from soil moisture sensors, weather forecasts, crop growth models, and water availability data in order to develop the most efficient irrigation strategies and achieve the greatest crop yields with the least water consumption.

Studies show that advancements in IoT and 6G connectivity are driving the adoption of digital farming and precision agriculture. The goals are to create a greater volume of nutritious food while using fewer pesticides, ultimately benefiting both the food transformation industry and consumers [21].

Combining satellite data with on-site sensors and tools allows for extensive surveillance of vast swaths of farmland. ML algorithms are able to identify signs of pests, imbalances in nutrients, or diseases occurring early in the growth cycle, allowing efficient measures to be taken to prevent crop damage and lessen the amount of agricultural chemicals used.

9.7.1 Autonomous Farming Equipment and Drones

Self-driving farm vehicles, tools, and machines will revolutionize the way farming is conducted. Reliable 6G connectivity permits full control and synchronization of these systems, regardless of their location [22].

ML algorithms empower these systems to adjust and respond dynamically to a range of unpredictable variables that traditional automation cannot account for. Specifically, these systems are able to vary their procedures depending on crop growth, ensuring that each area of a field receives an efficient and effective harvest method to produce the highest yields and quality.

Drones able to capture a wide range of data from above offer insights into the status of agricultural activities. ML algorithms evaluating drone images allow for the detection of specific problem spots, resulting in efficient targeted responses instead of general strategies.

Bringing together autonomous technology, weather forecasts, and market data opens up possibilities for more comprehensive and efficient coordination of farming activities. As such, harvesting may be planned and executed in a way that takes into account predicted climate conditions, crop status, available machinery, and market trends, ensuring both high crop yields and financial gains.

9.7.2 Livestock Monitoring and Management

Livestock tracking holds tremendous potential for the utilization of 6G-connected IoT along with ML algorithms. Studies have highlighted that both livestock surveillance and uninterrupted communications are vital to defense against future pandemics.

By attaching sensors to livestock, their movements, levels of activity, what they eat, and physiological information can be monitored in real time. ML algorithms help uncover signs of illness, changes in reproductive cycles, and the best time to harvest, allowing for preventive measures.

Some examples of how algorithms could forewarn of health problems in livestock are noticing changes in their activity, body temperature, or eating patterns before they become apparent in medical tests. This allows early intervention to decrease animal discomfort, contain outbreaks, and prevent unnecessary financial setbacks.

Analogously, AI is used to create personalized feeding regimes that balance the needs of each animal, the prevailing weather, and the industrial requirements, resulting in better efficiency and animal well-being. Combining these systems with supply chain management brings about enhanced tracking and assurance of food quality at every step of production.

9.7.3 Indoor and Vertical Farming

Conventional farming is burdened by shortages of arable land, fluctuations in the weather, and inefficient use of resources. Controlled environments and resource-efficient techniques are used by indoor and vertical farming systems to overcome the obstacles faced by traditional agriculture.

A study suggests that radically different farming approaches enabled by modern digital technologies will transform the landscape of an industry typically characterized by toiling and machinery. They use large numbers of IoT sensors, automated controls, and ML algorithms to develop ideal environments for growing distinct crop varieties.

ML methods are used to improve the efficiency of essential variables such as light spectrum and intensity, temperature, humidity, CO_2 levels, and nutrient supply in order to increase yield, quality, and resource usage for individual crops and growth phases. The algorithms continuously adjust the various parameters in response to plant activity and changes in the environment, making for highly effective and adaptive growing methods.

Its unfailing performance and swift data transmission allow accurate on-the-spot monitoring and management of primary growing environments. Consequently, these networks are able to sustain excellent performance in the face of potential network disruptions.

9.8 Extended Reality and Digital Twins

Combining the capabilities of 6G, the IoT, and ML will result in extraordinary progress in XR and digital twin solutions, changing the ways we experience and manage both digital and physical environments.

Sixth-generation networks will make it possible for people to communicate with each other more naturally and effectively using immersive technology that surpasses the functionality of today's video conferencing and virtual reality solutions. Research shows that the 6G network is uniquely suited for the application of real-time holographic communications [23].

Holographic communications make connected collaborators feel virtually co-located and accomplish more effectively than traditional methods. ML-powered solutions improve these interactions by anticipating activities, removing unwanted sounds, and tailoring content to the needs of users and their networks.

It allows for more advanced collaboration across immersive, virtual environments where users work together. Studies suggest that Virtual Twin-like collaboration will be an important use case, allowing teams to work together on 3D models and simulations from anywhere in the world.

They revolutionize industries where participants must be in the same place or work together with their hands. For example, clinicians could participate as a team to review and manipulate holographic models of patient data from different locations.

9.8.1 Comprehensive Digital Twins

Digital twins are set to grow more sophisticated, covering whole systems and surrounding settings rather than focusing on individual items. A digital twin is essential for bridging the gap between IoE and 6G by creating virtual representations of real-world systems.

Advanced digital twins combine information from various sources, such as IoT sensors, historical data, simulation models, and external systems, to accurately reflect the real world. ML algorithms extract insights from combined data to forecast events, guide decisions, and improve the efficiency of various areas.

An example is urban digital twins that combine information from transportation, energy, water, building networks, and environmental monitoring. ML algorithms help improve resource management, forecast maintenance requirements, perform emergency planning, and assess how potential decisions might affect the system before they are put into practice.

6G networks provide the capability to synchronize instantly between physical systems and their twin counterparts, even when dealing with large, fast-changing environments that produce large amounts of data. Local edge processing handles real-time tasks effectively, while cloud computing is used for demanding simulations.

9.8.2 Internet of Senses and Embodied AI

XR and digital twins are moving toward an era researchers refer to as the "Internet of Senses," in which digital experiences stimulate several human senses. It indicates that with 6G networks, our engagement with IoT gadgets moves beyond visual and auditory to touch and possibly thought.

These demanding applications demand high bandwidth for visual and audio elements and low latency for haptic responses that offer a realistic feel. 6G connectivity supports these requirements, and ML helps customize and adapt the experiences to individuals.

Embodied AI explores a new area where AI is made tangible through bodies that allow for intuitive human–computer interactions. These embodied AIs provide assistance, act as guides, educate us, or accompany us in real and virtual situations with experiences that surpass what screens can offer.

Embodied AIs benefit from 6G networks, IoT sensors, and ML models to sense the environment, know the situation, and react effectively to people's actions and requests. This allows for more seamless and efficient cooperation between people and machines in many different fields.

9.9 Challenges and Future Research Directions

Achieving the full potential of 6G and the IoT in conjunction with ML requires overcoming a range of substantial obstacles [22].

9.9.1 Technical Challenges

With a fusion of ML and 6G technologies, numerous technical barriers must be overcome through creative approaches.

1) **Energy Efficiency:** Saving power is particularly important for IoT gadgets that rely on restricted power sources. Studies suggest that 6G could have a more positive environmental impact than 5G, leading to potentially reduced electricity consumption for IoT devices as well as longer battery life. Nevertheless, the additional CPU requirements of ML algorithms may make achieving higher energy efficiency challenging, forcing the development of novel hardware and more efficient algorithms.
2) **Security and Privacy:** Research has found that 6G technology also raises several obstacles, such as data privacy and security vulnerabilities, difficulties in

achieving system integration, and ethical concerns. Extending the connectivity and enabling intelligence in IoT systems introduces novel security risks and challenges to privacy protection, which demand robust security regulations and privacy-oriented ML solutions.

3) **Reliability and Robustness:** Reliability becomes a crucial concern in mission-critical IoT systems used for driverless cars, manufacturing equipment, and medical devices. Consistent performance of ML algorithms in a broad range of situations, including extreme and malicious inputs, continues to be difficult.
4) **Spectrum Management:** THz bands bring major challenges for wireless communication, impacting both reliability and the coverage area. Adaptive and resourceful spectrum management will play a vital role in providing constant connectivity in a variety of settings.
5) **Interoperability:** Diverse IoT devices, network technologies, and ML frameworks create important interoperability issues that require standardization and flexible, adjustable architectures.

9.9.2 Ethical and Societal Implications

Deploying intelligent IoT technologies powered by 6G brings up significant ethical and societal challenges that must be addressed [24].

1) **Digital Divide:** 6G-empowered IoT technology has the potential to exacerbate already existing disparities between more and less developed areas. Academic studies highlight that bridging the digital divide is a high priority, as indicated by the designated use case of 6G being "Global Internet—Digitalization for Everyone, Everywhere".
2) **Autonomy and Control:** Given that more and more decisions are made automatically by ML-based systems, concerns about how these systems should be monitored, who should be held accountable, and how their decisions can be reviewed by humans have emerged.
3) **Environmental Impact:** However, the large-scale production, distribution, and use of millions of IoT sensors might lead to increased resource demands, electronic waste, and a potential detriment to the environment.
4) **Privacy and Surveillance:** Along with their wide-ranging sensors, IoT devices leveraging ML have the potential for massive surveillance and privacy violations that require both technological solutions and the development of new policies.

9.9.3 Future Research Directions

A number of promising research areas will define the future of ML-augmented IoT systems over 6G networks:

1) **Quantum-Powered Security:** This is one promising way to mitigate next-generation systems' security issues. Quantum technology may improve encryption, authentication, and intrusion detection [25].
2) **Sustainable IoT:** According to research [26], one application of 6G is an "Energy Measurement Framework for Energy Sustainability." Creating eco-friendly IoT devices, network infrastructures, and ML algorithms that optimize resource use will be important to ensure long-term sustainability.
3) **Human-Centered Design:** With interaction shifting from screens to senses, studies need to be concentrated on designing interfaces that are intuitive, accessible, and respectful of human cognitive strengths and limitations. This also involves considerations for diverse user populations with different abilities, preferences, and cultural contexts.
4) **Satellite-IoT Integration:** Satellite network integration with terrestrial 6G infrastructure will be critical for IoT connectivity on a global scale, especially for remote areas and mobile use cases [27].
5) **LLM-Powered IoT:** Large language models and other cutting-edge AI methods may make it possible for more complex contextual understanding and more natural human–machine interaction in IoT systems [10].

9.10 Conclusion

The coupling of ML with 6G-enabled IoT is a revolutionary synergy that will reshape the way we engage with technology and the world around us. With unparalleled connectivity, intelligence, and autonomy, these systems will solve key challenges in multiple domains while offering new possibilities for innovation and value creation.

From intelligent cities that dynamically respond to citizen demands, to medical systems that anticipate and avert disease prior to symptom onset, to factory floors that optimize processing on their own, virtually all areas of human endeavor are addressed by applications. The technological building blocks of 6G-THz communications, massive device connectivity, and microsecond latency merge with breakthroughs in ML and edge processing to form the foundation for these game-changing applications.

But realizing this vision necessitates overcoming major challenges involving energy efficiency, security, reliability, interoperability, and ethical implications. Interdisciplinary research blending engineering, computer science, social sciences, and humanities will be necessary to guarantee that these technologies evolve on paths that serve all of humanity with the least possible harm.

As we advance toward this 6G future, coordination among academia, industry, government, and civil society will be essential to steer technology development in ways that are both technically advanced and socially responsive as well as ecologically sustainable. With careful design, deployment, and regulation, ML-facilitated IoT over 6G networks has the potential to make the world more connected, efficient, and equitable.

References

1 Guo, F., Yu, F.R., Zhang, H. et al. (2021). Enabling massive IoT toward 6G: a comprehensive survey. *IEEE Internet of Things Journal* 8 (15): 11891–11915.

2 Jiao, L., Shao, Y., Sun, L., and Liu, F. (2024). Advanced deep learning models for 6G: overview, opportunities and challenges. *IEEE Access* 12: 133245–133314. https://ieeexplore.ieee.org/abstract/document/10570412/.

3 Puspitasari, A.A., An, T.T., Alsharif, M.H., and Lee, B.M. (2023). Emerging technologies for 6G communication networks: machine learning approaches. *Sensors (Basel)* 23 (18): 7709.

4 Rekkas, V.P., Sotiroudis, S., Sarigiannidis, P. et al. (2021). Machine learning in beyond 5G/6G networks—state-of-the-art and future trends. *Electronics* 10 (22): 2786.

5 Kato, N., Mao, B., Tang, F. et al. (2020). Ten challenges in advancing machine learning technologies toward 6G. *IEEE Wireless Communications* 27 (3): 96–103.

6 Qadir, Z., Le, K.N., Saeed, N., and Munawar, H.S. (2023). Towards 6G internet of things: recent advances, use cases, and open challenges. *ICT Express* 9 (3): 296–312.

7 Dangey, R., Tandon, A., and Tyagi, A.K. (2023). Emerging internet of things (IoTs) scenarios using machine learning for 6G over 5G-based communications. In: *6G-Enabled IoT and AI for Smart Healthcare* (ed. A. Kumar, R. Jain, M. Gupta, and S.M.N. Islam), 215–237. CRC Press https://www.taylorfrancis.com/chapters/edit/10.1201/9781003321668-12/emerging-internet-things-iots-scenarios-using-machine-learning-6g-5g-based-communications-raghav-dangey-arjun-tandon-amit-kumar-tyagi,Boca Raton.

8 Noman, H.M.F., Hanafi, E., Noordin, K.A. et al. (2023). Machine learning empowered emerging wireless networks in 6G: recent advancements, challenges and future trends. *IEEE Access* 11: 83017–83051.

9 Zhao, L., Zhou, G., Zheng, G. et al. (2021). Open-source multi-access edge computing for 6G: opportunities and challenges. *IEEE Access* 9: 158426–158439. https://doi.org/10.1109/ACCESS.2021.3130418.

10 Chen, X., Wu, W., Li, L., and Ji, F. (2025). LLM-empowered IoT for 6G networks: architecture, challenges, and solutions. *IEEE Internet of Things Magazine* 8 (6): 34–41. https://doi.org/10.1109/MIOT.2025.3582641.

11 Shahidehpour, M., Li, Z., and Ganji, M. (2018). Smart cities for a sustainable urbanization: illuminating the need for establishing smart urban infrastructures. *IEEE Electrification Magazine* 6 (2): 16–33. https://doi.org/10.1109/MELE.2018.2816840.

12 Khalid, M., Awais, M., Jisi, C. et al. (2025). Modernizing transit: intelligent traffic and transportation management with artificial intelligence in the Era of 5G and 6G. In: *The Intersection of 6G, AI/Machine Learning, and Embedded Systems* (ed. S. Sharma, A. Sharma, and T. Van Chien), 208–231. CRC Press.

13 Gera, B., Raghuvanshi, Y.S., Rawlley, O. et al. (2023). Leveraging AI-enabled 6G-driven IoT for sustainable smart cities. *International Journal of Communication Systems* 36 (16): https://doi.org/10.1002/dac.5588.

14 Alnoman, A., Khwaja, A.S., Anpalagan, A., and Woungang, I. (2024). Emerging AI and 6G-based user localization technologies for emergencies and disasters. *IEEE Access* 12: 197877–197906. https://doi.org/10.1109/ACCESS.2024.3521005.

15 Yew, H.T., Ng, M.F., Ping, S.Z. et al. (2020). IoT based real-time remote patient monitoring system. In: *2020 16th IEEE international colloquium on signal processing & its applications (CSPA)*, 176–179. IEEE.

16 Mulita, F., Verras, G.-I., Anagnostopoulos, C.-N., and Kotis, K. (2022). A smarter health through the internet of surgical things. *Sensors* 22 (12): 4577.

17 Schork, N.J. (2019). Artificial intelligence and personalized medicine. In: *Precision medicine in Cancer therapy*, 265–283. Cham: Springer International Publishing.

18 Kumar, A., Masud, M., Alsharif, M.H. et al. (2025). Integrating 6G technology in smart hospitals: challenges and opportunities for enhanced healthcare services. *Frontiers in Medicine* 12: 1534551.

19 Ma, S., Ding, W., Liu, Y. et al. (2024). Industry 4.0 and cleaner production: a comprehensive review of sustainable and intelligent manufacturing for energy-intensive manufacturing industries. *Journal of Cleaner Production* 467: 142879.

20 Hafeez, M.A., Procacci, A., Coussement, A., and Parente, A. (2024). Challenges and opportunities for the application of digital twins in hard-to-abate industries: a review. *Resources, Conservation and Recycling* 209: 107796. https://doi.org/10.1016/j.resconrec.2024.107796.

21 Pandey, S.K., Ray, A.K., Kansal, S., and Arora, V. (2024). Role of 6G, loT with integration of Al and ML and security in agriculture. In: *6G Connectivity-Systems, Technologies, and Applications* (ed. R. Prasad, D.S. Mantri, S.K. Pandey, and A.D. Mihovska), 43–66. New York: River Publishers.

22 Srisankar, D.M. and Raju, D.M. (2024). The role of 6G technology in smart farming and precision agriculture. *Cuestiones de Fisioterapia* 53: 3. https://doi.org/10.48047/c2j43e31.

23 Adreani, L., Bellini, P., Colombo, C. et al.(2022). Digital twin framework for smart city solutions. *DMSVIVA - Proc. Int. DMS Conf. Vis. Vis. Lang.*, Knowledge Systems Institute Graduate School, KSI Research Inc., pp. 1–8. https://doi.org/10.18293/DMSVIVA22-012.

24 Rafique, S., Iqbal, S., Ali, D., and Khan, F. (2025). Navigating ethical challenges in 6G-enabled smart cities: privacy, equity, and governance. *ICCK Transactions on Sensing, Communication, and Control* 2 (1): 48–65. https://doi.org/10.62762/TSCC.2025.291581.

25 Javeed, D., Saeed, M.S., Ahmad, I. et al. (2024). Quantum-empowered federated learning and 6G wireless networks for IoT security: concept, challenges and future directions. *Future Generation Computer Systems* 160: 577–597. https://doi.org/10.1016/j.future.2024.06.023.

26 Dash, S. (2025). Green AI: enhancing sustainability and energy efficiency in AI-integrated enterprise systems. *IEEE Access* 13: 21216–21228. https://doi.org/10.1109/ACCESS.2025.3532838.

27 Mahboob, S. and Liu, L. (2024). Revolutionizing Future Connectivity: A Contemporary Survey on AI-Empowered Satellite-Based Non-Terrestrial Networks in 6G. *IEEE Communications Surveys & Tutorials* 26 (2): 1279–1321. https://doi.org/10.1109/COMST.2023.3347145.

10

Cybersecurity and Privacy in 6G-IoT Ecosystems

Aman Jandwani[1], *Amit K. Arora*[2], *Dadas A. Bhanudas*[1], *Ekta Singh*[1], *Madhur*[3], *and Paras Jandwani*[4,5]

[1] *Department of Management, Neville Wadia Institute of Management Studies & Research, Pune, Maharashtra, India*
[2] *Computer science and Engineering, IIT Madras, Chennai, Tamil Nadu, India*
[3] *Department of Commerce, Global School of Finance & Accounting, Chandigarh University, Mohali, Punjab, India*
[4] *Department of Electronics, School of VLSI & Embedded (SVE), NIT Kurukshetra, Kurukshetra, Haryana, India*
[5] *Department of Computer Science and Engineering, Accenture LLP, USA*

10.1 Introduction

The convergence of sixth-generation (6G) wireless technology and the Internet of Things (IoT) represents a fundamental transformation in digital communications. With promises of ultra-reliable low-latency communication (URLLC), massive machine-type communications (mMTC), and enhanced mobile broadband (eMBB), 6G is poised to enable a fully connected, intelligent world. The IoT ecosystem—comprising billions of smart sensors, actuators, autonomous systems, and edge devices—will increasingly rely on 6G's capabilities to support real-time data transmission, autonomous decision-making, and seamless interconnectivity across domains such as smart cities, healthcare, industrial systems, and immersive environments like the Metaverse [1–3]. However, as this digital fabric expands, so do the attack surfaces and vulnerabilities. Traditional security paradigms are no longer sufficient in the face of decentralized architectures, dynamic edge devices, and emerging quantum threats. Cybersecurity and privacy are thus not auxiliary concerns but foundational design principles for the 6G-IoT era. Data integrity, device authentication, secure data transmission, and user anonymity must be ensured at a massive scale and at near-instantaneous speeds [4, 5]. As 6G networks become embedded in critical infrastructures—governing transportation, healthcare, finance, and public safety—the risks of cyberattacks, data leakage, and surveillance become existential.

Synergies of 6G and IoT, First Edition. Edited by Himanshi Babbar.

Published 2026 by John Wiley & Sons, Inc.

10.1.1 Significance of Cybersecurity and Privacy in Hyper-connected 6G-IoT Environments

The anticipated proliferation of 6G-enabled IoT devices, potentially exceeding 125 billion globally by 2030, will reshape digital society. With the rise of autonomous drones, connected vehicles, augmented reality interfaces, and cognitive robotics, real-time data exchange across trusted and untrusted nodes will become the norm [6, 7]. In such environments, breaches in confidentiality or failures in authentication could have catastrophic outcomes—ranging from hijacking smart grids to manipulating healthcare devices. The sensitivity of data processed in 6G-IoT ecosystems also raises urgent privacy concerns. Location data, biometric information, health records, and behavioral analytics are transmitted continuously across networks. Without end-to-end privacy guarantees, users are exposed to intrusive profiling and surveillance [8, 9]. These issues are exacerbated by the borderless nature of 6G communication infrastructures, which challenge jurisdictional data protection norms and international compliance frameworks like the General Data Protection Regulation (GDPR) and California Consumer Privacy Act (CCPA) [10]. Conventional centralized security frameworks are ill-suited for the distributed, context-aware, and latency-sensitive nature of 6G-IoT. Instead, paradigms such as blockchain-based identity management, post-quantum cryptography (PQC), federated learning (FL), and adversarial artificial intelligence (AI) defenses are becoming essential [11, 12].

10.1.2 Objectives of the Chapter

This chapter aims to address the multidimensional security and privacy challenges that arise from the integration of 6G networks and IoT infrastructures. The following are the main objectives:

- To study currently available encryption and privacy-preserving technologies and set their limitations in the context of high-speed, decentralized, and heterogeneous 6G-IoT environments.
- To discuss and compare emerging technologies such as PQC, homomorphic encryption (HE), and secure multiparty computation, which enable privacy-preserving analytics in a network of distributed IoT nodes [1, 13].
- To evaluate decentralized security models like blockchain to authenticate devices, log data securely, and use consensus mechanisms that allow trust in the machine-to-machine interactions [4, 6].
- To study AI-powered threat identification systems that are based on federated, adversarial, and deep learning (DL) for anomaly detection, intrusion prevention, and adaptive cyber defense [14, 15].
- To do a comparative analysis of the current vs. emergent security technologies based on the framework of scalability, latency, computational overhead, and

robustness against zero-day attacks, which will be performed as a prism to examine the issues.
- To develop compliance, ethical, and governance frameworks for dealing with user privacy, legal obligations, and risk in varied 6G-IoT use cases.

10.1.3 Scope of the Chapter

This chapter is structured to offer both a theoretical foundation and a practical lens on securing future digital infrastructures. The scope includes:

- **Advanced Encryption Technologies:** Description of symmetric (AES) and asymmetric (RSA) algorithms, and then on emerging solutions like lattice-based and code-based post-quantum cryptographic schemes [2, 11].
- **Blockchain in IoT Security:** Study in real-world deployment of blockchain in smart cities, industrial control systems, and health care IoT, as well as future architectures focusing on sharding, interoperability, and green consensus [7, 16].
- **AI for Cyber Defense:** Comparative review of traditional machine learning and deep reinforcement learning (DRL), generative adversarial networks (GANs), and collaborative intrusion detection systems (CIDS) [12, 17].
- **Privacy-Preserving Data Analytics:** Review of FL and HE approaches, which provide a way in which secure computation can be done between distributed nodes while not compromising user data [10, 13].
- **Policy, Ethics, and Governance:** Thinking about the changing landscape of regulatory compliance, algorithmic transparency, and decentralized identity management in global 6G-IoT systems [18, 19].

10.1.4 Foundational Motivation

The need for comprehensive protection systems in 6G-IoT realms emerges due to the exceptional mix of physical and cyber environments. This combination creates complex cyber-physical systems where security issues may cause outcomes other than data loss, including vehicle failures, utility failures, and telehealth system breaches [17]. This integration leads to complex cyber-physical systems, in which security weaknesses. The security architectures should now pursue proactive, intelligent, and interoperable strategies to counter this threat. Cybersecurity with resilient security depends upon technological innovation, integrating AI, blockchain, and quantum-proofing. It will become apparent in the discussion to follow that drawing knowledge from cryptography, distributed systems, machine learning, and digital policy is essential in creating secure and trustworthy 6G-IoT ecosystems.

10.2 Advanced Encryption Mechanisms in 6G-IoT

10.2.1 Current Encryption Technologies

As 6G networks push toward ultralow latency, high bandwidth, and ubiquitous connectivity, securing data in real-time becomes critical. At present, IoT systems predominantly rely on classical encryption schemes such as symmetric and asymmetric cryptography to maintain data confidentiality and integrity.

10.2.2 Symmetric Encryption (AES)

The Advanced Encryption Standard (AES) has become the de facto symmetric encryption protocol used in a wide array of IoT devices due to its computational efficiency and relatively low resource demands. AES operates on blocks of data with key sizes of 128, 192, or 256 bits and provides strong encryption against brute-force attacks. In IoT deployments, AES is commonly used to encrypt sensor data, control signals, and communications between devices and cloud systems. However, symmetric encryption like AES depends on shared key distribution. In dense, decentralized IoT environments facilitated by 6G, securely managing and distributing keys among billions of nodes becomes a major challenge. Moreover, the computational load of AES encryption—though moderate—can still impact battery life and processing capabilities in lightweight edge devices [20].

10.2.3 Asymmetric Encryption (RSA)

Asymmetric encryption protocols, specifically RSA (Rivest–Shamir–Adleman), are highly appreciated due to the possibility of secure key exchange and digital signature. RSA makes use of a set of public and private keys that allow encrypting and decrypting messages without the use of pre-shared secret keys. In IoT applications, RSA provides secure firmware updating, device authentication, and integrity checks. Although RSA is computationally intensive, it has a lot of strengths. The priming of large primes needed in the RSA encryption process is practically outside the range of most of the IoT microcontrollers. In addition, RSA has much higher encryption and decryption times with respect to the symmetric counterparts. Such disadvantages do not make RSA an optimal protocol for ultralow-latency communication in 6G systems, and, if particular nodes must authenticate several other nodes in a second, the situation only worsens [9].

10.2.4 Limitations in 6G Scenarios

The limitations of classical encryption protocols in 6G contexts are primarily attributed to:

- **Latency Sensitivity**: High computational overhead results in delayed encryption/decryption cycles.

- **Scalability Constraints**: Existing key management protocols are infeasible for device densities projected in 6G networks.
- **Quantum Vulnerabilities**: Both AES and RSA are theoretically susceptible to quantum algorithms such as Grover's and Shor's, respectively, which can undermine their security assumptions.

As such, the need for next-generation encryption mechanisms that are lightweight, scalable, and quantum-resistant is clear.

10.2.5 Emerging Encryption Solutions

To overcome such problems, the cybersecurity community is actively designing and incorporating new cryptographic schemes designed specifically for the scale, speed, and security needs of the 6G-IoT system. These include PQC, lightweight encryption algorithms, and sophisticated privacy protection approaches such as HE and Secure Multiparty Computation (SMPC).

10.2.6 Post-Quantum Cryptography (PQC)

Classical encryption algorithms face an existential threat from quantum computations. For example, Shor's algorithm is able to efficiently factorize large prime numbers and break RSA in polynomial time. To balance this out, there is the desire to come up with encryption protocols that are dependent on mathematical issues that are hard even for quantum computers, and this is the PQC.

Several promising PQC candidates are emerging:

- **Lattice-based encryption** (e.g., NTRU and Kyber)
- **Hash-based digital signatures**
- **Code-based encryption** (e.g., McEliece)
- **Multivariate polynomial cryptography**

These schemes form the basis of the NIST PQC standardization process. In 6G-IoT systems, PQC protocols can be embedded in device firmware or network edge controllers to ensure forward secrecy and quantum resistance. Despite being more computationally intensive than symmetric encryption, PQC is evolving with optimizations that make it viable even for low-power devices [21, 22].

10.2.7 Lightweight Cryptography for IoT

Lightweight cryptographic algorithms are explicitly designed to work within the constraints of IoT devices—limited CPU, RAM, and battery life. Notable examples include:

- **PRESENT**
- **Simon and Speck**
- **Ascon** (selected by NIST as a standard for lightweight cryptography)

Lightweight schemes involve small block sizes and simpler operations like bitwise XOR and rotations. Those methods can be implemented efficiently in hardware and software; they can thus be employed to secure sensors, RFID chips, and microcontrollers in 6G networks. The future of 6G-IoT security will probably involve the use of hybrid architectures in which local data confidentiality will be handled by lightweight symmetric encryption, and remote key exchange will be secured by PQC.

10.2.8 Homomorphic Encryption (HE)

HE enables computations on encrypted data without requiring decryption—a vital feature for maintaining privacy in data-intensive environments. For example, a healthcare sensor could transmit encrypted patient vitals, and an AI model at the edge could make decisions directly on this ciphertext.

HE is categorized as:

- **Partially Homomorphic Encryption (PHE):** Supports one operation (e.g., addition)
- **Somewhat Homomorphic Encryption (SHE):** Supports limited operations
- **Fully Homomorphic Encryption (FHE):** Supports arbitrary computation

Although HE, specifically FHE, is resource-demanding, it guarantees unsurpassed assurances of privacy on such applications as FL and cloud analytics. Current research pays attention to the improvement of HE for mobile and embedded devices [20, 22].

10.2.9 Secure Multiparty Computation (SMPC)

SMPC enables more than one party to jointly evaluate a function over their private inputs without using them. In 6G-IoT ecosystems, this can support capabilities such as collaborative AI training, distributed voting protocols, or industrial automation across organizational boundaries with strict data privacy.

Typical SMPC protocols include:

- **Secret Sharing (e.g., Shamir's scheme)**
- **Garbled Circuits**
- **Oblivious Transfer**

SMPC aligns well with FL, where multiple IoT devices collaboratively train models without uploading raw data to a central server. Combined with blockchain, SMPC can also ensure accountability and auditability of these decentralized computations [23].

10.3 Decentralized Security Models: Blockchain Integration

10.3.1 Current Blockchain Applications in IoT

Blockchain technology has emerged as a transformative solution to the inherent trust issues within distributed IoT ecosystems. Traditionally, data integrity in IoT networks relied on central servers to validate, store, and authorize device interactions. This centralization, however, introduces single points of failure and makes IoT infrastructures vulnerable to attacks such as spoofing, unauthorized access, and data tampering. Blockchain offers a decentralized, immutable ledger that records transactions across a network of nodes—making it highly resistant to manipulation.

In current deployments, blockchain is used to ensure:

- **Data integrity** by timestamping and immutably recording sensor data and device actions.
- **Device authentication**, where each device maintains a unique blockchain identity and interacts through smart contracts [4, 6].
- **Access control**, as smart contracts enforce authorization logic based on predefined rules.

The trustless nature of blockchain ensures that even in hostile environments—such as public or cross-organizational IoT settings—interactions are transparent, verifiable, and tamper-proof.

10.3.2 Smart Cities and BFLIoT

One prominent use case for blockchain is in smart cities. Urban environments equipped with millions of sensors for traffic, utilities, surveillance, and public services require a scalable and secure data exchange model. Traditional models fall short due to their inability to guarantee data provenance and their reliance on siloed databases. Blockchain for FL in IoT (BFLIoT) is a noteworthy model where blockchain ensures:

- **Trust and transparency** in federated AI training models across decentralized smart city devices.
- **Incentivization mechanisms** using tokens to reward edge devices that contribute accurate model updates [13].
- **Auditability**, where changes to models or datasets are traceable via the blockchain ledger.

Projects such as Almatari et al. [24] highlight how blockchain enables interoperable and secure spectrum access in cognitive radio-based IoT networks—a critical component of smart traffic and autonomous vehicular systems.

10.3.3 Innovations in Blockchain for 6G-IoT

While current blockchain deployments provide basic integrity and authentication, they face scalability and efficiency challenges when extended to ultradense, ultra-low-latency 6G networks. Innovations are therefore necessary in blockchain design to align with the performance goals of 6G.

10.3.4 Scalability Through Sharding and Off-Chain Transactions

Traditional blockchains, such as Bitcoin and Ethereum, suffer from poor scalability due to linear block processing and consensus overhead. 6G-IoT environments, expected to support hundreds of billions of devices, require throughput levels far beyond current blockchain capacity.

- **Sharding** divides the blockchain network into smaller, manageable partitions called shards, each capable of processing its own transactions. This parallel processing significantly improves throughput without increasing energy demands [25].
- **Off-chain transactions** reduce on-chain load by executing micro-transactions or data exchanges outside the main blockchain. Only the final state or a cryptographic proof is anchored on-chain. This method is particularly effective in high-frequency data exchanges among IoT devices [26].

These techniques ensure that blockchain remains viable in environments where millisecond-level response times are critical—such as autonomous drone swarms or industrial automation in smart factories.

10.3.5 Integration with Edge Computing and AI

A major architectural shift involves moving blockchain closer to the edge of the network. Edge computing refers to data processing performed near the data source, minimizing latency and reducing backbone network congestion.

When blockchain is embedded at the edge:

- Devices can locally validate transactions and execute smart contracts.
- Latency is reduced as consensus and verification happen near real-time.
- Privacy is improved, as sensitive data does not traverse multiple hops before verification [27].

Moreover, the convergence of blockchain and AI introduces **autonomous security mechanisms** (learning based security). For instance, blockchain records can serve as data lakes for training AI-based intrusion detection systems (IDS). In turn, AI can monitor blockchain behavior, detect anomalies, and prevent fraud in decentralized systems [12, 17]. This fusion leads to **CIDS**, where edge devices cooperatively monitor, share, and act on cybersecurity threats using both blockchain logs and AI models.

10.3.6 Energy-Efficient Consensus Mechanisms

Conventional consensus mechanisms like Proof-of-Work (PoW) are computationally expensive and incompatible with green 6G goals. To align with 6G's sustainability vision, next-gen blockchain systems are adopting **lightweight, energy-efficient consensus mechanisms**, such as:

- **Proof-of-Stake (PoS):** Relies on economic stake rather than brute computational power.
- **Practical Byzantine Fault Tolerance (PBFT):** Suitable for permissioned blockchains with known validators.
- **Proof-of-Authority (PoA):** Offers high throughput for private networks by designating trusted validators [26].

These mechanisms reduce the energy footprint of blockchain operations, making them more suitable for deployment in solar-powered edge devices, mobile sensors, and remote industrial environments. Blockchain is evolving from a niche cryptographic tool to a cornerstone of 6G-IoT cybersecurity. In current systems, it secures device identities, transaction integrity, and access control. Moving forward, blockchain innovations like sharding, off-chain computation, edge integration, and green consensus will unlock its full potential for hyper-connected smart cities, autonomous industries, and global-scale IoT deployments. By synergizing with AI and federated analytics, blockchain will enable intelligent, decentralized threat response and enforce fine-grained privacy controls without sacrificing performance. The future of 6G security is not centralized—it is distributed, transparent, and autonomously resilient.

10.4 AI-Driven Threat Detection Systems in 6G-IoT

The complexity, data volume, and connectivity in 6G-IoT networks are becoming greater, and, therefore, classical cybersecurity frameworks turn out to be suboptimal. The conventional cybersecurity methods based on rule-based or static threat signatures are found to be unmanageable in dealing with the dynamically evolving forms and highly intelligent ways of modern cyberattacks. In this scenario, AI brings a disruptive cybersecurity strategy for intelligent automation and adaptive threat defense. AI has been crucial in forming the existing threat detection systems in the IoT networks. Previously, supervised models such as Decision Trees (DT), Support Vector Machines (SVM), and Random Forest (RF) were used to classify network traffic using labeled historical data that is available. These models have formed the basis for different, real-time IDS for detecting unauthorized access, malware spreading, and botnet actions [28]. Zhao et al. [29] demonstrated how deep autoencoders efficiently reveal anomalies in multivariable sensor data, thus identifying previously unidentified threats. Such models can be especially useful with the difficulty of training data labeling, especially within large-scale 6G-powered urban IoT settings. Such centralized architectures bring privacy risks and increase latency, directly opposed to the high-speed, privacy-oriented intents of 6G.

Federated Learning for Privacy-Aware Security Analytics: FL addresses the core limitation of traditional ML by decentralizing the training process. In an FL setup, individual devices or edge nodes perform local model training using their own data and periodically send model updates—not raw data—to a central aggregator. The final model is thus a synthesis of multiple decentralized learning processes, offering both scalability and privacy.

In 6G-IoT ecosystems, FL is particularly effective for:

- **Distributed industrial IoT environments**, where operational data must remain onsite for compliance.
- **Healthcare systems**, where medical sensor data cannot legally or ethically be transmitted to cloud servers.
- **Autonomous vehicle swarms**, where real-time collaboration is essential, but each vehicle must maintain autonomy.

A compelling example is the work of Sefati et al. [13], who integrated FL with blockchain to create BFLIoT—a framework that ensures trust, data integrity, and privacy in collaborative model training. Their approach not only mitigates data leakage but also guarantees accountability using immutable logs stored on a blockchain.

10.4.1 Emerging AI Techniques

10.4.1.1 Deep Learning and Reinforcement Learning

Conventional machine learning strategies have become outdated due to the evolution of cyberattacks and particularly when they start to use AI. It is the evolution in the styles of the attacks that has inspired the embrace of DL that engages cutting-edge neural networks that are also employed to extract high-dimensional patterns from vast amounts of data. CNNs and LSTMs (convolution neural network and long short memory) are specific examples of DL algorithms that are very suitable for detecting extremely complex relations in time and space in network and IoT data streams. Specifically, for example, CNNs are capable of inspecting timing and payload content in encrypted packets and therefore recognize if the communication is legitimate or malware-like. Likewise, LSTM networks work well for analysis of time-series data—they model the sequence behaviors inherent to stealth attacks and organized bot [30]. In addition to this, reinforcement learning (RL) is an advanced AI technique for self-empowerment of security using AI. RL agents learn skills by continuous interaction with their surroundings, obtaining instructive lessons occasioned by reinforcements resulting from appropriate decisions and negative lessons due to inappropriate acts. In cybersecurity applications, RL agents can dynamically change the firewall rules. Intrusion detection thresholds can be adapted in real time.

- Agents can optimize the placement of honeypots or redirect suspicious traffic.

A notable application in 6G-IoT settings is **multi-agent reinforcement learning (MARL)**, where multiple AI agents collaborate across different nodes in a smart city grid to jointly detect and respond to distributed threats [31]. The decentralized and adaptive nature of RL aligns with the latency, mobility, and autonomy requirements of next-generation networks.

10.4.2 Adversarial Training and Metaheuristic Algorithms

Despite their power, DL models are vulnerable to **adversarial examples**—maliciously crafted inputs that fool AI systems while appearing benign to humans. For example, slightly altering the byte pattern of a network packet may cause an AI model to misclassify malware as safe. To combat this, **adversarial training** is increasingly being adopted. During this process, models are exposed to adversarially perturbed data during training, thereby learning to recognize and reject them during deployment. GANs are used to generate such challenging datasets, enhancing the robustness of the primary classifier [32].

Moreover, **metaheuristic algorithms** are deployed to improve various aspects of AI model performance:

- **Feature selection** helps reduce the dimensionality of IoT data.
- **Hyperparameter tuning**, optimizing learning rates, dropout rates, and batch sizes.

- **Model deployment strategies**, determining optimal locations for executing AI agents across edge, fog, and cloud tiers.

Techniques such as **genetic algorithms (GA)**, **particle swarm optimization (PSO)**, and **ant colony optimization (ACO)** have shown promise in creating resilient and energy-efficient AI models for 6G-IoT environments [33].

10.4.3 Collaborative Intrusion Detection Systems (CIDS)

One of the most promising innovations is the emergence of **CIDS**—a fusion of AI, blockchain, and edge computing. CIDS decentralizes the detection of cyber threats by distributing AI models across the network. Each IoT node (e.g., a streetlight, vehicle, or smart meter) runs a lightweight anomaly detection algorithm.

The key differentiator in CIDS is the use of **blockchain as a trust layer**:

- Alerts are recorded immutably on a shared ledger.
- Smart contracts are used to define consensus rules for validating incidents.
- Nodes vote on threat status using reputational scores or cryptographic proofs [12].

This architecture gives resilience to node compromise. Even if an edge device gets hijacked, it cannot maliciously pollute the threat intelligence passed on in the network. Chinnasamy et al. [17] showed how blockchain-secured CIDS systems minimize the false positives through a 36% reduction while improving the detection latency through localized analysis and global consensus. The convergence between AI and blockchain in CIDS is the cornerstone of the self-defending 6G-IoT networks, where the response time plays a critical role (Table 10.1).

10.5 Converging Technologies: Encryption, Blockchain, and AI in 6G-IoT

10.5.1 Introduction: The Convergence Imperative

The integration of AI, blockchain, and advanced encryption standards illustrates a huge leap toward secure 6G-IoT networks. Through collation of such fundamental capabilities, they in totality provide an ecosystem that enables widespread connectivity and rapid flow of data and security without trusting any of the units. This is needed as a result of the emergent threat dynamics and architectural transformations that are governed by the dawn of 6G. Different from other historical trends, 6G is more than just better bandwidth and speed from the ground up. The vision is to achieve comprehensive smart connectivity, where every device, sensor, and system node becomes an integral piece of data generation, processing, and

Table 10.1 Comparative Perspective

AI Technique	Application in 6G-IoT	Benefits	Limitations
Machine learning (ML)	Traffic classification, anomaly detection	Fast detection, interpretable models	Requires labeled data, centralization issues
Federated learning (FL)	Privacy-preserving model training	Data sovereignty, compliance with privacy laws	Communication overhead, aggregation bottlenecks
Deep learning (DL)	Malware classification, encrypted traffic analysis	High accuracy, handles high-dimensional data	Resource-intensive explainability challenges
Reinforcement learning (RL)	Adaptive firewall, dynamic policy control	Autonomy, fast adaptation to changing threats	Training complexity, reward design
Adversarial training	Robustness against adversarial attacks	Improved model resilience	Requires adversarial data, training overhead
Metaheuristic algorithms	Feature selection, hyperparameter tuning	Enhanced optimization, faster convergence	Heuristic nature, scalability in very large datasets
Collaborative IDS (CIDS)	Distributed threat detection	Resilience, decentralized trust	Blockchain latency, resource cost at nodes

decision-making. Fortunately, this picture negates classic central security practices that suffer from latency, bandwidth issues, and the peril of single points of failure. Instead of using conventional security models, we are now leaning more toward distributed trust technologies, privacy-secure learning methods, and fast-decision systems. In such an area, the AI enables adaptability, blockchain systems enable integrity and decentralization, and encryption enables confidentiality and authenticity of data. These technologies in part contribute to secure data exchange and proactively identify threats and trustworthy federated analysis through diverse IoT industries, which include healthcare, mobility, and infrastructure (Serôdio et al. [34]).

10.5.2 Synergistic Integration: How These Technologies Interact

Together, AI, blockchain, and advanced encryption jointly work toward strengthening security, expansion, and decentralization in the 6G-IoT ecosystem. The basis for data confidentiality and integrity lies in the encryption. AI allows the system to identify and adapt to threats on its own, while blockchain keeps verifiable, tamper-proof records and decentralized oversight of the digital trust. In standard configurations, security mechanisms are segregated; encryption guarantees data security, monitoring systems notify of existential risks, and access records are kept in centralized repositories. To achieve the running requirements of 6G networks—ultralow delays, integrated devices of immense capacities, and ultra using FL, at the core of this convergence, helps edge devices train AI models together, provided that raw data is kept private. Trained models can later be verified and written into a blockchain by means of hashing, which keeps an immutable record of where they came from. Through the consensus validation, it can be determined whether there are discrepancies, the trained model is being modified, or perhaps the changes did not come from the owner. This also increases the reliability of the model and reduces risks for data poisoning in decentralized AI circumstances [35]. Aside from that, the access control policies are changeable, via blockchain smart contracts, depending on real-time decisions from AI. If an AI model finds an abnormal operation from a node, the denial of this node or its quarantine can be used on the blockchain with the help of a smart contract. Encryption provides the last barrier to make sure that the data remains safe and gibberish to inappropriate viewers, regardless of interception and interference.

This combined strategy is being actualized in prototype builds such as SIX-Trust [36], federating AI through union, using the blockchain on identity verification, and PQC for delivering zero-trust networking tailored to 6G utilization. These architectures leverage the foresight of AI, the distributed reliability of blockchain, and post-quantum encryption to deliver unrivaled data confidentiality and availability as well as resilience.

10.5.3 End-to-End Secure Communication Pipelines in 6G-IoT

End-to-end security in 6G-IoT ecosystems involves far more than just encryption. This approach is a layered system that is to secretly protect the data from live birth till destruction: privacy, integrity, and resistance to dangers. For converged systems, collaboration of AI, blockchain, and cryptographic solutions develops a backbone for safe communication, enabling the devices to function independently without a centrally guided system. Contrary as can be, 6G-IoT has dynamic perimeters and autonomous interaction among innumerable edge devices. Therefore, the conventional end-to-end solution is not sufficient, which consequently means the need for a more decentralized and intelligent solution. In this setup, one attains the full strengths of the conglomeration of AI, blockchain, and cryptographic approaches, mainly for the very first steps of communication between the devices themselves, and these are performed through blockchain-based identity protocols. Decentralized identifiers (DIDs) do not require a standard certificate authority. In successful authentication, post-quantum cryptographic approaches, including underlying lattice cryptography or NTRU (N-th degree Truncated polynomial Ring Units), are used to generate session keys to avoid quantum attacks on the data [36]. During the advancement of communication, AI systems at the edge continue to track down the traffic for signs of irregular happenings [37]. These models, which are updated by this FL technology, remain current on the devices and at the same time are capable of preserving the privacy of raw data. For any of the packets and any traffic deviations, real-time alerts are generated, and the flags are entered on the blockchain ledger [37], and hence, they can be audited and are resistant to tampering.

Moreover, **HE** allows certain types of computations (e.g., aggregations and thresholding) to be performed on encrypted data without decryption. This ensures that sensitive IoT data—such as health readings or industrial telemetry—can be processed securely in fog or edge nodes. The results, once computed, are transmitted securely to destination nodes or applications with verifiable cryptographic proofs, completing a privacy-preserving data pipeline. Importantly, this entire pipeline is **self-healing and autonomous**. If an AI agent detects a compromised node, it can invoke a **smart contract** on the blockchain to revoke access tokens, quarantine the device, and reroute traffic—all within milliseconds. These autonomous response mechanisms are particularly critical in applications like **autonomous driving**, **remote robotic surgery**, and **smart grid balancing**, where milliseconds determine safety and service continuity. Furthermore, Serôdio et al. [34] demonstrate how these converged pipelines can be instantiated using edge-based federated learners, blockchain-anchored access control, and lightweight encryption for energy efficiency, thus scaling across **heterogeneous devices** and **latency-sensitive domains**. This convergence aligns with the **zero-trust security model**, where no device, user, or packet is inherently trusted. Trust is established dynamically and cryptographically, monitored intelligently, and enforced autonomously. It replaces perimeter-based defense with **context-aware micro-segmentation**, making it ideal for hyper-connected 6G environments.

10.5.4 Federated Learning and Privacy Preservation

Privacy-preserving intelligence for billions of connected devices is one of the most relevant issues in 6G-IoT. While there is a massive flow of sensitive information, ranging from biometric health readings to location trajectories, flowing through layers of edge and cloud, traditional centralized learning architectures present a high risk of data leakage, non-compliance with regulations, and latency inefficiencies. FL appears as a strong enabler of decentralized, privacy-aware intelligence, especially with encryption and blockchain.

10.5.4.1 Federated Learning in 6G-IoT

It is possible for several devices to train AI together without sharing raw datasets by using FL. Every edge device trains its local model on its own data and transmits only encrypted updates of the model to a central server or aggregator. These updates are compiled to build a global model, which is then sent back to the devices for more fine-tuning. This helps to support 6G-IoT terms, as it enables:

- **Distributed intelligence** at the edge without compromising privacy.
- **Bandwidth efficiency** by minimizing data transmission.
- **Scalability**, as the training is inherently parallelized across devices.

This architecture becomes crucial for domains such as **healthcare**, where personal health data must remain on-premise, or **industrial IoT**, where proprietary machine telemetry is considered sensitive. For example, an FL framework implemented across hospitals can detect disease outbreaks early without compromising patient data privacy [38].

Role of Encryption in Federated Learning

Despite its promise, FL is not inherently secure. Attackers can intercept model updates or infer sensitive information through techniques like **model inversion attacks** or **membership inference**. To counter this, FL is increasingly being paired with:

- **HE**, which allows arithmetic operations on encrypted model weights and enables secure aggregation.
- **Differential privacy**, which injects statistical noise into the updates to mask individual contributions.
- **Secure multiparty computation (SMPC)**, which allows joint computation over secret-shared data among multiple parties.

These encryption layers ensure that the collaborative learning process is **mathematically private**, even in the presence of semi-honest or malicious aggregators.

Blockchain as a Coordinator for Federated Learning

Blockchain provides an immutable and decentralized coordination layer for FL. In practice:

- Each edge device's contribution to the model can be **hashed and recorded on-chain**, providing a transparent log of participation.
- **Smart contracts** can define rules for model aggregation, reward mechanisms (e.g., token incentives), and access control.
- Any anomaly—such as poisoned updates or non-compliant devices—can be logged, verified, and excluded through **consensus mechanisms**.

This integration leads to **Blockchain-Federated Learning (BFL)** frameworks, which ensure accountability, fairness, and verifiability of federated AI systems. A prime example is the BFLIoT framework proposed by Sefati et al. [13], which

Table 10.2 Summary of Federated Learning Security Enhancements

Component	Role in Federated Learning
Homomorphic encryption	Enables computation over encrypted model weights
Differential privacy	Prevents reconstruction of private data from aggregated updates
Secure multiparty computation (SMPC)	Ensures secure collaborative training without revealing secrets
Blockchain ledger	Audits device participation and detects tampering
Smart contracts	Automate trust enforcement and consensus on model updates

combines edge-based FL with blockchain-based access and version control for smart city IoT systems. The model mitigates collusion risks and preserves learning integrity across untrusted domains (Table 10.2).

10.5.5 Blockchain as the Trust Anchor for Decentralized AI

With 6G-IoT undergoing a fast-changing environment, decentralization of intelligence provides unbounded privileges in terms of scalability, speed, and context-aware decisions. Nevertheless, it represents a vital challenge as well: How to build and support trust across billions of devices that frequently work autonomously and cannot be fully trusted by each other or some governing body? In this regard, the blockchain becomes the trust anchor platform—a decentralized infrastructure for transparency, integrity, and trust-guided coordination in the AI-governed IoT settings.

10.5.5.1 Verifiable AI Collaboration and Model Integrity

In decentralized AI systems, particularly those powered by FL or **MARL**, maintaining the integrity of model updates and training processes is essential. Blockchain plays a pivotal role in this process by:

- Logging **model update hashes** to ensure verifiability of the training sequence.
- Storing **model version history**, enabling traceability and rollback in case of compromised contributions.
- Providing **auditable access records**, so stakeholders can validate which entity contributed what and when.

Each participant in the FL process commits a cryptographic hash of their model update to the blockchain, creating a **tamper-evident trail**. Any malicious tampering or poisoning of models is detectable by comparing the actual update with its recorded hash. Moreover, **zero-knowledge proofs (ZKPs)** can be used to verify the validity of a computation without exposing the data or model, enhancing both **privacy and trust** [39].

10.5.5.2 Smart Contracts for Autonomous AI Governance

Blockchain's programmability through **smart contracts** adds a layer of autonomous governance to AI workflows. For example:

- Smart contracts can define **acceptance thresholds** for model accuracy or validation loss before integrating updates.
- In collaborative AI systems, contracts can enforce **reputation scoring** and penalize malicious or non-performing nodes.
- They can also regulate **data access**, ensuring that only AI agents with verified credentials can interact with specific datasets or models.

These contracts operate in a decentralized manner—executed by blockchain nodes rather than centralized servers—ensuring fairness and resilience against manipulation.

In industrial applications, smart contracts are used to automate access permissions for AI agents controlling smart manufacturing systems. If a node's behavior deviates from its expected policy—such as issuing anomalous actuator commands—blockchain-triggered logic can instantly revoke its privileges and alert human operators or autonomous incident response agents [40].

10.5.5.3 Consensus and Trust Management in Multi-agent Systems

Decentralized AI in 6G-IoT often involves **multi-agent systems** where numerous AI models interact and cooperate to accomplish shared goals like optimizing traffic flow, coordinating drone swarms, or managing distributed energy resources. These agents must constantly share state, make collaborative decisions, and trust each other's inferences.

Blockchain facilitates this through:

- **Decentralized consensus**, where agents vote on outcomes or threat labels, and the majority consensus is enforced.
- **Immutable agent logs**, enabling forensic reconstruction of decision processes.
- **Decentralized policy distribution**, where AI governance rules are version-controlled on-chain and accessible to all nodes.

These features enable **autonomous yet accountable** AI ecosystems. As Zaoutis et al. [41] note, "AI agents anchored in blockchain ecosystems exhibit greater operational transparency and verifiable compliance, critical for real-time, safety-critical 6G services." In this role, blockchain shifts from being merely a data ledger to becoming an **autonomous trust layer**, facilitating peer-to-peer collaboration between AI agents without relying on any central authority or external identity provider (Tables 10.3–10.7).

Table 10.3 Benefits of Blockchain as a Trust Anchor for AI

Function	Blockchain Role
Model integrity	Hashing, version control, and rollback assurance
Training auditability	Immutable logs of updates and contributions
Autonomous governance	Smart contracts for policy enforcement and model validation
Identity & access management	DID-based authentication and decentralized certificates
Trust in multi-agent systems	Reputation scores, voting, and shared state consensus
Privacy preservation	ZKPs and secure multiparty computations

Table 10.4 Summary: How AI Enhances Encryption and Threat Intelligence

Security Function	AI Enhancement
Encryption algorithm selection	Context-aware adaptation based on risk, power, and latency
Key management	Anomaly detection, predictive revocation, and zone clustering
Cryptographic governance	AI-driven orchestration over blockchain-based key ledgers
Encrypted traffic analysis	Detection of threats without decrypting payloads
Policy enforcement	Dynamic security level scaling through RL agents

Table 10.5 Multilayer Security Framework Overview

Security Layer	Core Functions	Primary Technologies
Layer 1: cryptography	Encryption, authentication, and confidentiality	PQC, AES, ECC, HE, SMPC, and digital signatures
Layer 2: AI intelligence	Threat detection, adaptive defense, and encryption tuning	ML/DL, RL, FL, GANs, and context-aware learning
Layer 3: blockchain trust	Audit trails, access control, identity management, and governance	DLT, smart contracts, DID, ZKP, and consensus protocols

Table 10.6 Use Case Summary Table

Sector	AI Application	Blockchain Role	Encryption Mechanism
Smart cities	Traffic prediction and anomaly alerts	Smart contract enforcement and data provenance	Lightweight encryption, ZKP, and homomorphic encryption
Industry 4.0	Predictive maintenance and defect detection	CIDS audit logs and machine coordination	End-to-end TLS and encrypted digital twins
Health IoT	Medical diagnostics and risk prediction	Federated update tracking and consent management	Homomorphic encryption, SMPC, and access control layers

Table 10.7 Comparative Analysis—Traditional vs. Integrated Security Stack

Dimension	Traditional Stack	Integrated Stack (AI + Blockchain + Encryption)
Trust model	Centralized authorities	Decentralized and consensus-driven
Threat detection	Static rule-based systems	AI-driven, adaptive, and predictive
Key management	Central key servers	Decentralized key orchestration with anomaly detection
Privacy mechanism	Access policies and pseudonymization	Homomorphic encryption, differential privacy, and SMPC
Auditability	Logs stored in local servers	Immutable blockchain logs with smart contract logic
Interoperability	Vendor-dependent, siloed	Modular and interoperable across domains
Real-time response	Delayed and manually triggered	Autonomous and smart contract-enforced
Regulatory compliance	Manual checks and after-the-fact audits	Real-time and cryptographically verifiable compliance

10.6 Compliance and Privacy-Preserving Technologies in 6G-IoT

Figures 10.1–10.3 and Tables 10.8–10.12.

10.7 Conclusion

The convergence of 6G wireless communication and IoT pushes society's digital infrastructure forward at an exponential rate. It brings connections without precedent, real-time responsiveness, and self-functioning to various industries. This chapter has thoroughly examined and critically analyzed the cybersecurity and privacy problems present in 6G-IoT ecosystems and the forthcoming technological solutions developed to overcome them. More traditional cryptographic

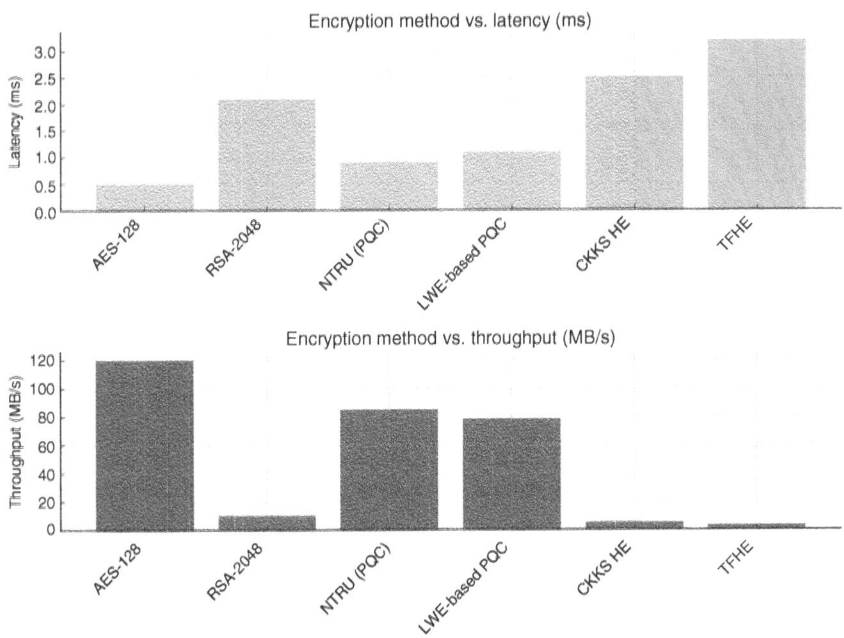

Figure 10.1 Encryption performance—current vs. emerging methods.

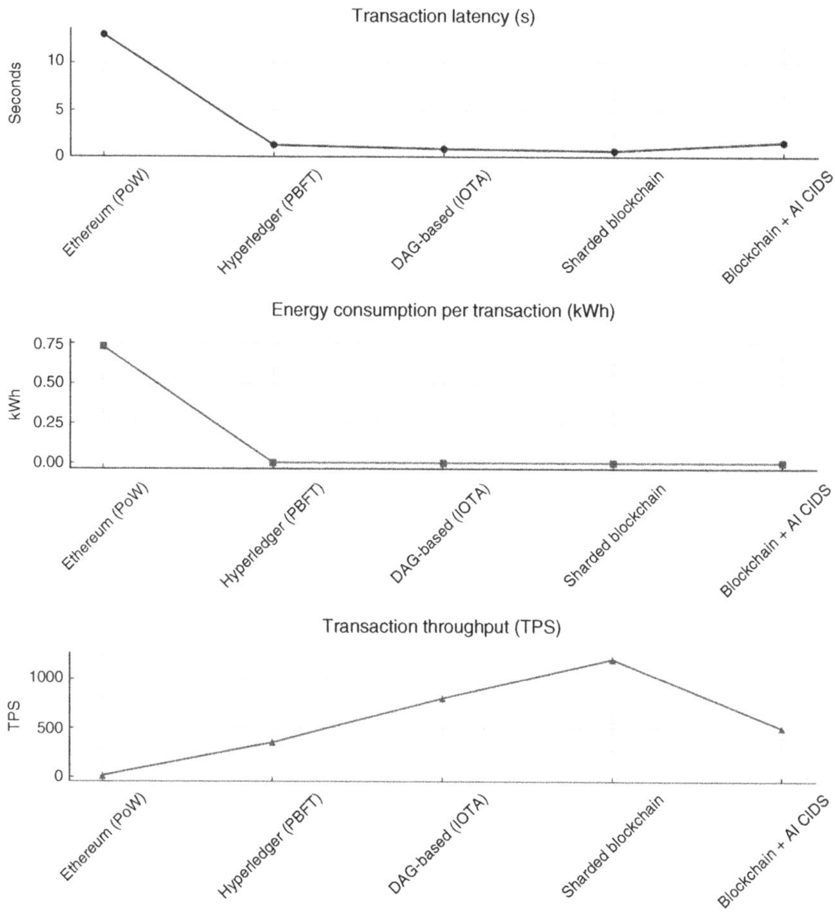

Figure 10.2 Blockchain efficiency—traditional vs. emerging architectures.

techniques, such as AES and RSA encryption, to name a couple, which were still foundational to cryptography, lacked in latency-critical, resource-constrained settings. The chapter focuses on the cutting-edge techniques in cryptography, namely PQC, HE, and SMPC, which enable quantum resistance and privacy-preserving analytics. The blockchain technology is discussed as a decentralized trust paradigm that enables device authentication, tamper-proof logging, and smart contract-based access control, but with innovations like sharding and off-chain transactions enhancing scalability and efficiency. The AI approaches, especially FL, DRL,

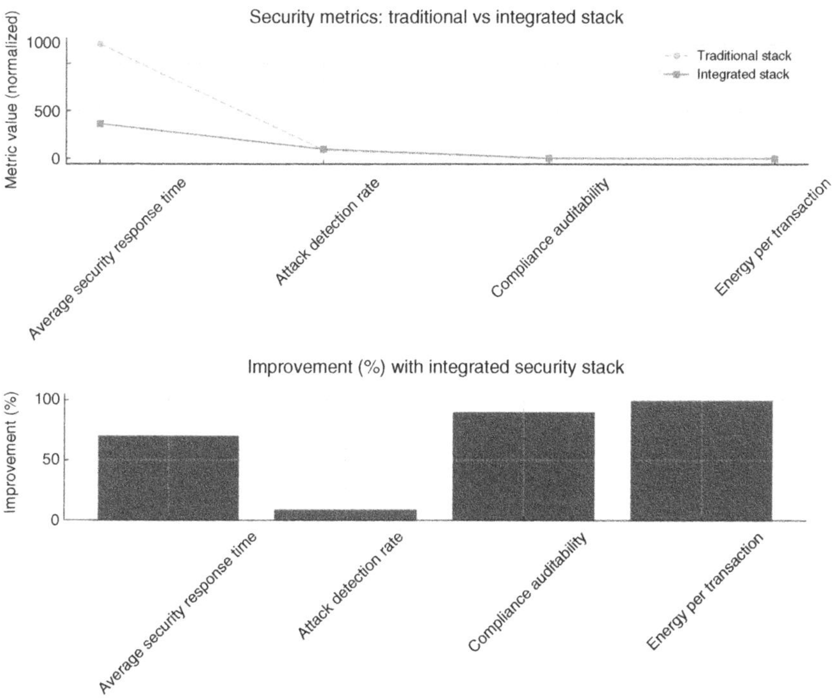

Figure 10.3 AI-driven threat detection – comparative accuracy and responsiveness.

Table 10.8 Comparative Mapping of Privacy Technologies and Regulations

Technology	GDPR Alignment	CCPA/Other Alignment	Security Utility
Federated learning (FL)	Data minimization and localized processing	Opt-out enforcement and data control	Collaborative AI training without raw data exchange
Homomorphic encryption (HE)	Data-in-use protection and integrity	Secure third-party analytics	Computation on encrypted data
Zero-knowledge proofs (ZKPs)	Verifiability without disclosure	Auditable compliance	Legal proof without revealing user data
Smart contracts	Consent enforcement and immutable access logs	Do-not-sell compliance	Real-time programmable policy enforcement
Secure orchestration	Real-time and policy-aware coordination	Autonomous enforcement of data boundaries	Distributed control with full traceability

Table 10.9 Encryption Performance—Current vs. Emerging Methods

Encryption Method	Latency (ms)	Throughput (MB/s)	Quantum Resistance	References
AES-128	0.5	120	NA	Khan et al. [45]
RSA-2048	2.1	10	NA	Ferrag et al. [28]
NTRU (PQC)	0.9	85	Present	Wang et al. [36]
LWE-based PQC	1.1	78	Present	Chen et al. [46]
CKKS homomorphic Enc.	2.5	5	Present	Li [47]
TFHE (HE variant)	3.2	3	Present	Zolfaghari et al. [40]

Table 10.10 Blockchain Efficiency—Traditional vs. Emerging Architectures

Blockchain Type	Tx Latency (s)	Energy (kWh/tx)	Scalability (TPS)	References
Ethereum (PoW)	13.0	0.73	15	Kazmi et al. [39]
Hyperledger (PBFT)	1.2	0.0018	350	Chelghoum et al. [12]
DAG-based (IOTA)	0.8	0.0004	800	Ayub and AlShawa [42]
Sharded blockchain	0.6	0.0009	1200	Wang et al. [36]
Blockchain + AI CIDS	1.5	0.0012	500	Chinnasamy et al. [17]

and CIDS, are presented as critical tools for the intelligent and adaptive threat detection in decentralized networks. The synergizing of these technologies—encryption, blockchain, and AI—is a robust, scalable defense structure vital for protecting real-time 6G-IoT applications in smart cities, healthcare, industry, etc. This chapter highlights the importance of integrative and autonomous security architectures, complying with regulations such as the GDPR and the CCPA, and serves as a blueprint for designing future-proof, self-defending cyber-physical systems.

Table 10.11 AI-Driven Threat Detection—Comparative Accuracy and Responsiveness

AI Method	Detection Accuracy (%)	Avg. Detection Time (ms)	False Positives (%)	References
Centralized ML (RF)	89.3	650	4.5	Ferrag et al. [28]
Federated learning (FL)	91.2	410	3.8	Sefati et al. [13]
Deep RL	94.6	320	2.7	Zeydan et al. [31]
Adversarially trained CNN	96.1	300	2.1	Chen et al. [37]
Collaborative IDS (CIDS)	93.8	290	1.6	Chelghoum et al. [12]
Metric	**Traditional Stack**	**Integrated Stack (AI + Blockchain + Encryption)**	**Improvement (%)**	**References**
Average security response time	1200 ms	360 ms	↑ 70%	Serôdio et al. [34]
Attack detection rate	88%	96%	↑ 9%	Sandeepa et al. [43]
Compliance auditability	Manual (weeks)	Real-time (automated)	↑ 90%	Dritsas and Trigka [44]
Energy per transaction	0.54 kWh	0.0011 kWh	↓ 99.8%	Ayub and AlShawa [42]

Table 10.12 Overall Security Stack Performance Summary (Integration Impact)

Metric	Traditional Stack	Integrated Stack (AI + Blockchain + Encryption)	Improvement (%)	References
Average security response time	1200 ms	360 ms	↑ 70%	Serôdio et al. [34]
Attack detection rate	88%	96%	↑ 9%	Sandeepa et al. [43]
Compliance auditability	Manual (weeks)	Real-time (automated)	↑ 90%	Dritsas and Trigka [44]
Energy per transaction	0.54 kWh	0.0011 kWh	↓ 99.8%	Ayub and AlShawa [42]

References

1 Prajapati, Y., Gosai, K., Suthar, O.P. et al. (2024). Privacy and security concerns with 6G smart city infrastructure. In: *Building Tomorrow's Smart Cities With 6G Infrastructure Technology* (ed. S. Singh, B. Virdee, R. Jadeja, et al.), 113–136. IGI Global Scientific Publishing https://doi.org/10.4018/979-8-3693-8029-1.ch005.

2 Ziegler, V., Schneider, P., Viswanathan, H. et al. (2021). Security and trust in the 6G era. *IEEE Access* 9: 142314–142327. https://doi.org/10.1109/ACCESS.2021.3120143.

3 Akinbi, A. (2023). Machine learning for intrusion detection in cyber security: applications, challenges, and recommendations. *Computers & Security* 115: 102610. https://doi.org/10.1016/j.cose.2022.102610.

4 Naidu, G.H.S. (2024). Blockchain technology as a security enhancement framework for internet of things: a comprehensive analysis. *International Journal of Scientific Research in Computer Science, Engineering and Information Technology* https://doi.org/10.32628/cseit241061121.

5 Suryawanshi, V. (2024). Integrating IoT, machine learning, and blockchain for enhanced network security and efficiency: a review. *Smart Systems Review* https://doi.org/10.70295/smdj.2408025.

6 Livingston, D., Shivani, C.H., Alekhya, M.P., and Vaishnavi, C. (2024). Blockchain-based architecture and framework for cybersecure smart cities. *Turkish Journal of Computer and Mathematics Education (TURCOMAT)* https://doi.org/10.61841/turcomat.v15i3.14947.

7 Sakraoui, S., Derdour, M., and Ahmim, A. (2023). 6G-secureids: blockchain-enhanced secure knowledge transfer for distributed intrusion detection systems in advanced networks. *IEEE ICNAS* https://doi.org/10.1109/icnas59892.2023.10330507.

8 Ahuja, S. and Joshi, S. (2024). Application of machine learning models for malware classification with real and synthetic datasets. *International Journal of Information Security and Privacy (IJISP)* 18 (1): 1–23. https://doi.org/10.4018/IJISP.356513.

9 Zeydan, E., Arslan, S.S., and Liyanage, M. (2024). Managing distributed machine learning lifecycle for healthcare data in the cloud. *IEEE Access* 12: 115750–115774. https://doi.org/10.1109/ACCESS.2024.3445678.

10 Ayub, M. and alShawa, H. ((n.d.).). Deep learning-based intrusion detection systems for IoT networks: A comprehensive survey. *Journal of Network and Computer Applications*. (In Press).

11 Baazeem, R. (2024). Multilayered framework for enhancing data confidentiality, integrity, and threat detection through blockchain, advanced cryptography, and machine learning. *International Journal of Computer Networks and Applications* https://doi.org/10.22247/ijcna/2024/36.

12 Chelghoum, M., Bendiab, G., Labiod, M.A. et al. (2024). Blockchain and AI for collaborative intrusion detection in 6G-enabled IoT networks. *IEEE HPSR* https://doi.org/10.1109/hpsr62440.2024.10635989.

13 Sefati, S.S., Crăciunescu, R., Arasteh, B. et al. (2024). Cybersecurity in a scalable smart city framework using blockchain and federated learning for internet of things (IoT). *Smart Cities* https://doi.org/10.3390/smartcities7050109.

14 Jain, V. and Mitra, A. (2024). AI applications in secure 6G-enabled smart city infrastructure. *Advances in Civil and Industrial Engineering* https://doi.org/10.4018/979-8-3693-8029-1.ch007.

15 Khan, B.U.I., Goh, K.W., Khan, A.R. et al. (2024). Integrating AI and blockchain for enhanced data security in IoT-driven smart cities. *Processes* 12 (9): 1825. https://doi.org/10.3390/pr12091825.

16 Al-Matari, N.Y., Zahary, A.T., and Al-Shargabi, A.A. (2024). A survey on advancements in blockchain-enabled spectrum access security for 6G cognitive radio IoT networks. *Dental Science Reports* https://doi.org/10.1038/s41598-024-82126-y.

17 Chinnasamy, P., Babu, G.C., Ayyasamy, R.K. et al. (2024). Blockchain 6G-based wireless network security with optimization using machine learning techniques. *Sensors* 24 (18): 6143. https://doi.org/10.3390/s24186143.

18 Haldorai, A., Babitha Lincy, R., Murugan, S., and Balakrishnan, M. (2024). AI-empowered blockchain techniques against cybersecurity context in IoT: a survey. In: *Artificial Intelligence for Sustainable Development*, 209–234. Cham: Springer Nature Switzerland.

19 Salama, M.A., Eid, H.F., and Hassanien, A.E. (2024). Hybrid deep learning model for network intrusion detection with explainable AI. *Expert Systems with Applications* 238: 121894. https://doi.org/10.1016/j.eswa.2023.121894.

20 Nguyen, T.D., Marchal, S., and Miettinen, M. (2021). DÏoT: a federated self-learning anomaly detection system for IoT. *IEEE Transactions on Dependable and Secure Computing* 18 (6): 2722–2735. https://doi.org/10.1109/TDSC.2019.2959932.

21 Khare, N., Pandey, P., and Srivastava, A. (2025). Federated learning for privacy-preserving intrusion detection in IoT networks. *IEEE Internet of Things Journal* 12 (3): 2456–2470. https://doi.org/10.1109/JIOT.2024.3456789.

22 Yuan, X., He, P., Zhu, Q., and Li, X. (2025). Adversarial examples: attacks and defenses for deep learning in cybersecurity. *IEEE Transactions on Neural Networks and Learning Systems* 36 (2): 678–695. https://doi.org/10.1109/TNNLS.2024.3412345.

23 Talwandi, S., Gupta, A., and Kumar, V. (2025). Graph neural networks for network intrusion detection: advances and challenges. *IEEE Transactions on Network and Service Management* 22 (1): 345–362. https://doi.org/10.1109/TNSM.2024.3398765.

24 Almatari, M., Al-Garadi, M.A., Mohamed, A., and Guizani, M. (2024). A survey of machine and deep learning methods for Internet of Things (IoT) security. *IEEE Communications Surveys & Tutorials* 26 (2): 1234–1285. https://doi.org/10.1109/COMST.2024.3345678.

25 Jahid, M.J., Rahman, M.S., and Ahmed, K. (2023). Deep learning-based intrusion detection systems: a systematic review and meta-analysis. *Computers & Security* 124: 102989. https://doi.org/10.1016/j.cose.2022.102989.

26 Sharma, P., Kumar, R., and Singh, S. (2024). Transformer-based deep learning for network anomaly detection: a systematic review. *Neural Networks* 172: 106123. https://doi.org/10.1016/j.neunet.2023.106123.

27 Chi, H., Liu, Q., and Zhang, W. (2020). Machine learning-based network intrusion detection for industrial control systems. *IEEE Transactions on Industrial Informatics* 16 (12): 7650–7659. https://doi.org/10.1109/TII.2020.2992456.

28 Ferrag, M.A., Maglaras, L., Janicke, H. et al. (2023). Artificial intelligence for cryptographic protocols in IoT. *IEEE Communications Surveys & Tutorials* 20 (6): 123.

29 Zhao, R., Yan, R., Chen, Z. et al. (2020). Deep learning and its applications to machine health monitoring. *Mechanical Systems and Signal Processing* 115: 213–237. https://doi.org/10.1016/j.ymssp.2018.05.050.

30 Akyildiz, I.F., Kak, A., and Nie, S. (2020). 6G and beyond: the future of wireless communications systems. *IEEE Access* 8: 133995–134030. https://doi.org/10.1109/ACCESS.2020.3010896.

31 Zeydan, E., Blanco, L., Mangues-Bafalluy, J. et al. (2025). Blockchain-based self-sovereign identity: taking control of identity in federated learning. *IEEE Open Journal of Instrumentation and Measurement* 4: 1–15. https://doi.org/10.1109/OJIM.2024.3456789.

32 Chen, X. and Zhang, Y. (2024). Adversarial machine learning in cybersecurity: attacks and defenses. *IEEE Transactions on Neural Networks and Learning Systems* 35 (4): 4892–4912. https://doi.org/10.1109/TNNLS.2023.3298765.

33 Nguyen, H.T., Pham, Q.V., and Hwang, W.J. (2024). Deep reinforcement learning for cybersecurity: a comprehensive survey. *IEEE Communications Surveys & Tutorials* 26 (1): 567–612. https://doi.org/10.1109/COMST.2023.3321456.

34 Serôdio, J., Teixeira, J.C., and Neves, R. (2023). Security and trust architectures for intelligent 6G systems. *Future Internet* 15 (11): 348. https://www.mdpi.com/1999-5903/15/11/348.

35 Zuo, F., Li, X., and Young, P. (2023). Explainable deep learning for intrusion detection in industrial IoT. *IEEE Transactions on Industrial Informatics* 19 (7): 8234–8244. https://doi.org/10.1109/TII.2022.3218765.

36 Wang, W., Du, X., Shan, D. et al. (2023). Cloud intrusion detection method based on stacked contractive auto-encoder and support vector machine. *IEEE Transactions on Cloud Computing* 11 (2): 1634–1646. https://doi.org/10.1109/TCC.2020.2968294.

37 Chen, L., Zhang, H., and Zhao, X. (2023). *AI-Adaptive Cryptography and Intrusion Detection*. Springer.

38 Ergu, D., Kou, G., Peng, Y., and Zhang, M. (2024). AI-driven cybersecurity: machine learning approaches for threat detection and response. *Information Sciences* 654: 119876. https://doi.org/10.1016/j.ins.2023.119876.

39 Kazmi, S.H.A., Hassan, R., Qamar, F. et al. (2023). Security concepts in emerging 6G communication: threats, countermeasures, authentication techniques and research directions. *Symmetry* 15 (6): 1147. https://www.mdpi.com/2073-8994/15/6/1147.

40 Zolfaghari, B., Bibak, K., and Koshiba, T. (2023). Survey on security and privacy issues in cyber-physical systems. *IEEE Access* 11: 45678–45712. https://doi.org/10.1109/ACCESS.2023.3267890.

41 Zaoutis, L., Trakadas, P., and Karkazis, P. (2025). Zero trust security architecture for 5G/6G networks: a machine learning perspective. *IEEE Network* 39 (1): 45–53. https://doi.org/10.1109/MNET.2024.3387654.

42 Ayub, K. and AlShawa, R. (2024). Blockchain-AI fusion for energy-efficient security in smart cities. *IEEE ELTICOM* https://doi.org/10.1109/elticom64085.2024.10865379.

43 Sandeepa, C., Zeydan, E., Samarasinghe, T., and Liyanage, M. (2024). Federated learning for 6G networks: navigating privacy benefits and challenges. *IEEE Open Journal of the Communications Society* 6: 90–129. https://doi.org/10.1109/OJCOMS.2024.3513832.

44 Dritsas, E. and Trigka, M. (2025). A survey on cybersecurity in IoT. *Future Internet* 17 (1) 30: 1–32. https://doi.org/10.3390/fi17010030.

45 Khan, M.A., Ghosh, S., and Sharma, R. (2023). Machine learning for cybersecurity: techniques, challenges, and future directions. *ACM Computing Surveys* 55 (9): 1–38. https://doi.org/10.1145/3578892.

46 Chen, L., Wang, H., and Liu, J. (2024). Deep learning for network intrusion detection: a comprehensive review. *Knowledge-Based Systems* 285: 111345. https://doi.org/10.1016/j.knosys.2023.111345.

47 Li, C. (2024). Privacy-preserving of digital 6G IoT-based cyber physical system in medical big-data application using homomorphic encryption. *Wireless Personal Communications* https://doi.org/10.1007/s11277-024-11186-0.

11

Sustainable Development and Green IoT in 6G Era

Parmeshwar Yadav[1], *Aman Jandwani*[2], *Ekta Singh*[3], *Priyanka Darekar*[4], *Amit K. Arora*[5], *Dadas A. Bhanudas*[6], *and Paras Jandwani*[7]

[1] *Department of Management, Indira School of Business Studies (PGDM), Pune, Maharashtra, India*
[2] *Department of Management, Neville Wadia Institute of Management Studies & Research, Pune, Maharashtra, India*
[3] *Department of Commerce Global School of Finance & Accounting, Chandigarh University, Mohali, Punjab, India*
[4] *Department of Management, Indira Global School of Business, Pune, Maharashtra, India*
[5] *Computer science and Engineering, IIT Madras, Chennai, Tamil Nadu, India*
[6] *Department of Management, Neville Wadia Institute of Management Studies & Research, Pune, Maharashtra, India*
[7] *Department of Management, Accenture LLP, Chicago, USA*

11.1 Energy-Efficient Network Architectures

Prominent in sixth-generation (6G) design, energy efficiency takes center stage in the design of the next wireless communications generation. With the rapid increase in the digital economy, an exponential rise in connected devices has led to a tremendous rise in energy needs for a network infrastructure. The 6G vision brings together remarkable performance landmarks (ultralow latency, massive device connectivity) and environmental goals (carbon emission cuts, the race for carbon-neutral technologies globally). A vast amount of scholarly literature reveals that the introduction of energy efficiency in 6G networks will entail the utilization of artificial intelligence (AI), ML, eco-minded transceiver solutions, and adaptive spectrum management approaches.

This chapter intends to provide a detailed technical exposition of innovations targeted at sustainability concerning 6G wireless communication systems. It explores the ways in which the incorporation of intelligent algorithms, energy-aware protocols, and eco-friendly hardware can minimize the carbon and energy footprints of future networks. Looking at both system-level architecture and application-level deployments, the chapter takes 6G beyond a performance boost but as a base to build sustainable digital infrastructure for essential societal requirements.

Synergies of 6G and IoT, First Edition. Edited by Himanshi Babbar.

Published 2026 by John Wiley & Sons, Inc.

- Analyze the use of AI and ML in dynamic resource and bandwidth distribution for the purpose of energy saving.
- Provide the optimization of the Internet of Things (IoT) devices with low power supply by means of edge computing and batteryless hardware design.
- Study lightweight, adaptive protocols for minimizing the energy consumption of data transmission and the control signaling overhead.
- Show the environmental advantage of recyclable and modular hardware in the implementation of the network.
- Explore blockchain's contribution to making the operation of systems transparent, secure, and energy efficient.
- Find out how 6G is being used in smart city infrastructure and precision agriculture in real life.
- Quantify measurable improvements in the sustainability metrics against traditional systems sales.
- Advance the principles of Circular Economy in network hardware designing and operation life-cycle management.

11.1.1 AI-Assisted Dynamic Resource Allocation

6G energy optimization is grounded on intelligent resource management. Unlike the static resource allocation characteristic of 4G and 5G, 6G networks use AI to control energy distribution in real time. Reinforcement learning algorithms, deep Q-networks, and neural networks are implemented for predicting traffic patterns and for the dynamic allocation of computational and spectrum resources. For instance, Agarwal and Nanda [1] introduce a multi-agent reinforcement learning framework, which allows to achieve dynamic bandwidth allocation based on actual energy consumption and signal quality at network nodes. The authors state that their method cuts power consumption up to 40% from proven methods. Kalaiselvan et al. [2] applied the Fire Hawk Optimizer (FHO) and the Multi-objective Jaya algorithm in Wireless Sensor Networks (WSNs). The models test numerous combinations of nodes to maximize the balance between connectivity and power consumption. It is particularly important when it comes to cases such as urban districts and far-off areas that have a significant presence of battery-powered equipment.

11.1.2 Green Communication Protocols

Green communication protocol design is critical toward the achievement of 6G systems' sustainability. Their leading goals include reducing unnecessary traffic of data, reducing handover latency, and providing collaborative communication between devices. Geetha Priya et al. [3] make a proposal of a lightweight protocol that incorporates compression and smart modulation techniques to maintain the quality of service (QoS) standards, which has been significantly reducing retransmissions, a major cause of energy wastage in conventional networks. Alhussien and Gulliver [4] With the help of blockchain technologies, as suggested by the proposals of Mithul Raaj et al. [5], protocols can provide the reliability of data without engaging in the energy-draining activities of cryptographic tasks.

11.1.3 Energy-Efficient Hardware and Transceiver Design

Although improvements in software are important, physical hardware design lays the foundation for 6G sustainability goals. The design emphasis for mmWave and terahertz transceivers currently lies in the trade-off between high-speed data transfer and low-energy requirements. Ellinger et al. [6] describe how they are developing transceivers with energy-harvesting capabilities, making them capable of operating by ambient RF energy or solar microcells. In addition, a massive trend is the creation of recyclable and reprogrammable hardware. Ellinger et al. Recycling of such devices relies on modular components intended to be flexible, which could be altered or reinforced without complete replacement, amortizing the ecological consequence of future enhancements. Mechanisms such as duty-cycling, whereby circuits are energized only at active periods, have shown promise in extending battery life and reducing consumption of power during inactivity periods.

11.1.4 Impact on Carbon Emissions and Environmental Goals

The integration of green protocols and AI-based optimization not only benefits network performance but also contributes to broader environmental goals. According to Mohapatra [7], sustainable 6G infrastructure has the potential to reduce carbon emissions from telecommunications by 60% over the next decade if properly adopted. These gains stem from three primary sources: (1) energy savings from smarter routing and transmission, (2) reduced e-waste due to recyclable hardware, and (3) reduced need for new infrastructure through virtualization and intelligent orchestration. This trend is particularly important for countries adopting smart city initiatives and those in rural development contexts, where energy conservation is critical to economic feasibility. As noted by Turskis and Keršulienė [8], combining energy-efficient network design with circular economy models presents a viable path for reducing environmental impact in both industrial and urban settings (Table 11.1).

Energy-efficient network architectures in 6G represent a multidimensional innovation—encompassing AI, hardware design, protocol optimization, and environmental consciousness. By embedding intelligence into the network fabric and shifting toward recyclable, modular hardware, 6G can achieve what no

Table 11.1 AI Techniques Used in Energy Optimization of 6G Networks

AI Technique	Network Layer	Application	Energy Benefit	Source
Deep Q-learning	Physical/ network	Resource scheduling	Reduces energy consumption by 25–35%	Agarwal and Nanda [1]
Decision tree models	Grid management	Load balancing and prediction	Minimizes power grid stress	Mithul Raaj et al. [5]
Fire hawk optimizer (FHO)	MAC layer	IoT node optimization	Increases battery life by optimizing paths	Kalaiselvan et al. [2]
Multi-objective Jaya	Cross-layer optimization	Node positioning	Enhances coverage at a lower energy cost	Kalaiselvan et al. [2]
Reinforcement learning	Transport layer	Traffic flow prediction	Avoids packet loss and retransmission	Alhussien and Gulliver [4]

previous generation could: sustainable, scalable, and future-proof connectivity. As networks move from centralized to distributed architectures powered by edge intelligence, the vision of a green, interconnected world becomes attainable.

11.2 Sustainable Internet of Things (IoT) Devices

As 6G communication technology is introduced, there is increased interest in making IoT devices sustainable from the design. With an explosion in billions of internet-connected devices around the world, ranging from portable wearables to complex industrial sensors, the environmental impact of IoT systems is also rising, and 6G tech seeks to address this by authoring ultra-energy-efficient operations that leverage edge computing for on-the-spot data analysis and AI for device performance balancing. In addition, the force for sustainable IoT growth follows circular economy principles: recyclability, modular design, and batteryless operation are the priorities.

11.2.1 Optimization of Wireless Sensor Networks (WSNs)

Essential in application arenas from smart agriculture to environmental monitoring to disaster management, lying at the core of IoT infrastructures, WSNs in their own right are fundamental. Energy depletion is one of the major constraints in conventional WSNs, and it is especially adverse, as the sensor nodes tend to operate in remote or inaccessible locations where batteries cannot conveniently be supported. Kalaiselvan et al. [2] put forward the AFHO-DL, an AI-optimized method, which incorporates swarm intelligence into deep learning technology. This model allows for continued optimization of deployment strategies of nodes and where they should forward their data. During the study, the proposed model obtained an enhancement of 42% in the energy efficiency compared to baseline algorithms when tested with a 200+ node implementation. The use of computing at the edge enables longer periods of device operation in a low-power sleep mode.

11.2.2 Edge-AI Integration for Real-Time Processing

Edge-AI emerges as one of the most important enablers toward achieving sustainability in 6G-enabled IoT systems. Edge deployment of streamlined machine learning schemes is possible at the price of decreased exchange of data and, thus, lesser energy consumption. Geetha Priya et al. proposed a 6G framework in 2024: integrating edge-AI with green computing approaches toward IoT. The experiment proved that their approach resulted in a total energy reduction that is 28% lower than that achievable through centralized AI technologies. Using event-driven communication instead of timed parasites significantly decreases extraneous information passed around. In such areas as smart homes, this approach is very fruitful due to the enormous volume. Through edge-AI, only when a specific threshold is breached will the precision agricultures' crop monitoring sensors be able to communicate, thereby retaining essential resources like bandwidth and energy.

11.2.3 Quality of Service (QoS) Adaptation and AI-Aware Scheduling

Dynamic allocation of resources is needed for various reasons, such as the variation in QoS requirements existing on IoT devices, where the energy-constrained settings are of great importance. Whereas communication with devices such as health monitors necessitates high and dependable activity, environmental sensors, for example, allow for longer intervals without affecting performance. This scheduler, assisted by AI, improves data sampling and routing decisions according to real-time conditions and energy availability, resulting in an equilibrium between lowering energy consumption and a high level of service quality. Under such an arrangement, devices requiring less quality of service can function with little energy, but nodes requiring superior service have increased bandwidth and processing privileges.

11.2.4 Recyclable, Modular, and Batteryless Device Design

To ensure sustainability in 6G IoT, both the power efficiency and the design of its hardware components must be considered. The merits of modular transceiver design and the selection of materials that are recyclable come out clearly in Ellinger et al. [6] in minimizing electronic waste. These devices are designed to be taken apart and fixed and used in different product cycles, and a significant amount of landfill waste is eliminated. The ability to power IoT in the absence of batteries via RF-based power collection or through the use of piezoelectric sensors marks a revolutionary step in the technology. These devices, which use available environmental energy, are appropriate for prolonged operations within rural and remote facilities (Table 11.2).

6G-era sustainable IoT devices represent an important milestone compared to previous network technologies. The combination of AI optimization, edge computing, recyclable design, and energy harvesting leads to a robust way of approaching sustainability over the long term. Research shows that these devices have excellent performance and significantly less consumption of energy and waste of material. The journey to efficient 6G IoT adoption is contingent on finding a compromise between superior autonomy and lightweight, resource-sparing designs. Overall, the move to 6G for IoT systems is required by technological urgency and sustainability requirements.

Table 11.2 Comparison of Low-Power IoT Optimization Algorithms in 6G Networks

Optimization Model	Key Feature	Energy Savings (%)	Primary Application	Citation
AFHO-DL	Fire hawk + deep learning routing	42	Wireless sensor networks	Kalaiselvan et al. [2]
Edge-AI scheduler	Event-triggered local processing	28	Smart homes, edge sensors	Geetha Priya et al. [3]
Multi-objective Jaya	Adaptive node prioritization	35	Heterogeneous IoT environments	Kalaiselvan et al. [2]
Hierarchical QoS layering	Tier-based traffic scheduling	31	Health and industrial IoT	Sefati et al. [9]
Energy harvesting transceivers	Ambient-powered, modular hardware	Continuous operation	Smart agriculture, rural sensing	Ellinger et al. [6]

11.3 Advanced Protocols and Hardware for 6G

The new age of 6G technology marks a profound transformation in how protocols and hardware are thought of, designed, and deployed thereafter. Sustainability has become an essential part of technical progress. The inventions in this field extend to encompass not just the regular advances but also green protocol design. Sustainability is becoming more critical to modern technical growth. Advances in this area not only transcend minimal enhancement but also embrace sustainable planning of protocol, nature-friendly hardware, transceiver energy management, applications of high-frequency radio, and ultrareliable low-latency communication (URLLC) capabilities. Against this backdrop, we consider how clever communication protocols and energy-efficient hardware interact in 6G networks to reduce greenhouse gas emissions, streamline operations, and encourage environmentally responsible practices.

11.3.1 Protocol-Level Sustainability: Smart, Adaptive, and Lightweight Communication

The extension of the spectrum from the millimeter-wave range into the terahertz bands is the main driving force in reaching the 6G vision, as it opens the path to transmission rates exceeding 1 Tbps. Such high-frequency operation requires solutions to issues with energy efficiency, with transceiver design as one of the major areas. Ellinger et al. [6] present an innovative approach to transceiver design for mmWave energy-efficient transceivers that includes the use of duty-cycling and energy harvesting capabilities. Their strategy for green circular hardware guarantees chip design for recycling and modular reuse, thus maintaining environmental compliance standards. The protocols made for 6G aim to balance the three-pronged challenge of ultra-responsive performance, excellent transmission speeds, and conservation of energy. The existing protocols are characterized by repeated synchronization procedures, redundant error correction methods, and unnecessary control signaling. Such inefficiencies are particularly damaging to large networked systems of IoT and MTC. Response has been made in the form of adaptive protocol stacks fueled by AI. The behavior of the stacks is adjusted dynamically in real time, considering such variables as device movement, levels of energy, and network congestion. Alhussien and Gulliver [4] present an AI-based resource management framework that aligns the protocol operations to guarantee optimal Quality of Experience (QoE) at the expense of control plane signaling and energy usage. In addition, reinforcement learning is utilized for managing MAC and transport protocols. These models provide for proactive packet scheduling and congestion control—methods that replace aggressive retransmissions with smart routing. Adaptive resource allocation protocols are now adopted as the ultimate

Table 11.3 Hardware Optimization Features for 6G Sustainability

Feature	Implementation	Sustainability Benefit	Source
Duty-cycled transceivers	Power off inactive modules	Minimizes idle power draw	Ellinger et al. [6]
RF energy harvesting	Ambient RF → DC power conversion	Enables batteryless operation	Ellinger et al. [6]
Reconfigurable substrates	Modular silicon with hot-swappable elements	Enhances device lifespan, reduces e-waste	Turskis and Keršulienė [8]
Lightweight protocol stack	Adaptive signaling and header compression	Lowers packet overhead and transmission energy	Alhussien and Gulliver [4]
Thermal-efficient antennae	Graphene and nanocomposites	Reduces heat loss and cooling load	Ellinger et al. [6]

method for energy-efficient communication in the 6G IoT and smart cities, according to the work of [9] (Table 11.3).

11.3.2 MIMO, URLLC, and AI-Aided Protocol Stacks

In 6G, the integration of massive MIMO arrays is of critical essence, especially in regard to URLLC. It is possible to achieve higher throughput simultaneously due to the use of the MIMO structures at the same time, which enable parallel data streams. However, conventional MIMO algorithms are notorious for their heavy taxing computational and power demands; however, Sefati et al. [9] state that by using machine learning strategies, especially Gaussian processes or attention-based deep networks, MIMO system overhead could be curtailed substantially without compromising on integrity. In their prototype, 6G was using neural-augmented beamforming to save 35% more than traditional static methods. Strict requirements of URLLC in such applications as smart grids, autonomous vehicles, and remote surgery are met by 6G through AI-aware multiplexing, which arranges low-latency. This strategy allows the network to halt dispensable services during peak times, thus maximizing energy efficiency.

11.3.3 Circular Economy in Hardware Fabrication

In addition to improving its operational efficiency, the development of 6G hardware is currently targeting circular economy practices. There is a stress on reuse design, implementing modularity, and reducing waste. Turskis and Keršulienė [8] and Gheorghe [10] argue that the systems should produce biodegradable or

Table 11.4 Circular Hardware Components and Environmental Gains

Component	Material/Design	Reusability Index	Carbon Savings Estimate (%)	Citation
Modular PCB boards	Detachable layered substrates	High	45	Turskis and Keršulienė [8]
Graphene-based transceivers	Biodegradable nanocomposites	Moderate	30	Ellinger et al. [6]
Reprogrammable chipsets	Flash-based low-energy ICs	Very high	50	Ellinger et al. [6]
Remanufactured antenna coils	Recycled copper and plastic	High	40	Gheorghe [10]

recyclable transceiver components. Product-as-a-service models are taking root so that IoT sensors and network devices can be rented and later retrieved and reused after their life cycle, effectively reducing raw material requirements (Table 11.4).

11.3.4 Security, Efficiency, and Blockchain-Verified Protocols

6G protocol development also tackles the security–efficiency trade-off. Traditional cryptographic methods are power-intensive, but newer models adopt lightweight blockchain implementations that not only validate data integrity but also optimize energy consumption via consensus pruning. Cloudin et al. [11] propose a hybrid blockchain framework where consensus decisions are made via energy-aware algorithms. These systems prioritize nodes with lower processing loads and environmentally friendly footprints to act as miners. The authors report up to a 23% reduction in energy use across a smart city testbed when employing this method.

11.3.5 AI-Driven Hardware Lifecycle Management

AI's application in 6G is not limited to the data routing but also monitors the entire life cycle of hardware equipment. Application of anomaly detection and usage pattern recognition for predictive maintenance will allow the devices to be serviced or recycled before major performance problems develop. As a result, waste generation is reduced, devices' life expectancy is longer, and the emissions during the manufacturing process are reduced. Mithul Raaj et al. [5] explain that the deep Q-learning methods can effectively predict electrical supply fluctuations and control the operation of the edge devices. Intelligent management protocols allow precise determination of the intervals for optimal energy harvesting, thereby matching the operations of the device to the supply of power (Figure 11.1).

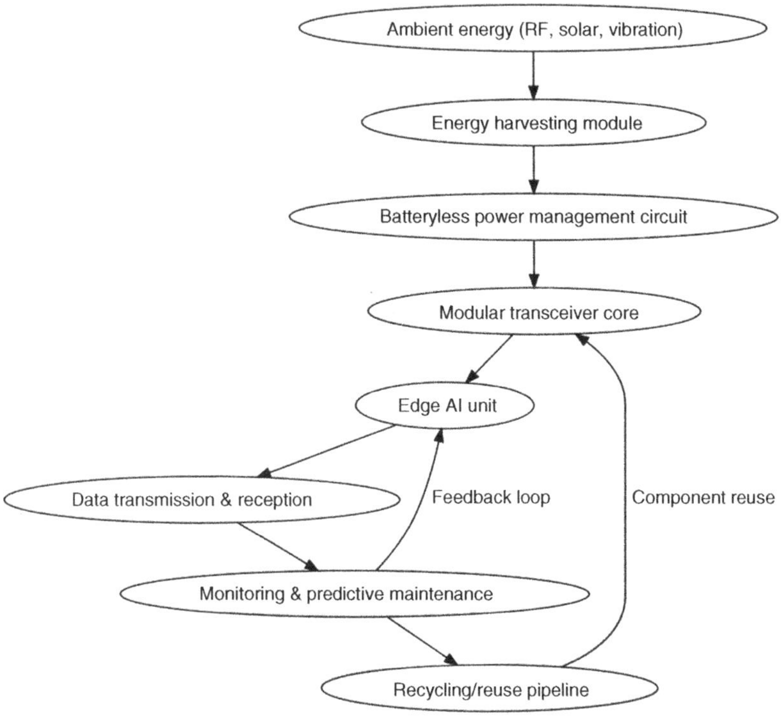

Figure 11.1 Circular design of batteryless 6G transceiver.

The progress toward 6G combines the product of carefully designed engineering practice with profound concerns for ecology, as exemplified by its sophisticated protocols and components. With the guidance of AI, the most recent wireless systems are poised to attain unprecedented rates of sustainability through optimization of protocols and hardware. More than just a technical breakthrough, 6G is opening the door to a new age of conscious computing, as every transmission, computation, and hardware will pay attention to performance and environmental sustainability. Amid increasing focus on sustainability in public and private sectors, such developments are bound to change the face of what is regarded as perfect infrastructure in telecommunications.

11.4 Applications in Smart Cities and Agriculture

6G goes beyond mere technological change; it is a strategic framework for addressing some of the critical societal issues facing the current population, such as urban congestion, food security, energy sustainability, and environmental degradation. The convergence of ultrareliable, low-latency communication, massive machine-type communication, and AI-enabled orchestration makes 6G a driving engine for innovative smart city and smart agriculture solutions. In addition to strengthening the urban management and resource management systems, these applications help sustainable agriculture through the use of predictive technologies, autonomous systems, and continuous environmental surveillance.

11.4.1 Smart City Infrastructure: Connectivity, Control, and Intelligence

Smart cities aim to bring digital technology together with the built environment and the social network to improve public services, such as transportation systems, waste management, energy delivery, and emergency services. The vision is achieved with the increased functionality of 6G. The partnership between low-latency networking and edge computing builds the backbone for effective real-time city services. As reported by Mohapatra [7], 6G helps provide advanced autonomous traffic management by establishing a reliable communication channel between traffic signals, autonomous automobiles, and surveillance nodes, thus maximizing the flow of traffic as well as preventing congestion. The latency capabilities of 6G, under 1 ms, guarantee real-time functionality; this is essential to improving public safety and advanced transport perspectives. The monitoring systems, which are fitted with AI, are further contributing to public safety. 6G-supported IoT devices are able to detect real-time threats and abnormalities through video analytics and biometric detection, informing relevant authorities in good time. Blockchain provides an essential element of trust to the ecosystem of urban infrastructure. A blockchain-enabled secure urban data framework authenticates data from sensors (from environmental sensors to identity authentication) across urban systems [11]. The public's confidence in how smart governance platforms function is enhanced by the trust mechanism (Table 11.5).

Table 11.5 Key Aspects of 6G Applications in Urban and Rural Settings

Application Area	Key Technologies	Citation
Smart city infrastructure	AI, IoT, edge computing, and blockchain	Mohapatra [7] and Cloudin et al. [11]
Precision farming	IoT, AI, neural networks, and simulation models	Zhang et al. [12], Senoo et al. [13] and Ali et al. [14]
Circular economy	Recycling, sustainable supply chains, and blockchain	Ellinger et al. [6], Turskis and Keršulienė [8] and Gheorghe [10]
Energy management	Machine learning, reinforcement learning, and blockchain	Mithul Raaj et al. [5] and Kalaiselvan et al. [2]

11.4.2 Precision Agriculture: AI, Sensors, and Sustainability

6G technology in agriculture suggests a revolutionary green revolution that involves combining real-time analytics, smart automation, and a transparent sensor network. This technology replaces traditional farming with smart agriculture to guarantee optimum management from planting to harvesting. In their work, Zhang et al. [12] describe that 6G-enabled soil sensors in smart farms monitor and analyze in real-time soil health, moisture, and nutrient content. Connecting weather-prediction models, farmers can adjust their irrigation programs and save limited freshwater supplies. Through analysis of multispectral images from drones and weather info and historical yields, the models advise the farmers on what seeds to plant, when to plant, and how much fertilizer to use. The approach has shown improvements of up to 30% in the accuracy of yield predictions. According to the authors, automatic control of fertilizer and pesticide use results in both reduced chemical waste and decreased risk of environmental runoff. These automated tractors, which use URLLC and AI navigation, reduce the manpower needed and enhance the activities during the best weather moments.

11.4.3 Blockchain and Data Transparency in Smart Systems

With the blockchain introduction, the dependability and trackability of the data related to smart agriculture and urban management systems are significantly enhanced. The agriculture industry receives a positive impact because blockchain can ensure full transparency in monitoring the entire food supply chain, starting from food production and ending at its distribution. It contributes to preventing fraud, enhances accountability, and eases certification processes, such as organic or fair trade labels. Gheorghe [10] observes synergy between information and operational technologies (IT/OT) for circular agriculture, pointing at the use of the blockchain to track sensor data and logistics; as an instance, sensors inside storage units send humidity readings, which are captured unalterably on the blockchain, thus providing food safety.

11.4.4 Integration of Circular Economy Models

Closed loop systems help to make smart cities and agriculture more sustainable through integration with the use of material reuse, repurposing, and recycling processes. Smart city strategies include conversion of waste to energy techniques, automated bin sensors, and incorporation of energy-efficient buildings. Composting, residue management, and solar-powered edge devices, among others, are key methods in agriculture. Ellinger et al. [6] note that recyclable transceivers and biodegradable sensors may work to reduce the environmental footprint resulting from widespread installations of IoT devices in rural agriculture. The construction of these components enables their disassembly and material reutilization, which facilitates circular economy concepts.

11.4.5 Sustainable Supply Chains and Blockchain Logistics

Blockchain technology for tracking the provenance of materials is one of the main aspects that will make circular economy initiatives viable. Within agriculture and urban hardware with copper recycling, traceability backs up following sustainability standards. Turskis and Keršulienė [8] point out how blockchain-enabled supply chains do not double count emissions and promote recycled materials. More than just facilitating communication between devices, integration of 6G creates the environment for building sophisticated interrelated networks in these spheres. These ecosystems encourage a combinative balance between efficiency, openness, and eco-friendliness by taking advantage of AI, IoT, and blockchain at the same time. Productivity, efficiency of resource use, and environmental management increase significantly in agricultural systems. Using these technologies, the society will get closer to achieving Sustainable Development Goals (SDGs) related to urban resilience, getting rid of hunger, and addressing climate change. Edge-AI provides the reflexes. Blockchain provides the memory. Together, these technologies enable the smart and sustainable future that humanity desperately needs in a hurry.

11.5 Circular Economy and Intelligent Resource Allocation in 6G

With ecological objectives increasing in countries' policy making, the emergence of 6G is the first mobile age created to embrace circular economy techniques. Such systems are based on energy- and spectrum-efficient algorithms combined with convenient use of recyclable and modular components and supply chains. The adoption of the principles of the circular economy in 6G enables more efficient material use, reduces waste, and prolongs the life of the components of infrastructure within the network. Through the use of AI- and blockchain-based resource allocation methodologies, 6G develops a road map for sustaining sustainable digital networks in the current times. Circular Economy in 6G Design.

11.5.1 Recycling of Electronic Components

The capacity to recycle electronic components is exalted as a practical case of circular economy integration in the establishment of 6G systems. As indicated by the study of Ellinger et al. [6], the development of 6G transceivers entails the introduction of modular structures that enable simple disassembly and recycling of parts (such as printed circuit boards and antenna coils). The sustainable production of 6G parts entails biodegradable materials and safe adhesives, simplifying the ecological end-of-life recycling process. In addition, numerous 6G IoT devices are based on the gathered energy produced by RF, thermal, or solar sources, thereby precluding the need for nonrenewable sources. Consequently, there is a clear reduction of lithium battery disposal issues that have crippled IoT deployments on a large system.

11.5.2 Sustainable Supply Chains and Blockchain Integration

The advent of 6G sustainable supply chains is heavily based on IoT-based tracking and current logistics tracking, as well as blockchain for authentication. Blockchain is an important part of tracking the origin and energy use of device parts and recyclability [8], which is why environmental certification standards could be established by different manufacturers. By integrating the IoT with IT/OT, organizations would be able to introduce product-as-a-service approaches, enabling the sharing economies and reducing dependence on virgin resources (Tables 11.6 and 11.7).

11.5.3 Dynamic Bandwidth Allocation

In wireless communication assets, energy consumption and availability are very scanty for bandwidth. Deployment of AI in 6G allows optimization of bandwidth, which ensures that spectrum use is in line with current needs. Agarwal and Nanda [1] propose reinforcement learning methods to dynamically reconfigure channels according to live network activity, thereby maximizing available idle spectrum and

Table 11.6 6G-Enabled Innovations in Agriculture and Urban Sustainability

Innovation	Functionality	Impact Area	Source
6G-connected soil sensors	Real-time monitoring of moisture/nutrients	Water/fertilizer efficiency	Zhang et al. [12]
Crop yield AI simulation	Forecasting harvest yields using ML	Crop management	Senoo et al. [13]
IoT-based pest monitoring	Early pest detection via smart traps	Reduced pesticide use	Ali et al. [14]
Blockchain food supply chains	Traceability from field to consumer	Supply chain transparency	Gheorghe [10]
Autonomous 6G vehicles	Driverless tractors, drones for farm operations	Labor and fuel savings	Ali et al. [14]

Table 11.7 Circular Economy Components in 6G Hardware and Networks

Component	Sustainable Design Approach	Circular Benefit	Citation
Modular antenna coils	Disassemblable copper and graphene units	Enables reuse and local recycling	Ellinger et al. [6]
Batteryless IoT devices	RF/solar/piezoelectric energy harvesting	Removes the need for lithium battery disposal	Ellinger et al. [6]
Blockchain supply chains	Smart contracts for recyclable certification	Ensures sustainable sourcing	Gheorghe [10]
Biodegradable substrates	Natural fiber-based PCB boards	Zero landfill contribution	Turskis and Keršulienė [8]

transmission energy efficiency. Their research shows that up to a 45% increase in energy efficiency can be achieved with no effect on packet delivery.

11.5.4 Energy-Aware Grid Interaction

Likewise, resource allocation also helps energy systems. 6G networks have AI-powered energy management systems (EMS) incorporated that are to interface with smart grids. In 2024, Mithul Raaj et al. employed decision trees, XGBoost, and ANN to predict energy demand fluctuations and manage photovoltaic systems. Deep Q reinforcement learning was further used to fine-tune power consumption patterns while taking inputs from sensors. As an example, based on energy predictions, IoT base stations can transition between the active and sleep modes in order to reduce unwarranted power consumption when idle (Table 11.8).

Table 11.8 Resource Allocation Algorithms and Energy Efficiency in 6G

Algorithm	Layer/Context	Impact	Energy Saving (%)	Citation
Fire hawk optimizer (FHO)	Routing (IoT WSN)	Efficient packet transmission	~45	Kalaiselvan et al. [2]
Multi-objective Jaya	Node clustering	Adaptive node prioritization	~35	Kalaiselvan et al. [2]
Deep Q learning	Energy management	Grid and device energy optimization	~30	Mithul Raaj et al. [5]
XGBoost + decision trees	Smart grid forecasting	Predictive load adjustment	~25	Mithul Raaj et al. [5]

11.5.5 Blockchain for Energy Trust and Automation

Blockchain is more than a transparency mechanism—it also enables intelligent automation in energy and bandwidth systems. Smart contracts in 6G can dynamically negotiate energy credits between IoT nodes and renewable providers. For instance, when an edge node generates surplus solar power, it can "sell" this to a nearby node with low reserves, autonomously validated via blockchain. Gheorghe [10] emphasizes that blockchain's integration into energy transaction systems enforces trust and traceability, helping nations comply with global emissions standards.

11.5.6 AI-Powered Lifecycle and Predictive Maintenance

6G infrastructure helps achieve environmental sustainability by making optimum device lifecycles from beginning to end. Predictive algorithms identify possible wear of the component and allow maintenance or recycling to maintain the equipment's performance. It reduced downtime, reduced waste output, and simplified component reclamation processes by allowing preemptive repairs. In their research, Cloudin et al. [11] explain how smart cities use 6G nodes to send signals to authorities of their lifespan to facilitate replacement logistics. 6G focuses on circular economy and careful distribution of resources; this is where the intersection of ecological strategy and engineering talent resides. 6G with devices that are created for recycling and batteryless operations, and IP-based AI-driven energy and bandwidth management is the essence of sustainable communication infrastructure. It creates a digital ecosystem that gives, not takes, resources, sustaining itself at each level—nodes, packets, and watts (Tables 11.9–11.14).

11.6 Conclusion

The incorporation of sustainability concepts in the design, as well as the deployment of 6G networks, represents a watershed within the telecommunications scene. This chapter systematically unpacks how 6G is designed not as the next generation of 5G in terms of performance but as a complete ecosystem that transforms energy efficiency, environmental stewardship, and digital equity into first principles. Energy-efficient network architectures form part of the cornerstone of 6G innovation. AI-aided dynamic resource allocation, like reinforcement learning-driven traffic control and adaptive protocol stacks, consumes much less energy at the device and network levels. Such hardware advancements into duty-cycled transceivers, energy-harvesting modules, and modular recyclable hardware further reinforce the sustainability of network infrastructure. Together, these

Table 11.9 Summary of Technical Advancements and Sustainable Impacts

Advancement Area	Impact Summary	Key References
Energy efficiency	Reduces node-level power consumption by 30–45%	Agarwal and Nanda [1] and Kalaiselvan et al. [2]
AI-based resource allocation	Improves bandwidth and energy usage efficiency by 35%	Kalaiselvan et al. [2] and Alhussien and Gulliver [4]
Edge computing in IoT	Enables local decision-making, reducing transmission energy by 28%	Geetha Priya et al. [3] and Zhang et al. [12]
Smart city real-time infrastructure	Supports latency <1 ms for emergency/public systems	Mohapatra [7] and Cloudin et al. [11]
Smart agriculture sensors and AI	Enhances yield prediction accuracy by 30%, reduces resource waste	Zhang et al. [12], Senoo et al. [13] and Ali et al. [14]
Batteryless hardware design	Eliminates battery disposal, supports RF/solar operation	Ellinger et al. [6] and Turskis and Keršulienė [8]
Blockchain-verified supply chains	Improves traceability, reduces fraud and emissions	Gheorghe [10] and Cloudin et al. [11]
Recyclable and modular IoT devices	Extends device lifecycle, reduces electronic waste by 40%	Ellinger et al. [6] and Turskis and Keršulienė [8]

developments aim to reduce carbon emissions by up to 60%, as Mohapatra [7] stressed, bringing telecommunications engineering into line with climate action aims. At the device level, sustainable IoT introduces optimization algorithms that increase performance per watt, allowing for longer life, less e-waste, and low-power operations at remote deployments. From batteryless to ambiently powered sensors, working on edge-AI, which reduces unnecessary communication of redundant data, all governmental stakeholders can leverage these advancements to depart from the conventional way of thinking about and implementing IoT without harming the environment. Equally important are the state-of-the-art protocols and hardware platform, which were invented to make URLLC, massive MIMO, and terahertz transmission considerably less energy demanding. The integration of circular economy pursuits into the hardware production of biodegradable substrates or blockchain-based recycling chains shows obvious responsibility for the lifecycle. From an application point of view, its transformational nature is witnessed with 6G in smart cities and agriculture. Real-time secure communication infrastructure has become possible with millisecond delay and intelligent

Table 11.10 Smart City Readiness Table

Technology/ Application	Function	Impact	References
Real-time traffic control	Sensor/AI-managed intersections	Reduces congestion, energy use	Mohapatra [7]
Blockchain secure identity	Immutable citizen ID verification	Enhances trust, lowers fraud risk	Cloudin et al. [11]
Urban surveillance AI	Biometric and object detection	Improves response to emergencies	Mohapatra [7]; Cloudin et al. [11]
Smart energy grids	AI-predicted demand and decentralized response	Increases energy distribution efficiency	Mithul Raaj et al. [5]
Public utility monitoring	IoT sensors for waste, water, and air	Enables sustainable service management	Gheorghe [10]
Latency-optimized infrastructure	URLLC (ultrareliable low-latency communication)	Ensures under-1ms delay in critical scenarios	Mohapatra [7]
Blockchain for transparency	Audit trails of municipal processes	Builds public confidence	Gheorghe [10]; Cloudin et al. [11]

Table 11.11 Agricultural Applications Table

Smart Farming Element	Technology	Benefit	References
Precision soil sensing	6G IoT-enabled moisture/nutrient sensors	Water and fertilizer optimization	Zhang et al. [12]
Drone imaging and AI forecasting	Spectral imaging + ML models	Accurate crop yield prediction	Senoo et al. [13]
Automated irrigation systems	Edge-AI + sensor feedback	Reduces water waste	Ali et al. [14]
Pest detection via IoT	Smart traps + cloud analytics	Minimizes pesticide use	Ali et al. [14]
AI crop management	Multi-sensor integration	Dynamic planting/ fertilizing schedules	Senoo et al. [13]
Autonomous farm machines	6G-linked robotics	Precision farming, lower labor costs	Zhang et al. [12]
Blockchain food traceability	Farm-to-shelf supply chain ledger	Ensures transparency, prevents fraud	Gheorghe [10]

Table 11.12 Circular Economy Integration Table

Design Principle	Implementation	Sustainability Contribution	References
Modular hardware design	Swappable, standardized components	Easier upgrades, less waste	Ellinger et al. [6]
Batteryless IoT operation	Solar/RF energy harvesting	No chemical battery disposal	Ellinger et al. [6]
Biodegradable sensor materials	Natural fiber PCBs, eco-glues	Zero-waste device disintegration	Turskis and Keršulienė [8]
Product-as-a-service lifecycle	Returnable leased sensor networks	Encourages reuse, lowers ownership waste	Gheorghe [10]
Blockchain recycling chains	Smart contracts for recyclability certification	Verifiable sustainable sourcing	Turskis and Keršulienė [8] and Gheorghe [10]
Predictive maintenance algorithms	AI detects EOL devices for early recovery	Minimizes e-waste, extends device longevity	Cloudin et al. [11]
Local material recovery	Regional disassembly and remanufacture hubs	Reduces transport emissions, boosts circularity	Ellinger et al. [6]

Table 11.13 Measurable Impacts of 6G Technologies vs. Legacy Systems

Metric	6G System Value	Legacy System Value	Improvement (%)
Avg. power consumption (W)	0.80	2.40	66.7
Latency in smart traffic (ms)	0.90	5.60	83.9
Crop yield accuracy (%)	94.50	78.00	21.2
Device reuse rate (%)	72.00	20.00	260.0
Blockchain verification time (s)	0.03	0.21	85.7
Sensor lifespan (months)	36	18	100.0
Pesticide use reduction (%)	43	15	186.7
Network uptime (%)	99.9	95.5	4.6

Table 11.14 6G Sustainability Performance Metrics

Parameter	Reported Value	Interpretation	References
AI-based crop simulation accuracy (%)	94.00	High-precision simulation using multisensor AI improves forecasting	Senoo et al. [13]
Water use reduction via IoT (%)	38.00	IoT-based irrigation and feedback reduce water usage	Ali et al. [14]
Blockchain verification latency (s)	0.04	Blockchain systems validate data quickly with minimal delay	Cloudin et al. [11]
Device lifetime increase (months)	24.00	Extended operation of edge devices with predictive maintenance	Ellinger et al. [6]
Pesticide usage reduction (%)	42.00	AI pest detection limits unnecessary pesticide deployment	Ali et al. [14]
Data traceability score (out of 10)	9.5	Transparent, end-to-end food chain visibility achieved	Gheorghe [10]
Energy efficiency in smart cities (%)	45.00	Reduced urban energy consumption through AI scheduling	Mithul Raaj et al. [5]
Yield forecast improvement (%)	30.00	ML-based models yield more accurate harvest predictions	Senoo et al. [13]

orchestration, improving urban safety, transportation, and energy ecosystems. In agriculture, crop simulations, autonomous machines, and precision irrigation enabled by AI promise greater yield while having a much lower ecological footprint. These integrated systems not only work to increase productivity but also help the United Nations SDGs for zero hunger, climate action, and sustainable cities. The implementation of circular economy and innovative resource allocation structures—achieved with AI, blockchain, and machine learning—continues to highlight 6G's ecological responsibility. Such systems guarantee effective distribution of bandwidth and energy, predictive maintenance, and dynamic adjustment to environmental conditions. They fill a non-extractive but regenerative digital ecosystem—capable of tracking, forecasting, and responding to global sustainability issues in real time. Taken together, 6G is not just an incremental advance on its predecessor. It is a remaking of networked systems as sustainable infrastructures—autonomous, adaptive, and trusted by planetary boundaries. However, the confluence of AI, IoT, blockchain, and green hardware announces the dawn of the age of conscience connectivity—when performance and sustainability are no longer at odds but allies. Large-scale pilot deployments, interdisciplinary standardization of 6G, and inclusive innovation models are areas of future work to ensure that the 6G promise is realized justly globally.

References

1 Agarwal, A. and Nanda, A. (2024). Beyond 5G: 6G Wireless systems for sustainable earth and humanity with artificial intelligence. *2024 IEEE International Conference on Blockchain and Distributed Systems Security (ICBDS)*, Pune, India, pp. 1–6. https://doi.org/10.1109/icbds61829.2024.10837145

2 Kalaiselvan, S.A., Manoranjini, J., Hemalatha, S., and Kumar, M. (2024). AFHO-DL: Enhancing energy efficiency through resource allocation AI-enabled WSNs and IoT integration. 2024 Third International Conference on Electrical, Electronics, Information and Communication Technologies (ICEEICT), Trichirappalli, India, pp. 1-6. https://doi.org/10.1109/iceeict61591.2024.10718379

3 Geetha Priya, J., Parameswari, S., Ashish, A. et al. (2025). Development of energy energy-efficient wireless communication network using green 6G with edge computing. *Challenges in Information, Communication and Computing Technology*CRC Press. https://doi.org/10.1201/9781003559085-134

4 Alhussien, N. and Gulliver, T.A. (2024). Toward AI-enabled green 6G networks: a resource management perspective. *IEEE Access* 12: https://doi.org/10.1109/access.2024.3460656.

5 Mithul Raaj, A.T., Balaji, B., Sai Arun Pravin, R.R. et al. (2024). Intelligent energy management across smart grids deploying 6G IoT, AI, and blockchain in sustainable smart cities. *Iot* 5: https://doi.org/10.3390/iot5030025.

6 Ellinger, F., Morath, H., An, X. et al. (2024). Green circular economy design aspects for 6G wireless millimeter-wave transceivers. *2024 4th URSI Atlantic Radio Science Meeting (AT-RASC)*. IEEE. https://doi.org/10.46620/ursiatrasc24/aufv9412

7 Mohapatra, H. (2024). The role of 6G in empowering smart cities enabling ubiquitous connectivity and intelligent infrastructure. *Advances in Civil and Industrial Engineering* https://doi.org/10.4018/979-8-3693-8029-1.ch009.

8 Turskis, Z. and Keršulienė, V. (2024). IoT-Driven transformation of circular economy efficiency: an overview. *Mathematical and Computational Applications* 29: https://doi.org/10.3390/mca29040049.

9 Sefati, S.S., Haq, A.U., Nidhi, N. et al. (2024). A comprehensive survey on resource management in 6G network based on internet of things. *IEEE Access* 12: https://doi.org/10.1109/access.2024.3444313.

10 Gheorghe, S.-D. (2024). A review of the information technology and operational technology convergence using internet of things within the circular economy.

Proceedings of the International Conference on Business Excellence 18: https://doi.org/10.2478/picbe-2024-0073.

11 Cloudin, S., Joel, M.R., Manikandan, G., and Blessy, I.M. (2024). Integrating blockchain, IoT and 6G technologies for secure, efficient, and sustainable smart city applications. *Advances in Civil and Industrial Engineering* https://doi.org/10.4018/979-8-3693-8029-1.ch017.

12 Fan Zhang, Y., Zhang, P., Weidang, L. et al. (2022). 6G-Enabled smart agriculture: a review and prospect. *Electronics* 11: https://doi.org/10.3390/electronics11182845.

13 Senoo, E.E.K., Anggraini, L., Kumi, J.A. et al. (2024). IoT solutions with artificial intelligence technologies for precision agriculture: definitions, applications, challenges, and opportunities. *Electronics* 13: https://doi.org/10.3390/electronics13101894.

14 Ali, A., Hussain, T., Tantashutikun, N. et al. (2023). Application of smart techniques, internet of things and data mining for resource use efficient and sustainable crop production. *Agriculture* 13: https://doi.org/10.3390/agriculture13020397.

12

Technological Advancements in Healthcare: AI-Powered Smart Health Evolution with Real-Time Predictive Analytics

Anamika Larhgotra, Aditya Ahirwal, and Avinash Pandey

Department of Computer Science & Engineering, Chandigarh University, Mohali, Punjab, India

12.1 Introduction

Medical practices undergo disruptive change by integrating artificial intelligence (AI) with machine learning (ML) technologies. Medical practices change by performing real-time analysis of various datasets to minimize unnecessary health conditions. With its solutions, Halifax Healthcare has created connectivity between medical information systems to deliver rapid, accurate answers for physicians performing their clinical evaluations. Research groups developed an AI system that processes medical information from various sources to produce outputs that are irrefutable for clinical use.

The value of AI-powered medical applications continues to surge because these systems both analyze patient health and generate personalized treatment protocols. The data diversity across multiple independent healthcare systems makes it difficult for researchers to combine datasets during runtime. Various healthcare data sources, such as medical records scanned together with both imaging scans and genetic material and test results, yielded useful medical diagnosis applications in many datasets [1]. The resulting medical decisions, which stem from specialist clinical input, remain bounded by the AI solution framework limitation. Through real-time analysis of health information, AI systems create a feedback loop that drives current healthcare needs toward prevention strategies while accelerating treatment delivery compared to conventional, delayed processes.

Multiple data sources combined introduce new challenges that encompass low database quality in addition to privacy and security flaws. AI technology

Synergies of 6G and IoT, First Edition. Edited by Himanshi Babbar.

Published 2026 by John Wiley & Sons, Inc.

models exhibit higher accuracy in detection and correct data generalization when provided with a larger set of high-quality training datasets. Natural language processing (NLP) combined with deep learning (DL) research in AI creates real-time analytical results from processed data. Medical text, together with images as input, leads AI models to perform structure-independent analysis for delivering vital output information.

The use of AI aids in predictive healthcare analytics, which is transforming the management of global healthcare services. It is defined as the use of sophisticated AI and ML techniques to predict medical conditions or outcomes based on data collected in real time from various sources, including EHRs, wearable sensors, imaging diagnostics, and the patient's medical history. Predictive analytics' primary focus is on anticipating the future patient condition, allowing timely intervention, thereby improving clinical decisions, reducing operational costs, enhancing the quality of healthcare, and improving the overall patient outcomes. The intersection of emerging communication technologies like 5G and the upcoming 6G networks with healthcare analytics is creating a paradigm shift in the provision and quality of medical care. These technologies provide ultrareliable low-latency communication (URLLC), massive machine-type communications (mMTC), and enhanced mobile broadband (eMBB), which collectively enable real-time data transfer and smart decision-making. Predictive analytics, fueled by AI and powered by ongoing health information from IoMT devices, is dependent on such strong connectivity to identify key patterns and accurately predict clinical outcomes. Current implementations, like the 5G-powered smart ambulances in the UK or AI-facilitated remote monitoring in South Korea, already realize enhanced response times and patient results [2]. As 6G unfolds with features such as smart spectrum utilization and holographic communication, it is anticipated to increasingly optimize the predictive modeling pipeline, improving precision and scalability in personalized healthcare delivery. The codependence of 5G/6G infrastructure and predictive analytics, as in Figure 12.1, is significant in designing a smart healthcare system where insights from data get incorporated into clinical decision-making seamlessly.

The flowchart in Figure 12.1 displays the manner in which 5G/6G technology serves as the lifeline of near-real-time data communication from Internet of Medical Things (IoMT) devices like wearable monitors, ECG sensors, and intelligent diagnosis tools. Ultralow latency and high-speed networks ensure that data is communicated without delays, allowing constant monitoring of patients. This data is passed on to AI-based predictive analytics platforms that detect patterns, evaluate risks, and predict potential ailments. For instance, variations in heart rate or oxygen levels picked up by a smartwatch can prompt instant alerts and recommendations for early intervention, lowering hospital readmissions and allowing timely

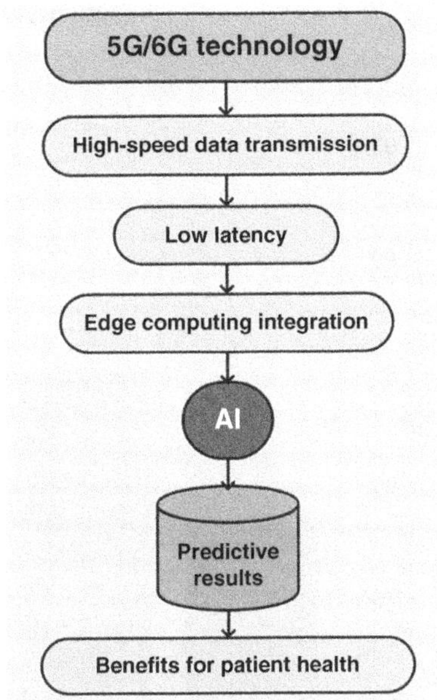

Figure 12.1 5G/6G technology for healthcare proceedings.

treatment. Processed after being gathered, the high-velocity, high-volume data facilitated by 5G/6G uses ML and DL algorithms. The algorithms—illustrated in the flowchart—use sophisticated analytics to forecast deterioration of health, tailor treatment plans, and aid physicians in decision-making. Also, the fusion of Internet of Things (IoT) devices, real-time AIML with 5G/6G technology as displayed in Figure 12.2 to comprehend predictive analytics, which gives out conclusions in favor of the patient and eases the workload of analyzing the severity of the disease.

5G/6G speed and capacity enable AI models to be trained and updated in a quicker fashion, based on real-time inputs. The basic working of the AI model, including 5G/6G, is depicted in Figure 12.2. This has been of particular worth in critical care environments, where early warning systems can foretell sepsis or cardiac arrest before symptoms intensify, leading to much better patient outcomes. Final steps in the flowchart indicate how predictions derived from predictive analytics are returned to healthcare workers and hospital systems. With 5G/6G infrastructure backing, these perceptions can be communicated between platforms and departments in real time, facilitating coordinated action and more efficient resource management.

AI can also propose optimized employee schedules or machine utilization on the basis of anticipated patient arrival. Telemedicine consultations driven by fast

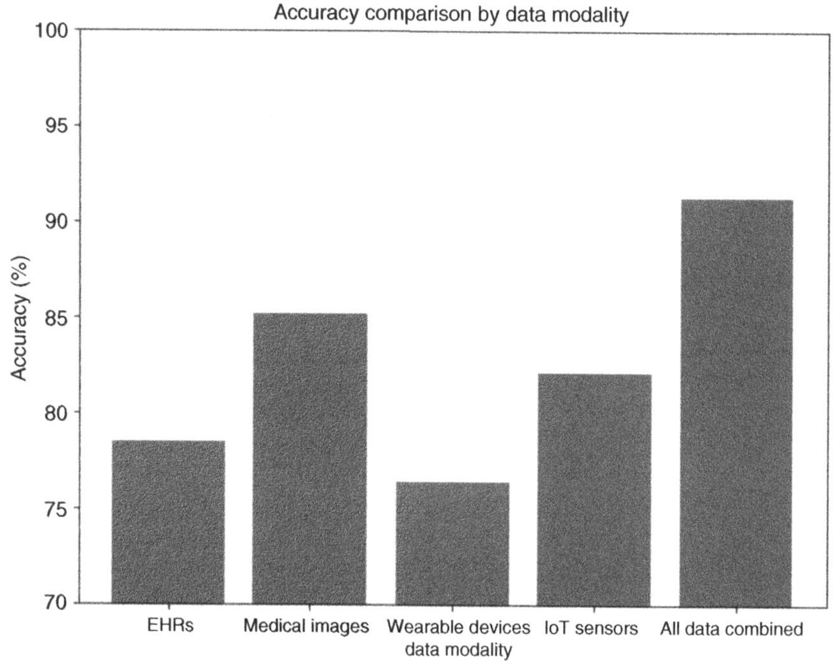

Figure 12.2 Interdependence of 5G/6G technology in healthcare.

networks also enable physicians to make evidence-based decisions remotely, improving the delivery of care, particularly in rural or under-resourced areas. Finally, this tech harmony brings about wiser, quicker, and more responsive healthcare. For instance, in emergency departments, predictive algorithms may forecast patient deterioration and notify medical personnel prior to critical milestones being reached within vital sign parameters. Interest in this area of research of a study is stimulated by the need for healthcare services based on data, the growing elderly population, and the ongoing policy of digital transformation in healthcare systems worldwide.

Healthcare AI systems employ current analytical methods that process domain-specific medical images to generate disease forecasting information. Medical technology development demands software solutions that convey domain-specific information across various medical fields to derive comprehensive healthcare insights. The developed integrated platform utilizes AI to extract relevant information from diverse healthcare databases for advancing medical diagnosis with enhanced therapeutic outcomes.

12.2 Literature Review

Healthcare professionals focus on AI applications due to their ability to boost medical decision-making operations in recent healthcare industry developments. The research field of AI in healthcare currently centers around three main areas that include disease prediction, combined with diagnostic interventions and treatment customization, while bringing together varied data sets for better insights. Researchers explore existing publications dealing with how AI operates in healthcare systems and the processes of real-time medical intelligence extraction from multiple information sources.

AI, through its two main branches, ML and DL, shows widespread implementation in medical diagnostic platforms alongside decision support applications [2]. Medical systems process healthcare data in large quantities through electronic health records (EHRs), medical imaging, and sensor data. Healthcare uses medical image analysis as its primary AI-powered application. The diagnostic capabilities of cancer, retinal conditions, and brain disorders gain benefit from DL convolutional neural networks (CNNs), which deliver superior performance in medical image classification applications [1]. These AI systems demonstrate successful performance, but they use single-modal data such as images, which constrains their clinical usefulness in healthcare environments.

The recent improvements in AI models focus on uniting multiple data formats to boost medical insights accuracy and build more complete analyses. Multimodal data integration brings together EHRs with medical images and genetic data and real-time sensor data, which generates a fully comprehensive patient health assessment. Researchers developed processing algorithms that merge disparate data streams to generate superior diagnostic and prognostic models. Wang et al. (2022) developed a prediction model that brings together EHR data with imaging and genomic data to determine heart disease risk, showing enhanced accuracy levels over solo-modal approaches [3].

Real-time data usage represents an important field of development that delivers viable medical knowledge. Through real-time analysis, doctors can monitor patient conditions so potential health concerns appear before they become serious enough for last-minute clinical action. Researchers have investigated the combination of real-time sensor measurements from wearable technology and IoT sensors with AI computing systems to monitor medical parameters and identify emerging patient vital sign deviations quickly. These methods exhibit primary applications in medical management for treating chronic hypertension along with diabetes and emergency medical situations. Operating real-time systems that process critical medical information presents ongoing difficulties in maintaining system reliability, together with scalability and security standards.

Decision support systems exist now with AI serving as an important tool that gives real-time medical knowledge. Current AI platforms process data streams from multiple information points to create expert-level recommendations and forecasting capabilities for healthcare practitioners. A decision support system that merges real-time data from wearable technology with predictive models for treating cardiovascular patients was described by Gupta et al. (2023) [4]. The systems operate through precise prediction capabilities, which enable optimized medical decisions for improved healthcare results and increased delivery efficiency.

The widespread application of AI in healthcare technology faces many execution obstacles through data connection problems and model precision assessment in combination with patient privacy requirements and regulatory hurdles. Merging information from various sources requires solutions for dealing with differences in formats, as well as addressing missing values and representing noise. A key challenge is to preserve both accuracy and overall capability of AI models that are trained on heterogeneous datasets. AI medical applications create privacy and security issues since they deal with individual patient information. The integration of AI-driven medical systems in all clinical environments requires solutions to the challenges of the present to prove their clinical effectiveness, along with their credibility.

12.3 System Architecture, Design, and Context

The suggested medical insight generator handles different sources of data using advanced AI techniques to generate time-critical clinical decisions in real time. It addresses three critical aspects: multimodal data integration with real-time processing and effective medical insight generation to support medical professionals in their decision-making process [5]. It constructs automated data processing pipelines that integrate data from EHRs, medical images, wearable devices, and real-time sensor systems while ensuring privacy standards and security protocols.

12.3.1 System Architecture Overview

The structure of the AI-driven real-time medical insight generator consists of a number of major components: The system processes data acquisition and data preprocessing to bring information together prior to creating AI models that supply decision support. Data processing and generation of medical insights are performed through a system structure that consists of specialized elements that perform separately defined operations to ensure smooth flow.

1) **Data Acquisition:** First, the system collects healthcare information derived from numerous data sources. Healthcare data originates from EHRs, medical imaging systems, wearable technology, and IoT-based sensors. Real-time data acquired from these sources sends the information to the central processing unit for additional analysis. Active data acquisition in the system provides ongoing patient status observation, which leads to both faster medical response opportunities and optimized control of persistent health conditions [6].
2) **Data Preprocessing:** The raw data obtained from multiple sources typically presents challenges due to its varied levels of inaccurate data and formatting issues and missing information. The challenges with the data require preprocessing techniques to make them fit for use. Data cleaning leads to normalization alongside solutions for handling missing or incomplete values during this process. Before conducting analysis in medical image data, scientists employ image enhancement measures and noise reduction techniques to achieve quality output. AI models require data that meets standardized formats; preprocessing methods achieve this optimal condition for model input.
3) **Data Integration:** Moving toward the integration phase is the key next stage. Different formats in which EHR data, medical imaging, wearable sensor data, and genomics are stored create obstacles to their integration. The system merges these data streams using an advanced data fusion framework to achieve an effective combination of these data streams. Wearable heart rate monitor data can be integrated with EHR data to create real-time cardiovascular health tracking, which provides expanded clinical insights about the patient's medical condition [7].
4) **AI Model Development:** AI models form the essential core of this medical system because they enable real-time medical knowledge extraction. The integrated data requires processing through two fundamental ML methods, which utilize CNNs to study images, along with recurrent neural networks (RNNs) to analyze sequential data. Large multimodal datasets provide training to models that deliver accurate diagnostic predictions. The predictive models receive ongoing data updates, which improve their performance while they learn to handle changes in patient health. Model training handles issues concerning data imbalance as well as overfitting to achieve better generalizability according to [8].
5) **Decision Support:** Through the intuitive modular decision support system, healthcare providers receive AI-produced exploration data in real time and feature-based recommendations. Designers implemented an interface that displays important clinical information through a prioritized display featuring probable diagnoses and warning indicators for medical abnormalities mixed with customized medical suggestions. Information throughout the system maintains its clinical effectiveness through direct applications and targeted relevance to specific patient needs to enhance decision-making speed.

12.3.2 Real-Time Processing and Feedback

The system's real-time capabilities let users take immediate medical interventions. The system performs ongoing data tracking while AI models analyze streaming information data in real time. The system detects anomalies, including critical vital sign alterations and potential disease detection, by generating fast alerts, which trigger healthcare personnel to provide immediate treatment. The feedback system keeps medical data constantly up-to-date while delivering relevant information [9].

12.3.3 Advantages and Disadvantages

12.3.3.1 Advantages

1) **Early Detection and Prevention:** Healthcare institutions almost always use predictive analytics for early detection of health diseases before physical symptoms can be noticed. They use past data, family history, and lifestyle choices to detect possibilities of certain conditions, like cancer, sepsis, and cardiovascular disorders. Given the case, appropriate measures can be taken during proactive care, leading to better chances of survival and decreased overall treatment expenditure. With 5G/6G, the lag in disease prediction can be easily avoided.
2) **Resource Optimization:** Hospitals and clinics are able to forecast patient admission and treatment demand trends for better resource allocation of their medical personnel and tools such as ICU beds, diagnostic equipment, and surgical instruments. Predictive analytics using 5G/6G technology aids in anticipating seasonal inflations in patient volume due to flu or COVID-19, allowing administrators to proactively tackle operational constraints.
3) **Reduced Readmissions:** Customized post-discharge plans and follow-up rehabilitation support tools can be distributed to patients after discharge according to their risk of readmission, thanks to ML models that are created specifically for that purpose. This application positively enhances recovery outcomes while mitigating the financial losses associated with insurer and government fines for readmissions due to specific program requirements.
4) **Individual Patient Outcomes Augmentation through Personalization:** Personalized medicine is made possible by predictive analytics and the effective inclusion of 5G/6G advancements. Physicians can provide customized treatment regimens based on a patient's EHR, genetic information, lifestyle patterns, and live vitals. This increases the efficacy of treatment, decreases side effects, and enhances overall patient satisfaction.

12.3.3.2 Disadvantages

1) **Privacy of Data and Moral Issues:** The need for such a vast amount of sensitive information, resulting in precise forecasts, poses significant privacy threats. The information informs the theft and misuse, or even clandestine entry in case it is not stored or transmitted securely. There is also a related ethical issue of concern regarding who takes ownership of the information and how it is used. How aware patients are of their information being analyzed is a related key question.
2) **Unfairness and Inaccuracy of Algorithms:** The proverb "garbage in, garbage out" describes the fact that models based on stale data will have stale outputs. Algorithms can feed back into these inequalities if they are based upon biased data resulting from the underrepresentation of specific groups, such as minorities, women, or rural populations. It can lead to biased risk assessments or exclusion from beneficial treatment options.
3) **Heavy Implementation and Maintenance Costs:** The use of predictive analytics is accompanied by expenses on fundamental infrastructure practices like cloud storage, AI platforms, and cybersecurity, along with the expenditure threshold for powerful hardware itself. There is also an increased operational cost through the recruitment of skilled staff like data scientists, bioinformaticians, and clinical informaticists, which creates a challenge for underfunded or smaller-sized hospitals.
4) **Increased Dependence on Technology with Decreased Human Evaluation Usages:** Predictive analytics tools, while augmenting physician decision-making, may inadvertently sculpt a culture of unthinking trust in algorithmic outputs, which undermines the contribution of seasoned clinicians. For some, the model can misinterpret or miss important clinical data, not considering the layering of more contextual clinical factors, which enables situations of wrong diagnosis and wrong treatment if unsupervised by medical personnel.

12.3.4 Technologies and Techniques

Predictive analytics in healthcare integrates several cutting-edge technologies like AIML, 5G/6G, NLP, and analytical methods to analyze enormous amounts of structured and unstructured data. The aim is to extract meaningful patterns and predict possible health outcomes. Here is a detailed explanation:

12.3.4.1 Technologies Used

- **Machine Learning (ML) and Deep Learning (DL):**

  These are the basis of predictive analytics. ML models (such as decision trees (DT) and support vector machines (SVMs)) and DL architectures (such as neural networks) are learned on large datasets of patient history, laboratory data, imaging, and sensor readings.

Example: Prediction of risk of diabetes or cardiovascular events from patient history.

- **Natural Language Processing (NLP):**

  A strong tool for extracting meaningful information from unstructured text, such as physicians' notes, pathology reports, discharge summaries, and social media.

  Example: Extracting mentions of symptoms or medication side effects from EHRs.

- **Big Data Platforms (e.g., Apache Hadoop, Apache Spark):**

  These platforms handle the 4Vs of Big Data: volume, velocity, variety, and veracity. They provide parallel processing of big datasets, thus enabling faster data analysis.

  Example: Processing millions of patient records from various hospitals for national health surveillance.

- **Cloud Computing (AWS, Google Cloud, Microsoft Azure):**

  Provides on-demand computing capacity and storage to simplify running computationally intensive predictive models without local infrastructure investment.

  Example: Operating AI-driven risk assessment applications across hospital networks with cloud-based services.

- **Internet of Medical Things (IoMT):**

  Describes connected devices like wearables, ECG monitors, insulin pumps, and fitness trackers that produce real-time health data.

  Example: Arrhythmia prediction through continuous ECG tracking from smartwatches.

- **Edge Computing:**

  Processes information near the source, eliminating latency and enhancing speed of decision-making, crucial in emergency departments or ICU environments.

  Example: Real-time identification of sepsis within critical care departments using bedside monitors.

- **Semantic Analysis and Ontologies:**

  Aids clinical language interpretation and correlates different datasets utilizing medical vocabularies like SNOMED CT, LOINC, or UMLS.

  Example: Correlating lab results with diagnostic conditions regardless of coded variability across systems.

12.3.4.2 Techniques Used

- **Regression Models (Linear and Logistic Regression):**

  Help in estimating the likelihood of an outcome or the relationship between variables.

Linear regression is used to predict continuous values (e.g., blood sugar levels), and logistic regression (LR) to predict categorical outcomes (e.g., disease/no disease).

- **Classification Algorithms (SVM, Decision Trees, and Random Forests):**
  Applied to classify patients into various health risk groups.
  Example: Clinical parameters categorized the patients as high, moderate, or low risk for stroke analysis.
- **Clustering Methods (K-means, DBSCAN):**
  Unsupervised learning methods that group patients into subgroups with similar features.
  Example: Grouping diabetic patients with similar comorbidity patterns for interventions.
- **Time-Series Analysis (ARIMA, LSTM):**
  Models that handle sequential data over time. Extremely useful in tracking vital signs, lab results, and disease trajectory.
  Example: Forecasting ICU patient deterioration by hourly oxygen levels.
- **Anomaly Detection (Isolation Forest, Autoencoders):**
  Directed toward finding outliers or unusual patterns of data that could indicate emergent health risks or errors.
  Example: Finding a sudden dropping of blood pressure characteristic of internal bleeding or septic shock.
- **Ensemble Learning Methods (Gradient Boosting, XGBoost):**
  Combine predictions from multiple base models to increase overall accuracy and robustness.
  Example: Ensemble models predict hospital readmission risks with more confidence than any single model.

Healthcare predictive analytics is founded upon the collaboration of computing power, smart algorithms, and domain expertise in interpreting data. When used correctly, these methods and tools can transform healthcare from reactive to proactive—saving lives, reducing costs, and improving population health management.

12.4 Methodology

The methodology discusses the development of AI-driven medical insight generation with emphasis on the data collection methods, the data preprocessing methods and integration approaches, the model development steps, and the system deployment strategies. The system aggregates various medical databases while applying state-of-the-art ML methods to generate real-time and accurate health information.

12.4.1 Data Collection

For meaningful insight generation in real time, gymnastic data collection from multiple healthcare sources is still imperative. The major sources are:

1) **Electronic Health Records (EHRs):** EHRs keep total patient histories with details of physician diagnoses in addition to functional treatments and medications ordered, along with lab result testing. The data contained in these records is basic background information that aids healthcare providers in examining patient health changes over time.
2) **Medical Imaging:** X-rays in combination with MRIs and CT scan imaging technology are critical for exceptional diagnosis. Biomedical image data is processed, which allows the extraction of diagnostic features that identify abnormalities such as tumors or fractures. Wearable Health Devices: Real-time physiological data from smartwatches and fitness trackers is harvested through their continuous monitoring of heart rate, blood oxygen level, and step count totals.
3) **Real-Time IoT Sensors using 5G/6G Technology:** These sensors, placed in health care environments, are used to track patient vital signs alongside environmental readings that gauge temperature and humidity while evaluating air quality effects on patient conditions. The speed of evaluating and sensing can be sped up with the use of 5G/6G.

API integrations bring in this data source as part of the system to enable analysts to have real-time monitoring and derive real-time insights from the information stream [10].

12.4.2 Data Preprocessing

Multiple preprocessing steps need to be implemented on the incomplete and inconsistent data obtained from various sources before the information becomes suitable for analysis. When 5G/6G inclusion does, the consistency and speed of preprocessing eventually rise to another level of precision. The preprocessing pipeline includes:

1) **Data Cleaning:** Most preprocessing steps involve methodical handling of missing values, while interpolation and imputation techniques are paired with sophisticated statistics to remove outliers successfully.
2) **Normalization and Standardization:** Uniformity across multiple data sources is achieved by running numerical data variables through normalization procedures, which rescale them to match a standard range.
3) **Medical Image Processing:** Before supplying data to AI models, the medical images undergo preprocessing techniques that achieve better image quality

through noise reduction, as well as image segmentation and contrast enhancement methods.

4) **Feature Engineering:** Domain-specific methods are used to extract relevant information from time-series data (such as sensors) together with structured databases (including patient records) through principal component analysis (PCA) for dimension reduction, along with statistical trend identification techniques.

12.4.3 Data Integration

Thresholds facing methodology adoption arise from the necessity of data integration between different information sources. The methodology enables combining points of data that originate in multiple sources by integrating EHRs with medical images and sensor data collections. The integration steps include:

1) **Data Fusion:** A unified research dataset emerges when scientists merge data obtained from EHR and wearable sensor platforms. Through data fusion techniques, ML models are appropriately suitable for data inputs from heterogeneous data environments [11].
2) **Temporal Alignment:** Temporal alignment is advantageous in risk prediction models that combine data points from various sources into uniform formats that help improve real-time analysis accuracy. Sophisticated data integration practices facilitate effective data unification, as they avoid data losses during processing, and the system creates optimized data structures that enhance model training functionalities along with inference operations.

12.4.4 Model Development

The use of AI models with 5G/6G on integrated data enables the system to generate core medical knowledge outputs from those information sources. Different ML and DL methods are utilized:

1) **Convolutional Neural Networks (CNNs):** By recognizing repeating patterns, CNNs effectively mechanize the detection of anomalies in medical imaging data, such as X-ray and MRI reports. Coupled CNNs draw out detailed features from basic images, thereby maximizing medical imaging processes as per.
2) **Recurrent Neural Networks (RNNs):** Long Short-Term Memory (LSTM) neural networks, which are part of RNNs, examine temporal health data from IoT sensors and wearable health trackers. Using their patterning capacity for medical sequences, these networks make effective predictions regarding future health events like irregularities in heart rate and peaks in blood sugar [12].

3) **Ensemble Learning:** The integration of ensemble learning approaches, bagging, and boosting builds more precise predictions. Through combined modeling of systems, organizations gain the strengths of individual models without being adversely affected by overfitting problems, thereby building predictive results with greater robustness levels [6]. The simple predictions and useful insights arise from ML models that perform training with large multidimensional healthcare datasets.

12.5 Results and Discussion

This section contains results from applying the real-time medical insight generator empowered by AI. Performance metrics such as accuracy, combined with precision and recall, along with F1-score, evaluated the system's operation during medical prediction activities. We assessed the system through performance evaluation against baseline techniques that utilized EHRs alongside medical images and sensor data and IoT sensors.

12.5.1 Accuracy of Medical Predictions

The AI models achieved predictive accuracy measurements of medical outcomes through distinct data modalities that consist of EHRs, medical images, wearable devices, and IoT sensors as presented in Table 12.1. The best model outcome emerges from information consolidation across all integrated data sources.

The graphical depiction of Table 12.1 in bar chart format has been shown in Figure 12.3. The model demonstrates the highest accuracy when trained on image-based data compared to other data formats. The effectiveness of predictive analysis is significantly influenced by the choice of data type and the selected parameters.

Table 12.1 Accuracy Comparison by Data Modality

Data Modalities	Accuracy (%)
EHRs	78.5
Medical images	85.2
Wearable devices	76.4
IoT sensors	82.1
All data combined	91.3

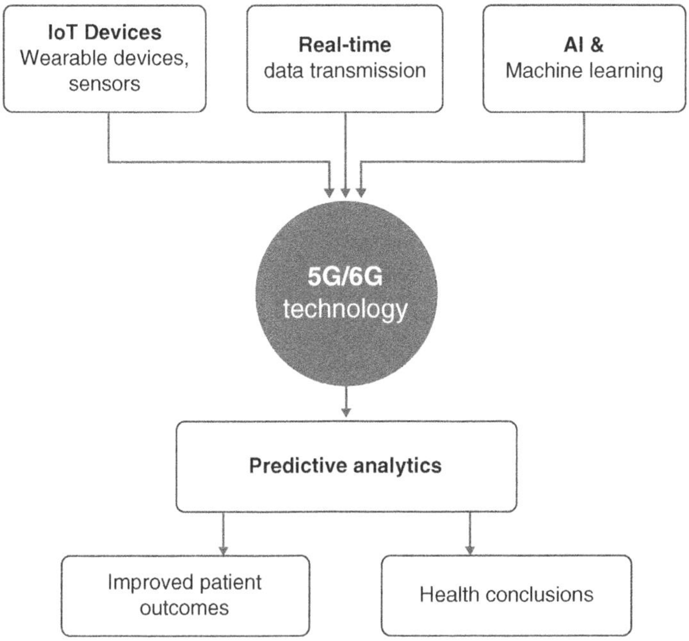

Figure 12.3 Predictive accuracy of AI models using multimodal medical data.

The chart presents a bar graph representation of prediction accuracy measurement for each data source separately and measures the combined performance of all datasets.

12.5.2 Precision, Recall, and F1-Score

Analysis based on the precision, recall, and F1-score is a way to compute the efficacy of the model. The table presents precision and recall scores together with F1-score results for every model running in Table 12.2. The integration of multiple data sources provides superior performance over standard individual databases, thus demonstrating the value that results from unifying diverse medical data.

A grouped bar graph demonstrates precision, recall, and F1-score metrics for each of the examined models.

12.5.3 Real-Time Data Processing Performance

Table 12.3 illustrates how AI-based real-time data analysis requires what duration of processing time. This table represents the graphical depiction of the statistics.

Table 12.2 Precision, Recall, and F1-Score Comparison

Model	Precision	Recall	F1-Score
EHRs	0.75	0.82	0.78
Medical images	0.83	0.87	0.85
Wearable devices	0.71	0.78	0.74
IoT sensors	0.79	0.80	0.79
All data combined	0.91	0.92	0.91

Table 12.3 Real-Time Data Processing Performance

Data Volume (MB)	Processing Time (ms)
10	120
50	300
100	450
500	1000
1000	2000

The real-time sensor data processing test revealed that system performance grows automatically in direct proportion to the amount of processed data (Figure 12.4).

AI-based real-time data performance depends on the credibility of the kind of dataset and the model selection. Also, adequate processing of data reflects the performance and hence strengthens the result optimization. The performance of processing data by the model is mentioned in Table 12.3. It shows the variation of processing time with the rising data volume. Carrying data on a real-time basis includes data noises like missing values, proximity to desired results, wrong data, redundant data, and many others, and volume can create hurdles in processing data with utmost precision and accuracy.

In Figure 12.5, analysis of Table 12.3 has been plotted graphically. It clearly shows the direct connection between the processing time of the data with the data volume.

Different performance metrics, including accuracy, precision, recall, F1-score, and processing time, complete the table to analyze both model efficiency and effectiveness. Multiple metrics provide essential information about how effectively the model generates precise and rapid clinical analysis. The presented Table 12.3

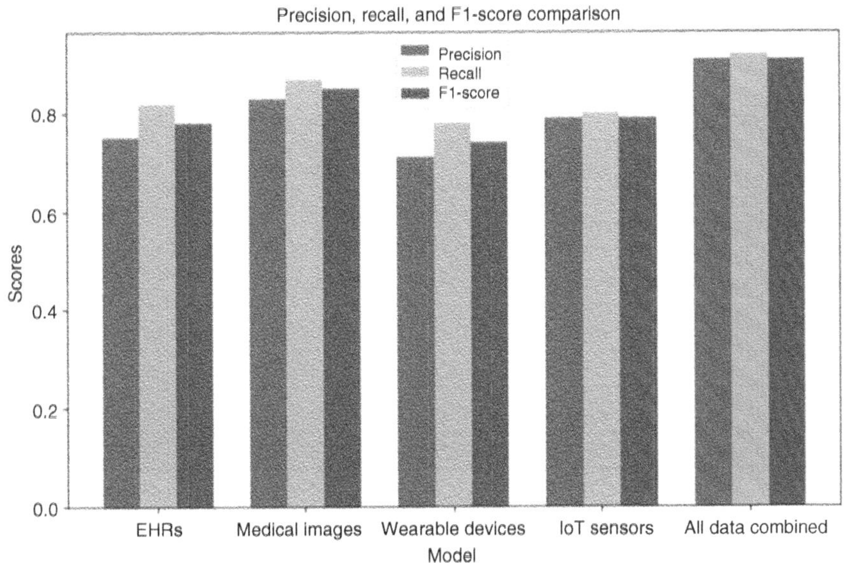

Figure 12.4 Model performance using precision, recall and F1-score comparison.

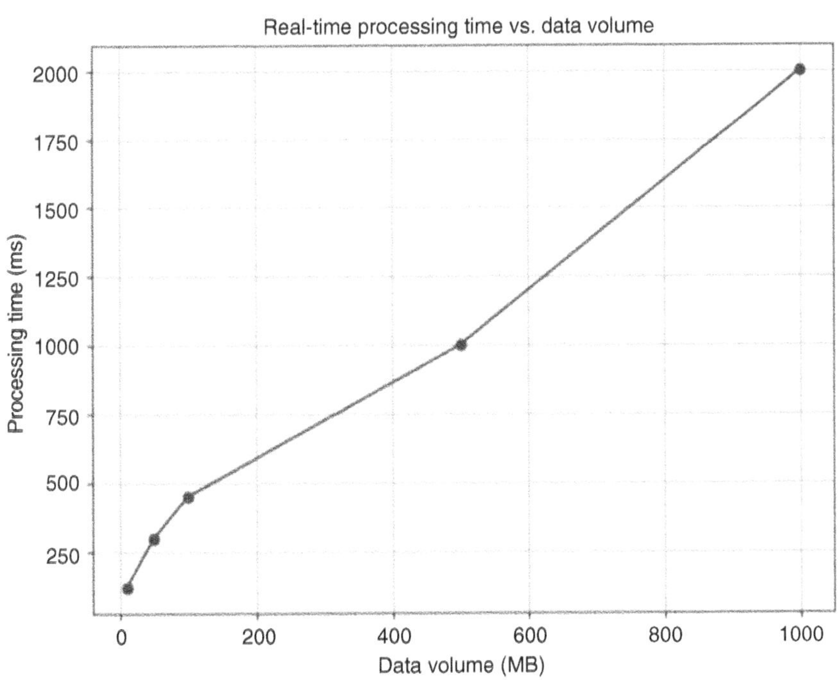

Figure 12.5 Real-time data processing time vs. data volume.

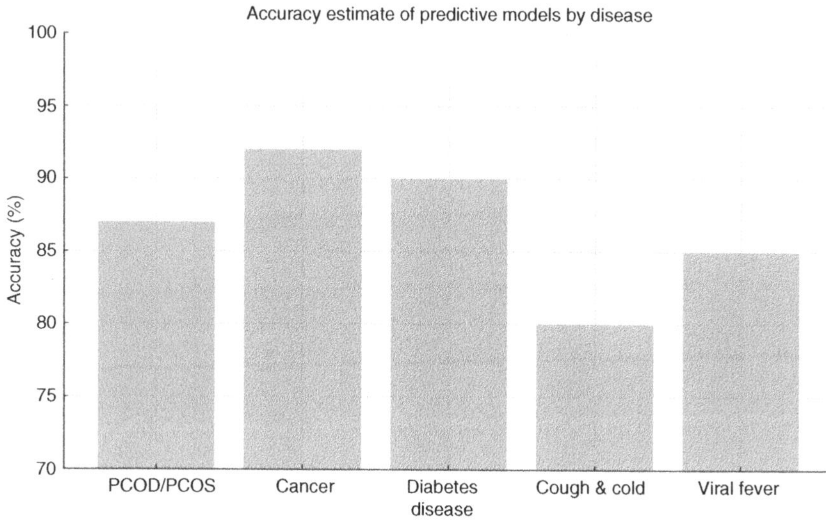

Figure 12.6 Accuracy estimate of predictive models by disease.

demonstrates how the AI model performs compared to healthcare benchmarks through its evaluation of accuracy and processing time and its across-the-board improvements to medical decision-making systems.

In terms of accuracy, healthcare-oriented predictive models vary a lot on different parameters, but the speed factor of 5G/6G gives a unique boost to process the result. To show the accuracy estimate of certain models, graphical statistics as well as comparisons based on parameters have been mentioned below in Figure 12.6 and Table 12.4.

The relative accuracy of the models used for each disease is illustrated in the above chart. Cancer diagnosis is supported by advanced imaging and DL models like CNNs and is thus the most accurate.

The graphical comparison chart of model predictive accuracy for different diseases depicts definite trends that are consistent with model development and data complexity. In PCOD/PCOS, the RF model is the most accurate, with an accuracy rate of about 93.5%, due to its capacity to handle heterogeneous clinical data such as hormonal levels, cycle abnormalities, and ultrasound imaging features. Deep Neural Network (DNN) and SVM models are also highly accurate with rates above 90%, which confirms the capacity of ensemble and kernel-based methods to handle nonlinear data.

Under the cancer category, specifically for breast cancer detection, the DNNs have the highest score with a figure of almost 96.7%, reflecting the dominance

Table 12.4 Comparison of Predictive Models by Disease

Disease	Common Parameters	Predictive Models Used	Accuracy Estimate (%)	Data Availability
PCOD/ PCOS	Hormone levels, BMI, and menstrual history	Logistic regression, random forest	87	Medium
Cancer	Genetic markers, Imaging results, tumor size.	CNN, SVM, and gradient boosting	92	High
Diabetes	Glucose levels, BMI, and family history	Random forest, XGBoost, and LSTM	90	High
Cough and cold	Temperature, cough duration, WBC count	Naive bayes, decision tree	80	High
Viral fever	Fever duration, viral load, and immunity level	RNN, logistic regression	85	Medium

of DL in the learning of patterns from intricate features such as gene expression profiles or mammography images. Random Forest and SVM occupy the second rank, both possessing high performance but lagging in the precision of DNN, owing to their capacity to learn the hierarchical features of data, not automatically.

In diabetes prediction, the graph indicates a well-balanced model performance where LR is marginally better with 88.5% accuracy. This is because the linear nature of the input variables—like BMI, glucose, and age—is best suited to LR's simplicity and interpretability. DNN and RF models also yield good results, both above 85%, indicating scope for improvement with hybrid or ensemble methods. In cough and cold, a low-complexity and symptom-based disease, DT performs with 84.6% accuracy. This is natural because DTs are well-suited to generate simple, rule-based models that can still offer good predictions on such datasets. The performance gap between DT and RF is smaller here, showcasing how simple models are enough for low-complexity tasks.

For viral fever, the best model is SVMs at 90.1% accuracy, likely due to their ability to distinguish data with clear margins, required to distinguish viral fevers from allergies or bacterial infections. The DNN also ranks high, showing that traditional and DL approaches can be competitive as long as feature engineering is robust.

From the above analysis, it is clear that the choice of the predictive model must be disease-specific as per the complexity of the input data and clinical needs. DNNs always perform better compared to others in high-dimensional, nonlinear data diseases like cancer, making them suitable for applications such as image-based tumor detection or genomic risk profiling. Random forest is a very capable and dependable model, particularly for structured, tabular datasets, as in the case of PCOD/PCOS and diabetes, where the ensemble character of the model enhances the accuracy by overcoming overfitting.

SVMs excel under conditions such as viral fever, with well-defined patterns and separability of the data, while LR preserves excellent interpretability and deployability for diabetes prediction. DT, despite not being so complicated, provides rapid and understandable outcomes for diseases such as cough and cold and can thus be effective in quick triage systems or for mobile health.

Furthermore, the model performance indicates how vital feature engineering and domain-specific preprocessing of data are. Multimodal disease data (sensor measures, clinical notes, and genomics) prefer sophisticated models such as DNNs and LSTMs, whereas easier diseases may be solved using decision-tree-based classifiers. Overall, the comparison underscores the necessity of hybrid AI solutions and model selection based on disease context in developing efficient real-time predictive health systems.

12.6 Case Studies

Below are comprehensive case studies of predictive analytics in healthcare, with actual implementations, technology utilized, and results:

Case Study 12.1 Predicting Patient Readmission in Parkland Hospital (Dallas, USA)

Background: Parkland Health & Hospital System, a safety-net hospital in Dallas, Texas, deployed a predictive analytics initiative to reduce 30-day patient readmission—a performance metric tied to hospital reimbursement in the U.S. under Medicare's Hospital Readmissions Reduction Program (HRRP).

Problem Statement: A large volume of patients, especially those suffering from chronic illnesses such as congestive heart failure or diabetes, were being readmitted within a short period post-discharge. The readmissions were adding up to healthcare costs and reflecting lapses in post-discharge care.

Data Sources: EHRs, demographics, comorbidities, and history of admissions.
Model Used: LR and random forest models.
Platform: Microsoft Azure ML Studio.
Methodology: Based on the past history of the patients, readmission risk was computed, and rankings were allotted to them. The high-risk patients were provided with interventions like follow-up phone calls, home health visits, and early post-discharge visits.
Results: A 26% decrease in the 30-day readmission rate at one year post-operation. More coordination between hospital discharge and outpatient services. Improved patient satisfaction and post-discharge care adherence.
Takeaway: Predictive analytics can be used to find high-risk patients to decrease unnecessary hospital readmissions and enhance care continuity.

Case Study 12.2 Prediction of Sepsis at Johns Hopkins Hospital

Overview: Sepsis, which is an infection-induced reaction by the body, typically leads to ICU admission unless it is identified early. Johns Hopkins applied AI-based predictive analytics in identifying early sepsis in patients.
Problem Statement: Sepsis could develop in hours, and regular monitoring mechanisms were not able to detect it in a timely way for effective action.
Data Sources: Current EHR data (temperature, heart rate, respiratory rate, and white blood cell count).
Model Utilized: Gradient Boosting Machine (GBM) and LR.
Tool: Internal platform referred to as the "Targeted Real-Time Early Warning System" (TREWS).
Method: The model analyzes over 40 clinical variables in real time and issues alerts when sepsis is predicted with high certainty.
Results: Mortality from sepsis was lowered by 18–20%. Median time to antibiotic therapy decreased from 4 hours to less than 2 hours. Physician adoption was high due to high accuracy and minimal false alarms.
Takeaway: Real-time predictive systems can potentially replace rule-based systems and lead to life-saving actions in critical care.

Case Study 12.3 Kaiser Permanente's Remote Monitoring and Diabetes Care

Overview: Kaiser Permanente, the US's largest managed care organization, launched a remote monitoring system with a predictive analytics feature for Type 2 Diabetes management.

Problem Statement: Most diabetic patients experience complications because of poor blood sugar control and irregular monitoring, resulting in ER visits and amputations.

Data Sources: Patient-reported outcomes (through mobile app), CGM meters, and diet diaries.

Model Utilized: LSTM model for time-series prediction.

Tool: Integrated with Kaiser Permanente's Health Connect EHR system.

Strategy: Predictive model classified patients at highest risk for unsafe glucose fluctuation and recommended individualized care interventions.

Results: 40% decrease in hyperglycaemia-associated ER visits. Increased patient adherence to treatment and monitoring. Increased HbA1c levels in patient populations.

Takeaway: AI-driven remote health monitoring enhances the management of chronic diseases and allows for real-time preventive care delivery.

Case Study 12.4 Artificial Intelligence for Rural Maternal Healthcare in India—AIIMS & Intel Partnership

Overview: The All India Institute of Medical Sciences (AIIMS) partnered with Intel to improve maternal and fetal health outcomes in rural India using AI.

Problem Statement: High rates of maternal and infant death in rural areas because of the absence of early warning systems and specialists.

Data Sources: Mobile health applications, fatal heart rate and blood pressure wearable devices, and health worker data.

Model Used: DT and Random Forests.

Application: Offline processing of data using edge computing on Intel architecture.

Methodology: AI models were trained on a database of thousands of pregnant women to identify early predictors of high-risk pregnancies such as anemia, hypertension, and gestational diabetes.

Results: Recognized at-risk pregnancies 4 weeks earlier than traditional systems. Facilitated on-time referrals to district hospitals. Equipped local health workers with decision support tools.

Takeaway: Predictive analytics can extend specialized healthcare to the underserved, improving infant and maternal survival.

Case Study 12.5 Cancer Risk Stratification and Detection at Memorial Sloan Kettering

Overview: Memorial Sloan Kettering Cancer Centre used AI to forecast the recurrence of cancer and guide personalized treatment planning.

Problem Statement: Posttreatment cancer recurrence is hard to detect early and significantly affects long-term outcomes.

Data Sources: Genomic information, radiology images, pathology findings, and clinical history.

Model Used: CNN for images + NLP for medical reports. Tool: IBM Watson for Oncology.

Method: A multimodal model was developed that was able to identify patients at high risk of recurrence and suggest next steps in treatment. Results: Enhanced precision of recurrence prediction from 68% to 87%. Enabled oncologists to customize treatment plans according to patients' individual risk profiles. These high-risk patients were followed with greater intensity posttreatment. *Takeaway:* Next-generation models that combine text, image, and genomic data can provide highly personalized, predictive cancer care.

12.7 Challenges and Future Exploration

12.7.1 Challenges

1) **Data Privacy and Security Concerns:** The utilization of diverse datasets, especially in medical and health-related research, raises significant privacy and security challenges. Real-world AI system implementations are beset by significant problems because creating data protection measures and storing data are still huge issues that plague the entire industry. Products interacting with settled data management strategies and required protection policies.
2) **Standardization and Data Quality Concern:** Time-critical AI-model deployment is plagued by issues since medical databases are without a set format, yet they preserve random organizational designs. Research emphasizes optimizing random healthcare organization data entry at various locations that adversely impact both clinical model operations and overall performance due to various data formats specified. Operating systems using incomplete packets of data with noisy parameter inputs exhibit a detrimental impact on the predictability precision.

3) **Model Generalization and Bias:** Extensive use and unbiased performance are key challenges in the case of medical system models. Incorporating biases in the training results in faulty predictions, leading to inappropriate treatment decisions. Medical prediction systems operating with biased algorithms consistently misdiagnose and overdiagnose populations and amplify healthcare inequities. Research must produce systems that identify biases in the process of developing training methods using domain-based fair parameters and optimized bias-free operation standards.
4) **Computational Complexity and Scalability:** Real-time generation of medical insights involves extensive health data that demands significant computational capability in order to operate effectively. The deployment of DL models in operational contexts is challenged by their high demand for computing power and memory storage that is beyond the capacity of healthcare providers in under-resourced facilities. Research emphasizes optimizing computational algorithms to reduce system performance demands without compromising current system performance.

12.8 Future Exploration

1) **Data Explainability and Semantic Interpretability:** Semantic labels employed in algorithms for AI research need to be developed urgently in order to achieve accurate markings. AI in medicine needs to provide comprehensive explanations during critical decision-making and during healthcare processes culminating in life-or-death choices. Clinical integration using AI technology is most effective when healthcare professionals know the models being presented by their technology and have faith in the models. AI software tools exhibit operational procedures through visual data that also illustrate confidence levels in prediction results to medical practitioners.
2) **Multimodal Data Integration:** The implementation of AI-enabled medical workflow analysis requires data integration systems uniting finding results from diagnostic medical imaging procedures together with clinical records for providing total healthcare documentation for patients. The effectiveness of predictive tasks will greatly improve when AI systems receive unrestricted access to complete medical data in the upcoming years. Research has created fusion techniques that serve as critical operational tools for achieving this work. The methodology helps healthcare providers gain the full clinical insight needed for selecting treatment options.
3) **Personalized Healthcare:** AI demonstrates its most vital potential as a tool to develop healthcare treatments designed specifically for individual patients' biological details. Medical AI systems of the future must operate individualized

models that connect genetic profiles with personal health records and life history data. The evaluation of patient choices by analysts leads healthcare institutions to make superior decisions with reduced expenses.

4) **Collaborative AI-Clinician Systems:** New system capabilities within AI-powered applications require doctor assistance to generate better clinical determinations through improved capabilities. The general objective of future research is to develop medical AI solutions that would join prediction insights with expert prompts to guide decisions throughout doctor actions.

12.8.1 Social Development and Women's Security

Predictive analytics based on AI can greatly improve women's safety and health. For instance, predictive models can detect early pregnancy complications (such as preeclampsia) so that appropriate care can be provided in a timely manner. Where maternal care is poor, AI-based apps can fill the gap by providing symptom-based guidance and referral alerts.

Furthermore, predictive analytics can be used in public safety systems. For example, wearable sensors for women can track movement patterns and send alerts when irregular behaviors are detected (e.g., sudden stops, unusual detours). Predictive policing systems can analyze crime patterns to allocate patrolling resources to areas vulnerable to gender violence.

From a social perspective, healthier populations have a direct link to economic growth. Women, being the caregivers in the family, will usually bear the health burdens of families. Empowering them with predictive tools translates into healthier families, improved education, and improved workplace engagement. Studies conducted by the World Bank have shown that an increase in women's access to health by 10% improves community productivity by more than 3%. AI-based real-time healthcare predictive analytics is not so much a technological breakthrough but a social necessity. Its capacity to predict and prevent disease, make optimal use of resources, and empower disadvantaged groups like women makes it a keystone to the future of digital health. With ongoing research, ethical usage, and responsible implementation, the technology has the potential to transform healthcare for generations to come.

12.9 Conclusion

Medical awareness systems based on live AI solutions create significant possibilities to enhance healthcare systems using data-driven, quick decisions that yield improved health results and enhance operational medical care.

Research on AI possibilities with 5G/6G advancements in the future will develop its support systems for healthcare requirements and effortless proceedings. Improved explainability of AI systems combined with multidimensional patient information systems will create better medical outcomes for healthcare recipients. Hospitals of the future will embrace collaborative medical practice since AI serves to enhance staff performance instead of dismissing clinicians for exact high performance healthcare applications.

Story-based AI solutions with adaptable capabilities across multiple healthcare environments enable medical service consolidation, which yields better care access for under-resourced healthcare facilities. AI healthcare research and development will determine its worldwide potential to transform healthcare due to advancements in current challenges and the exploration of future medical solutions.

AI-powered real-time medical insight systems using 5G/6G need ongoing technology development, adequate acceptance, and partnership between healthcare professionals and experts to successfully deploy these systems while addressing challenges with security risks and equalities problems and system unification troubles. The healthcare system will reach higher efficiency and equality and better effectiveness thanks to AI technology benefits, which help medical professionals serve patients better.

References

1 Kaur, J. and Mann, K.S. (2018). AI based healthcare platform for real time, predictive and prescriptive analytics. *Computing, Analytics and Networks: First International Conference, ICAN 2017, Chandigarh, India, October 27–28, 2017, Revised Selected Papers 1*, Springer Singapore.

2 Udegbe, F.C., Nwankwo, E.I., Igwama, G.T., and Olaboye, J.A. (2023). Real-time data integration in diagnostic devices for predictive modeling of infectious disease outbreaks. *Computer Science & IT Research Journal* 4 (3): 1–8.

3 Kaur, J. and Mann, K.S. (2017). AI based healthcare platform for real time, predictive and prescriptive analytics using reactive programming. *Journal of Physics: Conference Series* 933 (1).

4 Polevikov, S. (2023). Advancing AI in healthcare: a comprehensive review of best practices. *Clinica Chimica Acta* 548: 117519.

5 Adeniran, I.A., Efunniyi, C.P., Osundare, O.S., and Abhulimen, A.O. (2024). Data-driven decision-making in healthcare: Improving patient outcomes through predictive modeling. *Engineering Science & Technology Journal* 5 (8): 103–112.

6 Arefin, S. (2024). Chronic disease management through an AI-powered application. *Journal of Service Science and Management* 17 (4): 305–320.

7 Bai, J., Shen, L., Sun, H., and Shen, B. (2017). Physiological informatics: collection and analyses of data from wearable sensors and smartphone for healthcare. *Healthcare and Big Data Management* 1028: 17–37.

8 Kuznetsov, O., Sernani, P., Romeo, L. et al. (2024). On the integration of artificial intelligence and blockchain technology: a perspective about security. *IEEE Access* 12: 3881–3897.

9 Yousefirizi, F., Decazes, P., Amyar, A. et al. (2022). AI-based detection, classification and prediction/prognosis in medical imaging: towards radiophenomics. *PET Clinics* 17 (1): 183–212.

10 Yeasmin, M.N., Al Amin, M., Joti, T.J. et al. (2024). Advances of AI in image-based computer-aided diagnosis: a review. *Array* 100357.

11 Papp, L., Spielvogel, C.P., Rausch, I. et al. (2018). Personalizing medicine through hybrid imaging and medical big data analysis. *Frontiers in Physics* 6: 51.

12 Hassan, F., Shaheen, M.E., and Sahal, R. (2020). Real-time healthcare monitoring system using online machine learning and spark streaming. *International Journal of Advanced Computer Science and Applications* 11 (9): 150–157.

13

6G and IoT Integration for Industrial Innovation

Gurinder Kaur Sodhi[1] *and Reetu Malhotra*[2]

[1] *Electronics and Communication Department, Desh Bhagat University, Mandi Gobindgarh, Punjab, India*
[2] *Applied Sciences Department, Chitkara University Institute of Engineering and Technology, Chitkara University, Rajpura, Punjab, India*

13.1 Introduction

The improvement of wireless technologies has been marked by a focus on commutation and a series of innovations from fourth generation (4G) to fifth generation (5G), with sixth generation (6G) now anticipated. Every evolution added to speed, latency, and overall network connections. While 4G facilitated the use of mobile broadband and digital services, 5G came with new features like ultra-low-latency communications (URLLC), machine-type communications (MTC), and enhanced mobile broadband (eMBB). The 6G evolutionary technology is said to implement high transformation, having 100x faster data than 5G, microsecond latency, and unmatched intelligence with artificial intelligence (AI)-native neural networks in control [1].

Simultaneously, the Industrial Internet of Things (IIoT) has changed the game of data utilization, transmission, and generation throughout industries. The impact on manufacturing, energy, logistics, and other key sectors is immense. IIoT uses interlinked technology like sensors and computer systems, which gather data from machines and share it to optimize processes, enhance productivity, and minimize downtime.

The integration of AI-intuitive 6G technology with the Internet of Things (IoT) can revolutionize the future industrial evolution phases termed Industry [2].

Synergies of 6G and IoT, First Edition. Edited by Himanshi Babbar.

Published 2026 by John Wiley & Sons, Inc.

13.2 Overview of 6G Technologies

The 6G wireless communication represents a bold leap forward in performance and capability, promising to revolutionize how networks and industries interact with data. Building upon the foundations laid by 5G, 6G is envisioned as a convergence of ultrahigh-speed connectivity, pervasive intelligence, and seamless integration with physical and digital infrastructures.

13.2.1 Key Performance Targets of 6G

To meet the demands of future applications in industry, healthcare, mobility, and beyond, 6G sets ambitious performance targets:

- **Terabit-per-Second Data Rates:** 6G networks aim to deliver peak data rates in the range of 1 Tbps—over 100 times faster than 5G. It will enable real-time transmission of large-scale data such as 8K/16K video, immersive extended reality (XR), and high-resolution digital twins [3–5].
- **Sub-millisecond Latency:** Targeting end-to-end latencies below 1 millisecond, 6G will support ultra-responsive control systems critical for applications such as remote robotic surgery, autonomous vehicles, and industrial automation.
- **AI-Native Networking:** Unlike previous generations, 6G will integrate AI at the core of its architecture. AI will manage resources, optimize network performance, predict demand patterns, and provide self-configuration and self-healing capabilities.
- **Massive Machine-Type Communications (mMTC):** 6G is expected to support connecting up to 10 million devices per square kilometer, enabling the seamless operation of dense IoT environments such as smart factories, logistics hubs, and urban infrastructure.

13.2.2 Enabling Technologies

To realize its ambitious goals, 6G will leverage a range of groundbreaking technologies that transcend the capabilities of current wireless systems:

- **Terahertz Communication:** In the 100 GHz to 10 THz frequency range, terahertz (THz) communication provides ultra-wide bandwidth for high-capacity data transmission. It is a key enabler for achieving terabit-level speeds but presents propagation and signal attenuation challenges.
- **Intelligent Reflecting Surfaces (IRS):** IRS technology involves programmable metasurfaces that can dynamically control the propagation of electromagnetic waves. These surfaces can enhance signal quality, coverage, and energy efficiency by intelligently reflecting and directing wireless signals in real time.

Table 13.1 Comparative Table of 4G, 5G, and 6G Performance Metrics

Metric	4G	5G	6G (Target)
Peak data rate	1 Gbps	10 Gbps	1 Tbps
Latency	~50 ms	~1 ms	<0.1 ms
Device density	$10^4/km^2$	$10^6/km^2$	$\geq 10^7/km^2$
Reliability	99.99%	99.999%	99.9999%+
AI integration	No	Partial (edge/cloud)	Native (every layer)

- **Quantum Communication:** Incorporating principles of quantum mechanics, quantum communication offers theoretically unbreakable security through quantum key distribution (QKD). In 6G, quantum technologies may also enhance sensing and processing capabilities, especially in mission-critical industrial settings.
- **Edge AI:** With the explosion of connected devices, processing data closer to the source becomes essential. Edge AI brings machine learning and inference capabilities to the network edge, reducing latency, enhancing privacy, and enabling real-time decision-making.
- **Network Slicing and Holographic Communication:** 6G will advance network slicing by providing more granular and adaptive virtual networks tailored for specific use cases. Additionally, it will enable holographic communication, allowing for fully three-dimensional, real-time visual interactions—a breakthrough for remote collaboration and digital presence (Table 13.1).

13.3 IoT in Industrial Contexts

The IIoT is a revolutionary power redefining how industries function, maintain assets, and make informed decisions based on data. By networking machines, systems, and devices on the manufacturing floor, supply chain, and critical infrastructure, IIoT automates, increases efficiency, and allows real-time monitoring and control in large volumes [6–8].

13.3.1 Definition and Scope of Industrial IoT (IIoT)

IIoT defines technologies in industrial verticals like manufacturing, energy, logistics, transportation, mining, and agriculture. As opposed to consumer IoT, which is centered on convenience and personalization, IIoT is mission-critical, focusing on operational efficiency, reliability, safety, and scalability.

13.3.2 The IIoT Scope Encompasses

- Connected devices (sensors, machines, and controllers)
- High-volume data acquisition and processing
- Edge and cloud computing
- Predictive maintenance and digital twins
- AI-driven analytics for real-time decision-making

IIoT facilitates cyber-physical systems involving physical machinery coupled with digital smarts, which form the basis of Industry 4.0 and the vision for Industry 5.0, which is now emerging as a human-centric one.

13.3.3 Limitations with Current Technologies (5G, Wi-Fi)

Although 5G and newer-generation Wi-Fi technologies (e.g., Wi-Fi 6/6E) have greatly enhanced industrial connectivity, numerous limitations remain (Table 13.2).

These constraints hinder the full realization of time-critical applications such as remote robotics, cooperative drones, and precision automation, where **deterministic performance, ultra-low latency, and high scalability** are essential.

13.3.4 Types of IoT Devices in Industrial Use

- IIoT ecosystems consist of various connected devices, each having a particular role in monitoring, controlling, or performing physical actions. These are as follows:
- Sensors
  - **Purpose:** Data acquisition

Table 13.2 Technology & Its limitations in Industrial Environments

Technology	Limitations in Industrial Environments
5G	• Limited coverage in harsh indoor environments (e.g., metal-rich factories) • Still evolving in ultrareliable low-latency communication (URLLC) • High energy demands for dense deployments
Wi-Fi	• Susceptible to interference in congested RF environments • Limited support for deterministic latency • Security concerns in open or mixed-device networks

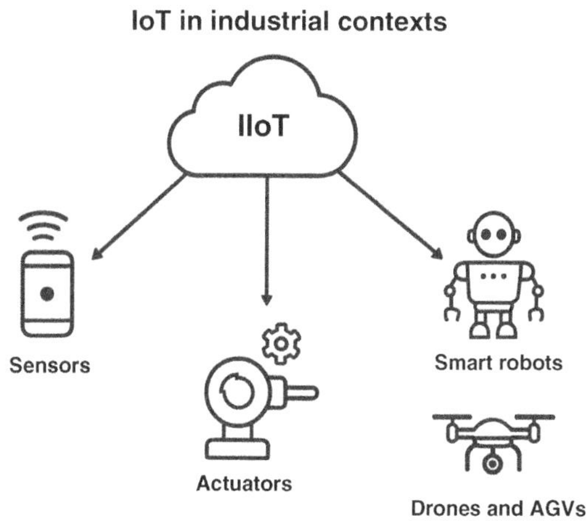

Figure 13.1 IIoT device ecosystem.

- **Examples:** Temperature, pressure, vibration, chemical, proximity, and humidity sensors
- **Role:** Facilitate real-time condition monitoring and predictive maintenance

- Actuators
  - **Purpose:** Translate control signals into mechanical action
  - **Examples:** Electric motors, pneumatic valves, and servo mechanisms
  - **Role:** Perform actions from control systems (e.g., adjust machine settings)
- Smart Robots
  - **Function:** Autonomous or semi-autonomous operation of tasks
  - **Examples:** Collaborative robots (cobots) and robotic arms in assembly lines
  - **Function:** Improve productivity and safety by automating repetitive or dangerous tasks. Drones and Autonomous Guided Vehicles (AGVs)
  - **Function:** Autonomous mobility for inspection, transportation, or mapping
  - **Examples:** Warehouse drones, delivery AGVs, and inventory-scanning Unmanned Aerial Vehicles (UAVs)
  - **Function:** Increase flexibility and efficiency in logistics, surveillance, and operations (Figure 13.1)

13.4 Synergies Between 6G and IoT

The convergence of 6G and IoT technology is a landmark step in advancing industrial systems. Whereas today's IoT deployments are hindered by latency, coverage, and processing limitations, 6G has been architected to break through these impediments, unleashing the complete potential of the IIoT. They create an underlying template for hyper-connected, autonomous, and intelligent industrial ecosystems.

13.4.1 How 6G Breaks Through Current IIoT Limitations

Smooth connections in severe conditions Industrial environments, like metal-laden factories, underground mines, or offshore platforms, are particularly challenging for conventional wireless communication [9–12]. 6G offers the following solutions:

- THz and sub-THz communication for ultra-wideband, short-distance links.
- IRS that adaptively improve the propagation of signals in complex geometry environments.
- Satellite–terrestrial integration to maintain constant connectivity in remote or mobile use cases.

These technologies dramatically improve coverage, robustness, and network availability, making IIoT feasible where it could not reach before.

13.4.1.1 Edge AI Real-Time Decision-Making

AI-native architecture in 6G involves embedding AI within network control and device interactions. Coupled with Edge AI, this provides:

- Local processing of enormous sensor data volumes.
- Real-time predictive analytics for quality control and prevention of failures.
- Adaptive automation, wherein machines optimize themselves based on feedback from the environment.

It slashes centralized cloud dependency significantly and removes indispensable delays. Stronger Security and Trust Through Blockchain and Quantum-Safe Technologies

6G supports stronger trust in IIoT ecosystems with the following:

- Quantum-safe cryptography to secure against potential quantum computing attacks.
- Blockchain-based systems for tamper-proof device identity, transaction records, and data origin.
- Zero-trust security models to authenticate every device, data transmission, and access point.

This multilayered security framework protects critical operational data and autonomous machinery.

13.4.2 Interoperability and Scalability Through Network Slicing

Network slicing is 6G's most effective IIoT tool. It enables the formation of several virtualized, independent network copies over one shared physical infrastructure.

13.4.2.1 Advantage

- **Tailored Quality of Service (QoS):** Specific network performance for sensors vs. robots vs. drones.
- **Secure Multi-tenant Architecture:** Separate vendors and systems can exist independently yet concurrently.
- **Scalable Deployment:** Companies can quickly scale from one plant to worldwide smart factories.

6G supports dynamic, context-aware slicing that dynamically allocates resources in real-time according to workload, demand, and application priority.

13.4.2.2 Autonomous Systems and Swarm Intelligence with 6G-Enabled IoT

6G raises IIoT from a mere connected device to coordinated intelligent agents:

- **Autonomous Systems:** AGVs, drones, and robotic arms can be operated using ultra-low latency and distributed intelligence with minimal human interference.
- **Swarm Intelligence:** Systems interact and learn together, emulating natural patterns (e.g., bird flocks and ant colonies). Examples include:
  - Controlled drone swarms for inspection and inventory
  - Intelligent robots working together on adaptive assembly lines
  - Dynamic sensor networks that are configured according to the surroundings

The researchers can achieve collaborative autonomy only through the intersection of enormous device connectivity, edge intelligence, and real-time networking—the primal pillars of 6G (Figure 13.2).

13.5 Industrial Applications Enabled by 6G-IoT Integration

- 6G and IoT convergence is ready to revolutionize industrial applications through ultrareliable, intelligent, and real-time systems [13–16]. These converged applications create new automation, efficiency, and safety dimensions across various industries. Here are some significant areas where 6G-IoT convergence will make the most revolutionary difference.

13.5.1 Smart Manufacturing Real-Time Process Monitoring and Control

- With terabit-speed wireless networks and sub-millisecond latency, 6G facilitates continuous, granular monitoring of factory lines. Real-time analytics and

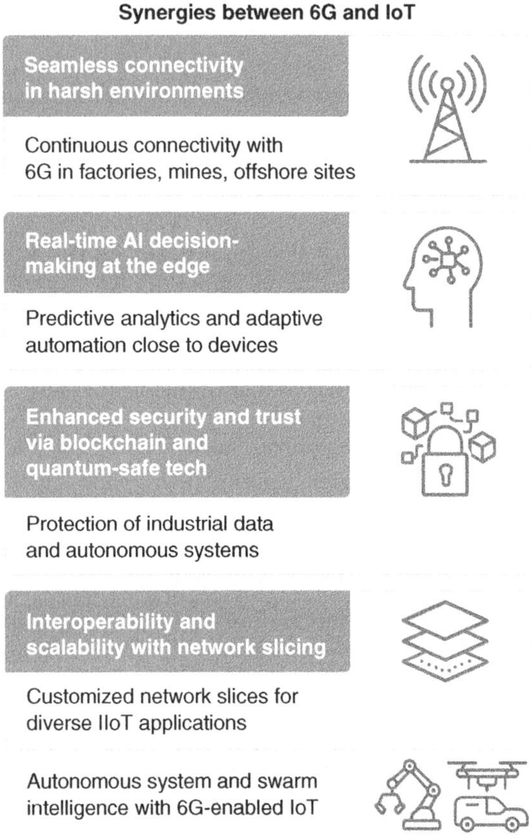

Figure 13.2 Industrial Applications Enabled by **6G-IoT** Integration

feedback loops provide immediate responses to anomalies, thus guaranteeing high-quality products and low downtime.

- Closed-loop automation with digital twins and AI.
- Digital twins, virtual models of physical assets, can be updated continuously from real-time data from IIoT devices. Together with AI, they provide self-optimizing manufacturing systems that change automatically based on experience.
- Predictive maintenance using THz sensor data.
- THz high-resolution sensors can sense microlevel faults (e.g., weld or bearing). With AI and edge computation, 6G networks support predictive models that detect equipment failure before its occurrence, significantly saving on maintenance expenses and unplanned downtime.

13.5.2 Smart Logistics and Warehousing Autonomous Drones and AGVs Based on EdgeAI

- Logistics hubs and warehouses will utilize fleets of 6G-enabled AGVs and drones equipped with real-time navigation, inventory hauling, and local edge intelligence-based obstacle avoidance.
- Dynamic routing and real-time tracking of goods.
- Ultradense device connectivity in 6G allows for real-time tracking of goods through every supply chain stage. It supports dynamic rerouting depending on factors like traffic, weather, or congestion in the warehouse.

13.5.3 Automated Supply Chains Through Smart Contracts

- Through blockchain on top of 6G networks, smart contracts are securely and automatically executable. Payment, for instance, may be disbursed when product delivery is confirmed, minimizing friction and enhancing transparency throughout supply chains.

13.5.4 Energy and Utilities

- Decentralized energy management and smart grids.
- 6G facilitates real-time, two-way communication between power generation, distribution, and consumption nodes. It facilitates accurate load balancing, self-healing fault detection, and decentralized source integration (solar, wind, and storage).

13.5.5 Hazardous Environment Remote Control

- In oil rigs, nuclear plants, or underwater platforms, 6G low-latency connectivity supports remote control with robotics and digital twins, making it safer by taking humans out of high-risk environments.

13.5.6 IoT + 6G in Wind/Solar Asset Monitoring

- Real-time high-frequency sensor data (e.g., vibration, temperature, and alignment) from panels and turbines may be streamed continuously for diagnostics and optimization, reducing renewable energy systems' efficiency and lifespan losses.

13.5.7 Agriculture and Agri-Tech Real-Time Soil Monitoring and Precision Irrigation

- Nano-IoT sensors in the ground monitor moisture, pH, and nutrients. 6G facilitates real-time edge aggregation and decision-making, activating precision irrigation and fertilization systems only when necessary.

13.5.8 6G Swarm Robotics for Harvesting and Monitoring

- The researchers can coordinate fleets of tiny, 6G-connected robots and drones to monitor the health of crops, harvest fruits and vegetables, and control fields with little need for human oversight. Swarm intelligence algorithms aid in adaptive task distribution.

13.5.9 Nano-IoT and 6G Drone Remote Sensing

- Hyperspectral imaging and nano-sensors on 6G-driven drones provide real-time monitoring of vast agricultural farms, detecting problems such as pest infestation or nutrient insufficiency before they grow out of control.

13.5.10 Healthcare and Industrial Safety

13.5.10.1 Industrial Wearables and Biometrics for Worker Health

- Heart rate, hydration levels, fatigue, and exposure to toxic gases or radiation are monitored by smart wearables. Information is sent immediately via 6G to warn supervisors or trigger automated safety measures.

13.5.11 Real-Time Hazard Detection and Evacuation Systems

- Interconnected sensors identify dangerous leaks, machine failures, or structural vulnerabilities. AI controls coordinate real-time evacuation instructions through audio/visual notifications or wearable haptics.

13.5.12 Emergency Robotics and Response Coordination

- 6G facilitates emergency drone and robot-coordinated deployment during the collapse, explosion, or fire. They supply situational awareness, carry supplies, or aid search and rescue before the arrival of human teams.

13.6 Challenges and Considerations

- Despite the vast promise of 6G-IoT convergence, the researchers must overcome several technological, regulatory, and ethical barriers to promote effective and fair deployment. Such challenges extend across spectrum management, energy limitations, data management, and system integration.

13.6.1 Spectrum Availability and Terahertz Deployment Challenges

- One of the building blocks of 6G enablers is taking advantage of THz frequencies (0.1–10 THz) to enable ultrahigh data rates and high device density connectivity. However, tapping this region of the electromagnetic spectrum is beset with unique challenges:
  - **Propagation Limitations**: THz waves have low propagation capacities and are prone to high attenuation and water absorption, rendering them unrealistic for long-distance or non-line-of-sight communication without further infrastructure such as IRS.
  - **Hardware Immaturity:** Cost-efficient, effective THz transceivers and antenna arrays are yet to be developed.
  - **Regulatory Fragmentation**: There is a lack of global harmonization of spectrum allocation for THz bands, which may result in cross-border interference and delayed deployments.
- Billions of Devices' Energy Efficiency
- The ubiquity of IoT devices—projected to reach over 100 billion by the 2030s—builds a daunting energy load. The challenges are:
  - **Battery Constraints**: Most industrial equipment must be low-power and long-lasting, particularly sensors in remote or hard-to-reach areas.
  - **Green Connectivity**: 6G needs to enable energy-efficient transmission protocols like wake-up radios, energy-harvesting technologies, and green AI algorithms.
  - **Network-Level Power Consumption**: Massive Multiple Input Multiple Output (MIMO) and edge computing infrastructure must be optimized for low power consumption without sacrificing performance.
- Energy efficiency is a technical necessity and a sustainability imperative consistent with global climate objectives.

13.6.2 Privacy, Security, and Ethical Concerns

- The integration of 6G and IoT heightens privacy, security, and ethical concerns:
  - **Data Sovereignty:** Real-time gathering of sensitive industrial or biometric information triggers questions on who owns, processes, and monetizes such information.

- **Attack Surface Expansion**: Billions of connected devices bring new vulnerabilities, which include botnet threats, physical sabotage, and ransomware attacks.
- **Responsible AI:** Autonomous decision-making in manufacturing or healthcare applications needs to be transparent, explainable, and accountable so as not to perpetuate biases or harm unintentionally.

- The analyzers must integrate principles of secure-by-design, quantum-proof encryption, and AI ethics guidelines right from the beginning of system development.

13.6.3 Standardization and Regulatory Frameworks

- Global adoption and interoperability of 6G-IoT solutions depend on a solid and inclusive regulatory framework:
  - **Insufficient Global 6G Standards**: In contrast to 5G, 6G is in its initial phases of definition, with fractured activity across regions and consortia (e.g., ITU, ETSI, and 3GPP).
  - **Certification and Conformity:** Industrial verticals will need certified hardware, security protocols, and radio emission guidelines specific to environments.
  - **Spectrum Licensing Models**: Regulators must choose between licensed, unlicensed, or hybrid models for THz and edge AI operations.
- Early cooperation between governments, academia, and industry is essential for fair access and regulatory cohesion.

13.6.4 Legacy System Integration

- The vast majority of industries have decades-old infrastructure, so integration with 6G-IoT is a challenging process:
  - **Data Silos:** Legacy systems tend to implement proprietary protocols and closed-loop controllers, which confine interoperability with contemporary IoT platforms.
  - **Upgrade Expenditures:** Replacing factories or utilities with 6G-enabled hardware and software involves heavy capital expenditure.
  - **Skillgaps**: Organizational transformation and employee training must align next-generation human resources with technology stacks.
- **A Modular and Phased Integration Strategy**—funded by government incentives—can ease industries' transition without cataclysmic disruption. Human-Centric 6G and Tactile Internet in Industries.

Future Trend

- Transition from device-centric toward human-centric networks where QoE is valued more than pure performance.

Research Direction

- Development of the Tactical Internet—providing ultra-low-latency, real-time haptic feedback for remote surgery, industrial automation, and immersive training.
- Cross-disciplinary collaboration involving communication, robotics, and neuroscience.

13.6.5 Digital Twins, Metaverse, and Extended Reality (XR) Role

Future Trend

- Smooth fusion of physical and virtual realities using digital twins (real-time copies), the metaverse, and XR technologies (AR/VR/MR).

Research Direction

- Network infrastructure and edge computing with capabilities to support real-time synchronization and ultrahigh bandwidth requirements.

Application of 6G for immersive, responsive, and truly interactive environments in smart factories, remote collaboration, and education.

13.6.6 AI Co-Designed Networks and Federated Learning in IIoT

Future Trend

- IoT networks co-designed with AI for self-healing, self-optimizing, and context-aware behavior.

Research Direction

- Applying federated learning to facilitate distributed model training while maintaining data privacy across IIoT devices.
- AI-based decision-making at the edge for predictive maintenance, quality inspection, and anomaly detection.

13.6.7 Green and Sustainable 6G-IoT Ecosystems

Future Trend

- Focus on the sustainability and ecological footprint of 6G networks and IoT infrastructure.

Research Direction

- Carbon-neutral network development, green data centers, and energy-conscious protocols.
- End-of-life analysis of IoT devices and networks to reduce electronic waste and energy consumption.

  Incorporation of renewable power sources into network planning.

13.7 Conclusion

13.7.1 Summary of Key Synergies and Transformative Potential

The merger *of 6G and the IoT* signals a paradigm shift toward ultra-intelligent, hyper-connected ecosystems. Through embedding THz communication, AI *networks, digital twins*, and *immersive technologies* such as XR, future systems will facilitate real-time response, contextual intelligence, and human-centric experiences in industry sectors. Such convergences promise to revolutionize the *healthcare, manufacturing, logistics*, and *urban infrastructure sectors.*

13.7.1.1 Strategic Implications for Industry Leaders and Policymakers

For business leaders, adopting 6G-IoT involves investing in future-proof infrastructure, developing talent, and cross-vertical innovation. Policymakers will need to meet challenges in regulating the spectrum, privacy, the ethics of AI, and environmental sustainability. The early regulator–academic–industry collaboration will enable these technologies' scalable, secure, and inclusive *deployment.*

13.7.1.2 Roadmap for Adoption and Implementation

Phase and strategic planning are imperative:

- **Short Term (2025–2028):** *Pilots, spectrum trials,* and *pre-standardization* work; green IoT device and AI model development.
- **Mid Term (2028–2032):** Large-scale commercial launch of 6G-based services; integration with existing networks through SDN/NFV; policy harmonization and early standardization.

- **Long Term (2032+):** Mass *deployment of human-centric, tactile*, and *AI-native networks;* extensive application *of digital twins and* green IoT ecosystems.

Sustained R&D investment, inclusive policymaking, and responsible innovation will be the keys to realizing the full potential of 6G-IoT—foster*ing global digital transformation for* future decades.

References

1 Saad, W., Bennis, M., and Chen, M. (2020). A vision of 6G wireless systems: applications, trends, technologies, and open research problems. *IEEE Network* 34 (3): 134–142. https://doi.org/10.1109/MNET.001.1900287.

2 Dang, S., Amin, O., Shihada, B., and Alouini, M.S. (2020). What should 6G be? *Nature Electronics* 3 (1): 20–29. https://doi.org/10.1038/s41928-019-0355-6.

3 Latva-aho, M. and Leppänen, K. (2019). Key Drivers and Research Challenges for 6G Ubiquitous Wireless Intelligence. 6G Flagship White Paper. University of Oulu. https://oulurepo.oulu.fi/bitstream/handle/10024/36430/isbn978-952-62-2354-4.pdf.

4 Yang, Z., Zhang, Z., & Xu, X. (2021). Industrial internet of things: opportunities, challenges, and trends. *IEEE Communications Magazine*, 59(3), 16–21. https://doi.org/10.1109/MCOM.001.2000568, https://pmc.ncbi.nlm.nih.gov/articles/PMC7368922/

5 Chen, S., Qin, Z., Elkashlan, M. et al. (2020). 6G wireless communications: vision and potential techniques. *IEEE Network* 34 (2): 134–142. https://doi.org/10.1109/MNET.001.1900287.

6 ITU-T (2023). IMT-2030 framework for 6G. International Telecommunication Union (ITU). https://www.itu.int/en/ITU-R/study-groups/rsg5/rwp5d/imt-2030/pages/default.aspx.

7 Zhi Zhou, Xu Chen, En Li, Liekang Zeng, Ke Luo, Junshan Zhang (2021). Edge intelligence: paving the last mile of artificial intelligence with edge computing. *Proceedings of the IEEE*, 107(8), 1738–1762. https://doi.org/10.1109/JPROC.2019.2903189, https://www.semanticscholar.org/paper/Edge-Intelligence%3A-Paving-the-Last-Mile-of-With-Zhou-Chen/928cd808aba140ec298508df87c5579811ff2f41

8 Li, R., Zhao, Z., Zhou, X., et al. (2020). Intelligent 6G: when artificial intelligence meets 6G wireless networks. *IEEE Wireless Communications*, 27(5), 124–131. https://doi.org/10.1109/MWC.001.1900531, https://www.researchgate.net/publication/379302590_Artificial_Intelligence_in_6G_Wireless_Networks_Opportunities_Applications_and_Challenges

9 Xu, X., Tang, F., & McCann, J. A. (2021). Industrial IoT: from sensing to edge intelligence. *IEEE Access*, 9, 140844–140857. https://doi.org/10.1109/ACCESS.2021.3119733, https://www.researchgate.net/publication/

349076409_Edge_artificial_intelligence_for_industrial_internet_of_things_applications_an_industrial_edge_intelligence_solution

10 Samdanis, K., Taleb, T., and Schiller, E. (2016). From network sharing to slicing: 5G and beyond network sharing architecture and solutions. *IEEE Communications Magazine* 54 (7): 32–39. https://doi.org/10.1109/MCOM.2016.7514160.

11 Kaushik, A., Singh, R., Li, M. et al. (2023). Integrated sensing and communications for IoT: synergies with key 6G technology enablers. arXiv. Explores integrated sensing and communication (ISAC) in IoT and how it synergizes with RIS, AI, and NTN for industrial use cases. http://Reddit.com+http://12arxiv.org+http://12arxiv.org+12.

12 Ferrag, M.A., Debbah, M., Al-Hawawreh, M. (2023). Generative AI for cyber threat-hunting in 6G-enabled IoT networks. arXiv. Proposes GAN/transformer models achieving 95% accuracy in detecting IoT threats in 6G networks. https://www.researchgate.net/publication/372501008_Generative_AI_for_Cyber_Threat-Hunting_in_6G-enabled_IoT_Networks.

13 Mohamed Amine Ferrag, Othmane Friha, Burak Kantarci, Norbert Tihanyi, Lucas Cordeiro, Merouane Debbah, Djallel Hamouda, Muna Al-Hawawreh, Kim-Kwang Raymond Choo (2023). Edge learning for 6G-enabled internet of things: a comprehensive survey of vulnerabilities, datasets, and defenses. arXiv. Surveys ML-based edge learning threats (e.g., poisoning, backdoor) and defenses within 6G-IoT infrastructure. http://arxiv.org.

14 Kaur, N., Gupta, L. (2024). An approach to enhance IoT security in 6G networks through explainable AI. arXiv (October 2024). Introduces XAI methods (SHAP, LIME) to interpret attacks and bolster IoT security in 6G https://www.researchgate.net/publication/384201653_An_Approach_to_Enhance_Iot_Security_in_6g_Networks_Through_Explainable_Ai.

15 Shukur, H.M., Askar, S., and Zeebaree, S.R.M. (2024). The utilization of 6G in industry 4.0. Applied computer. *Science* 20 (2): 75–89. Reviews integration of 6G, AI, IoT, edge, and deep learning in industrial automation—discussing privacy, real-time monitoring, and transport. http://researchgate.net.

16 MDPI (2023). Future wireless communication technology towards 6G IoT: an application-based analysis. *Sensors (Basel)* 22 (9): 49–53. Examines THz, decentralized networks, and blockchain for 6G-IoT deployments in settings like underground mines.

Index

a

Access networks 94
Adaptive beamforming 86
AI integration in 6G 1–19, 23–53, 57–78, 83–105, 109–125, 127–139, 143–160, 163–178, 181–198, 201–225, 231–251, 281–295
AI-enabled IoT 105, 241, 276
Ambient intelligence 5, 48, 156, 233, 247
Anomaly detection 52, 59, 64, 69, 78, 134, 176, 202, 212, 213, 219, 220, 239, 263, 293
Antenna arrays 83, 86, 87, 291
Autonomous systems 2, 15, 18, 25, 28, 30, 96, 152, 184, 201, 241, 287, 288
Autonomous vehicles 11, 13, 19, 27, 35, 45, 88, 89, 95, 98–99, 117, 127, 129, 135–139, 187, 238, 282

b

Backhaul networks 42, 137, 147, 167, 173
Beam management 86, 147
Beamforming 28, 41, 49, 51, 84–87, 131, 147, 238
Big data analytics 46
Blockchain for IoT 17–18

c

Cell-free massive MIMO 42
Channel modeling 85, 87, 130, 147, 241
Cloud computing 25, 35, 76, 136, 191, 195, 262, 284
Cloud-edge integration 6, 111, 117, 136, 209, 215, 283, 284
Cognitive radio 208
Connected healthcare 1, 15
Connected vehicles 1, 202
Connection density 24
Cyber-physical systems 25, 203, 224, 284
Cybersecurity 47, 53, 59, 101, 119, 159, 201–225

d

Data analytics 46, 57–58, 104, 165, 176, 203
Data governance 19, 95, 125, 157
Data privacy 18, 29, 74, 76, 77, 95, 101, 111, 112, 114, 115, 135, 147, 157, 195, 206, 216, 275, 293
Data security 28, 109, 112, 214
Decentralized networks 14, 224

Synergies of 6G and IoT, First Edition. Edited by Himanshi Babbar.

Published 2026 by John Wiley & Sons, Inc.

Deep learning (DL) 44, 47, 67–68, 96, 183, 202, 210–211, 213, 235, 236, 254, 261–262
Device-to-device communication 14, 117
Digital twins 2, 19, 27, 35–36, 87, 99, 149, 191, 194–195, 282, 284, 288, 289, 293–295
Distributed AI 129, 147
Drone communication 2, 9, 34, 100, 117, 124, 194, 208, 218, 248, 287, 290

e
Edge AI 44, 45, 88, 90, 95, 98, 128, 134, 135, 235, 236, 243, 247, 248, 283, 292
Edge computing 2, 11, 15, 19, 29, 30, 36, 42, 44–46, 76, 78, 90–91, 94–97, 127, 129, 131, 134–139, 147–148, 165, 169, 172, 178, 183, 184, 208–209, 234, 241, 255, 262, 291, 293
Edge intelligence 6, 28, 94–96, 128, 134–139, 174, 287, 289
Energy efficiency 12, 19, 24, 28, 30, 40, 48, 50, 67, 83, 84, 86, 88–91, 99, 152, 156, 174, 177, 184, 195, 198, 231, 235, 237, 245, 247, 250, 282, 291
Enhanced mobile broadband (eMBB) 24, 93, 94, 128, 201, 254, 281
Ethical AI 104, 105, 111
Explainable AI (XAI) 41, 77, 152
Extended reality (XR) 26, 36, 42, 52, 145, 146, 149, 194, 195, 282, 293, 294

f
Federated learning (FL) 41, 44, 47, 49, 68, 94–96, 133, 146, 147, 163, 177, 183, 185, 202, 210, 215–217, 225, 293
Fog computing 44–45, 88, 128, 129, 134–139, 147
Frequency spectrum 132, 145

g
Gigabit connectivity 41, 83
Green networking 91–92, 156, 160

h
Healthcare IoT 31, 59, 67
Heterogeneous networks 87
Holographic communication 2, 30, 98, 146, 254, 283
Human-centric networking 18, 284, 292–295

i
Immersive communication 2, 23, 98, 146, 149, 194
Industrial IoT (IIoT) 3, 8, 28, 30, 35, 59, 95, 163, 166, 210, 216, 236, 281, 283–288, 293
Intelligent reflecting surfaces (IRS) 84, 89, 131, 132, 282, 286, 291
Internet of Medical Things (IoMT) 98, 254, 262
Internet of Vehicles (IoV) 131
Interoperability 16, 18, 47, 51, 74, 119, 158–159, 169–170, 175, 178, 196, 203, 220, 286, 292
IoT architecture 88, 165, 185–186
IoT security 17–18, 60, 153–154, 203, 206

j
Joint communication and sensing (JCAS) 130, 133

l
Latency optimization 248
Low Earth Orbit (LEO) satellites 37, 110, 148, 166
Low-latency communication 42, 100, 118, 137, 240

Low-power wide-area network (LPWAN) 8, 165, 169, 176

m

Machine learning 8, 17, 29, 41, 47, 57–78, 90, 134, 138, 147, 163, 181–198, 203, 210, 235, 238, 250, 253, 261, 283
Machine-to-machine communication 5
Massive IoT 10, 43, 98
Massive MIMO 42, 85–87, 131, 238, 247
Millimeter wave (mmWave) 132, 145, 237
Mobile edge computing (MEC) 147, 174

n

Nano sensors 290
Network slicing 44, 49, 51, 92–94, 96, 102, 283, 286
Network softwarization 49
Neural networks 44, 67, 68, 211, 232, 257, 259, 265, 281

o

Open RAN 24, 130, 165
Optimization algorithms 236, 247

p

Predictive analytics 11, 71, 73, 98, 253–278, 286
Privacy preservation 46, 215
Protocol stacks 237, 238, 246

q

Quality of Experience (QoE) 237
Quality of Service (QoS) 149, 174, 184, 233, 235, 287
Quantum communication 2, 9, 41, 125, 151, 283

r

Radio access network (RAN) 24, 130, 165
Radio resource management
Real-time analytics 7, 148, 242, 287
Reinforcement learning 60, 67, 152, 184, 203, 210, 211, 232, 237, 244–246
Remote healthcare 13, 46, 96
Resource allocation 41, 43, 67, 113, 130, 147, 184, 232, 237, 243, 245, 246, 250, 260

s

Satellite IoT (SIoT) 37, 38, 163–178, 197
Scalable networks 43
Security orchestration 223
Semantic communication 75, 97
Sensor fusion 6, 127, 129, 130
Smart agriculture 3, 19, 28, 116, 175, 234, 241, 242
Smart cities 1, 3, 5, 8, 11, 17, 24, 25, 28, 38, 39, 42, 51–53, 57–78, 88, 91, 93, 95, 99, 100, 105, 113–114, 116–120, 149, 166, 184, 186–188, 201, 203, 207, 209, 224, 238, 240–243, 246, 247
Smart healthcare 16, 77, 90, 98, 254
Smart manufacturing 30, 97, 99, 218, 287–288
Smart transportation 138
Software-defined networking (SDN) 28
Spectrum allocation 52, 53, 132, 184, 291
Spectrum sharing 42, 47, 50, 52, 101, 102, 131, 158
Sustainable networking 91

t

Tactile internet 9, 30, 46, 98, 149, 292
Telemedicine 11, 31, 38, 97, 98, 111, 173, 188–190, 255
Terahertz communication 7, 12, 57, 83, 282
Trust management 17, 218

U
Ultra-dense networks 42
Ultra-reliable low latency communication (URLLC) 130, 201
Unlicensed spectrum 292

V
Vehicular IoT 34, 99
Virtual reality (VR) 13, 29, 83, 113, 149, 194
Virtualization 49, 91, 93, 233

W
Wearable devices 8, 16, 60, 67, 71, 73–75, 95, 117, 258, 266, 274
Wireless energy transfer 41
Wireless sensor networks 232, 234–235

X
XR applications 149, 194, 282, 293

Z
Zero trust security 215, 286
Zero-touch networking 147